SPECTROSCOPY

SPECTROSCOPY

Principles and Instrumentation

Mark F. Vitha
Drake University
Des Moines, IA
USA

Registered Office
John Wiley & Sons, Inc., 111 River Street, Hoboken, NJ 07030, USA

Editorial Office
111 River Street, Hoboken, NJ 07030, USA

For details of our global editorial offices, customer services, and more information about Wiley products visit us at www.wiley.com.

Wiley also publishes its books in a variety of electronic formats and by print-on-demand. Some content that appears in standard print versions of this book may not be available in other formats.

Library of Congress Cataloging-in-Publication Data

Names: Vitha, Mark F., author.
Title: Spectroscopy : principles and instrumentation / Mark F. Vitha.
Description: First edition. | Hoboken, NJ : Wiley, 2019. | Includes bibliographical references and index. |
Identifiers: LCCN 2018022525 (print) | LCCN 2018023111 (ebook) | ISBN 9781119436638 (Adobe PDF) |
    ISBN 9781119436607 (ePub) | ISBN 9781119436645 (hardback)
Subjects: | MESH: Spectrum Analysis–methods | Spectrum Analysis–instrumentation
Classification: LCC QD272.S6 (ebook) | LCC QD272.S6 (print) | NLM QY 90 | DDC 543/.5–dc23
LC record available at https://lccn.loc.gov/2018022525

Cover design by Wiley
Cover photo: Rich Sanders, Des Moines, Iowa. This image is used with the permission of the Des Moines Art Center, David Eppley, and Rachel Sussman

Set in 10/13pt Palatino by SPi Global, Pondicherry, India

Printed in the United States of America

10 9 8 7 6 5 4 3 2 1

CONTENTS

ABOUT THE COVER

The cover image is a detail of two works of art in the lobby of the Des Moines Art Center: *Set One, Set Two*, 2017, by David Eppley and *Sidewalk Kintsukuroi*, 2017, by Rachel Sussman. This book focuses on how light and matter interact and how we use those interactions to obtain chemical information. In the cover image, Eppley's work uses matter – colored vinyl tape – to evoke light that has been split into the spectrum. Eppley's carefully arranged, linear presentation breaks into jubilant strands and hexagonal shapes when the "light" interacts with Sussman's work, which is a ribbon of resin dusted with gold that fills a crack in the lobby floor. Thus, the theme of the interactions of light and matter is manifested in the combination of these two artworks. Sussman's work is based on a Japanese tradition of fixing broken ceramic pottery with lacquer dusted with gold and was installed for an exhibition titled "Alchemy: Transformations in Gold." Traditional kintsukuroi and Sussman's work are based on the idea that through golden repair, things are rendered more beautiful for having been broken. An interview with Sussman is available at https://www.desmoinesartcenter.org/blog/radio/alchemy-sussman and a video about Eppley's installation is available at https://www.youtube.com/watch?v=UFYfaKRo948. This image is used with the permission of the Des Moines Art Center, David Eppley, and Rachel Sussman. Photo Credit: Rich Sanders, Des Moines, Iowa.

PREFACE

If you have ever marveled at a sunset, a painting by Salvador Dali, glow-in-the-dark bracelets, fluorescent highlighters, the rich and varied colors of Kool-Aid, the reflections of clouds in a still lake, hair that is dyed bright pink, or how we can tell the strength of a cup of coffee based on how dark it is, you have marveled at light interacting with matter. My own interest in photography has made me more attuned to colors and to the properties of light, such as its softness or intensity, and more appreciative of what a remarkable achievement it is to be able to record light using sophisticated digital sensors. Everything we see – and much of what we don't – is a result of the interactions of matter with light. For example, matter absorbs, reflects, refracts, scatters, polarizes, and emits light. The focus of this book is on how we build instruments to take advantage of these interactions in order to measure the concentration of molecules in samples and to determine molecular structures of unknown compounds.

This book is a companion textbook to *Chromatography: Principles and Instrumentation*. Together, they cover two of the most commonly used analytical techniques and thus represent a significant majority of the instrumental analyses performed worldwide each day, with the noted exception of pH measurements. The books facilitate a modular approach to teaching and learning about chemical instrumentation, allowing instructors to integrate selected topics from this and other modular books, along with their own materials, into traditional instrumental and quantitative analysis courses, biochemistry courses, and analytical or integrated laboratory sequences.

The aim of each chapter is to connect the fundamental theory of how electromagnetic radiation affects matter to the design and use of instruments that make chemical measurements. Figures and problems are designed to help you develop mental images of how light in different regions of the electromagnetic spectrum interacts with matter and to understand the changes induced in the matter and in the electromagnetic radiation as a result of those interactions. Diagrams of instruments and their components connect the principles of spectroscopy to the design of instruments. It is through this connection of principles and instrumentation that we gain chemical information. Each chapter also provides case studies to illustrate how the techniques are used in practice, including analyzing works of art, studying the kinetics of enzymatic reactions, detecting explosives, and determining the DNA sequence of the human genome.

This book has five chapters. Chapter 1 provides an overview of the electromagnetic spectrum and the ways in which radiation interacts with, affects, and is affected by matter. Specifically, you will learn about the fundamental characteristics of electromagnetic radiation, such as wavelength, frequency, and amplitude. Variations in these characteristics change the way radiation impacts molecules in terms of making bonds vibrate with greater amplitude, promoting electrons to higher energy orbitals, and perhaps even breaking bonds. Phenomena such as reflection, refraction, diffraction, and polarization are also described in this chapter, as these phenomena impact virtually all spectroscopic techniques. This chapter also includes a discussion of scattering – a phenomenon that ultimately explains why the sky is blue and sunsets are red, illustrating just how familiar we already are with the effects of light interacting with matter.

Chapter 2 focuses on UV-visible spectroscopy and lays the groundwork for all the methods that rely on the absorption of electromagnetic radiation. It also introduces important instrumental components such as gratings and light sources, as well as common configurations of spectrophotometers. You will read about the quantized nature of the absorption process, with emphasis on how a molecule's immediate surroundings affect its absorption characteristics. The chapter concludes with a case study that shows how UV-visible spectroscopy was used to examine the kinetics of the enzyme myeloperoxidase.

The main topic of Chapter 3, luminescence, follows that of Chapter 2 in that molecules must absorb electromagnetic radiation before they can emit fluorescence or phosphorescence. In this chapter, you will learn about the molecular-level processes that occur after absorption but prior to the emission of light. The rates of these processes affect the intensity of emitted radiation and thus the qualitative and quantitative information we gain from them. Chapter 3 ends with a description of how fluorescence played a pivotal role in sequencing the human genome.

Chapter 4 deals with infrared spectroscopy, focusing primarily on the mid-infrared portion of the spectrum but including significant discussion of the growing applications that take advantage of the near- and far-infrared portions, too. Applications of infrared spectroscopy discussed in the chapter include determining the moisture content of biscuit dough and the long-range detection of explosives for security purposes.

The last chapter, expertly written by Dr. Jason Maher, explains the improbable phenomenon that lies behind Raman spectroscopy and how the fundamental principles of the technique dictate certain instrumental requirements that are not present in complementary techniques like infrared spectroscopy. You will also read about several applications of Raman spectroscopy, including the identification of pigments in artworks, the quantitation of heart medications in tablets, and the forensic identification of alkaloids such as heroin and morphine. I am grateful for Jason's contribution to the book and am certain readers will benefit from his excellent chapter.

It must be stated up front that in some chapters, particularly Chapter 2, much of the text and many of the figures come directly, or nearly so, from the textbook *Instrumental Methods of Analysis*, 7th edition, by H.H. Willard, L.L. Merritt, Jr., J.A. Dean, and F.A. Settle, Jr. It is in that textbook that the present book has its origins, and Frank A. Settle, Jr., Professor Emeritus, Washington and Lee University, who owns the copyright to that textbook, generously granted me permission to use material from it. Professor Settle also agreed that rather than using quotation marks or indents throughout the text to indicate passages from that work, acknowledging the use in this introduction is acceptable. I am grateful to him for this agreement, his support of my efforts, and his permission to use his and his colleagues' excellent work as a basis from which to build and develop this book.

As was the case with the publication of the chromatography book, this book is the result of the support, help, and input of a large number of people. Maura Lyons provided unwavering support throughout the writing and production of this book. I appreciate the calming effect she has and the model that she provides through her own scholarly pursuits. The education, love, and support provided by my mother continue to be the foundation upon which my professional efforts rest. How can one say "thank you" enough? Maria Bohorquez, Phani Chevru, Andrew Curtis, Teresa Golden, Gary Mabbott, Frank Settle, Emily Smith, and Matt Zwier all made improvements to this book. I greatly appreciate their advice and help. Much that is right in this text is due to their expertise and willingness to share it. Any errors present in the text are mine. Brian Gregory, Brian Lamp, Yinfa Ma, and Dave McCurdy also helped shape the chapters in this book. Megan Cohen at the Des Moines Art Center generously granted permission to use the image that appears on the cover of this book. The artists Rachel Sussman and David Eppley, who created the work that is pictured, and the photographer, Rich Sanders, who took the photo, also all generously gave me permission to use the image. I am happy to have their work on the cover of this book. I am also exceptionally appreciative of the time and help that dozens of people at instrument companies and publishing houses provided related to the acquisition of permissions to use many of the figures that appear in this textbook. It is easy to imagine that such requests come on top of people's regular responsibilities and busy work commitments, which makes me appreciate their willingness to accommodate permission requests even more. Thanks also to Drake University for providing funds to cover costs associated with writing this book.

FUNDAMENTALS OF SPECTROSCOPY

All instruments are designed to take advantage of some molecular property or behavior. For example, chromatography is based on the different strength of intermolecular interactions that molecules have with mobile and stationary phases. Electrochemistry is based on the ability of molecules to gain or lose electrons. In this book, we focus on the fact that atoms and molecules absorb and emit electromagnetic radiation (EMR). By measuring the amount and the characteristics of the EMR absorbed and emitted, we can measure the concentration of particular molecules present in a sample or gain structural information about them. You may already be familiar with several of the instrumental methods used to measure the absorption and emission of electromagnetic radiation, such as UV-visible, infrared (IR), and fluorescence spectroscopy.

In order to better understand the fundamental basis of these techniques, in this chapter we examine the properties of electromagnetic radiation and its effects on atoms and molecules. In subsequent chapters, we examine specific spectroscopic techniques. While all the techniques share common features, the specific instruments required and the information we gain from each are quite different and therefore require individual examination.

1.1. PROPERTIES OF ELECTROMAGNETIC RADIATION

Spectroscopic methods ultimately rely on measuring characteristics of electromagnetic radiation, which travels through space as a wave, as shown in Figure 1.1. As the name implies, it has two components, an electric field and a magnetic field, which are at right angles to one another. Figure 1.1 shows only a single wave with its electric field oriented along the x-axis, but in reality, most sources of electromagnetic radiation, like light bulbs and car headlights, emit radiation in which the electric field of the waves are randomly distributed around the x-axis. For now, however, we take a simplified view by focusing on only a single electromagnetic wave.

Spectroscopy: Principles and Instrumentation, First Edition. Mark F. Vitha.
© 2019 John Wiley & Sons, Inc. Published 2019 by John Wiley & Sons, Inc.

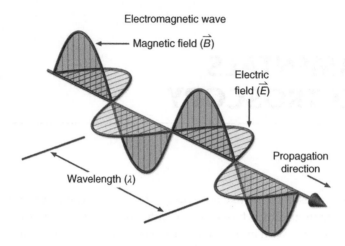

FIGURE 1.1 Diagram of a single electromagnetic wave propagating through space. The diagram indicates that an electromagnetic wave has both an electric field (*E*) and magnetic field (*B*) associated with it and that they are oriented at right angles to each other. It also indicates that the wavelength (*λ*) is the distance the wave travels during one oscillation of the electric and magnetic fields. Source: Reproduced with permission of Eric Clarke.

All electromagnetic waves have the properties of:

1. Speed
2. Amplitude
3. Frequency
4. Wavelength
5. Energy

Each of these characteristics is described below.

1.1.1.   Speed, c

Electromagnetic radiation in a vacuum travels at 2.998×10^8 m/s, commonly referred to as the speed of light, *c*. This speed only pertains to light traveling in a vacuum, though, because EMR slows down when it travels through matter such as air and water. We discuss the speed of light when it travels through matter in a later section.

1.1.2.   Amplitude, A

The amplitude, *A*, of a wave is the maximum length of the electric field vector, as shown in Figure 1.2. We seldom consider the amplitude of the wave because detectors are not fast enough to measure the magnitude of the electric field vector. Instead, we measure the radiant power, *P*, of a beam, which is proportional to the square of the amplitude. Radiant power is the amount of energy transmitted per unit time and is given by Eq. (1.1), where *E* is the energy of a photon and *ϕ* is the flux (i.e., the number of photons per unit time) [1]:

$$P = E\phi \tag{1.1}$$

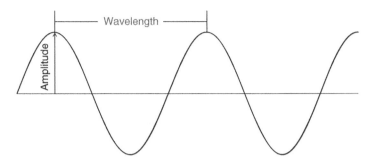

FIGURE 1.2   A side view of the electric field component of an electromagnetic wave as it propagates from left to right across the page. The amplitude is the displacement along the *y*-axis and the wavelength is the peak-to-peak distance between a single oscillation of the wave.

Although radiant power is commonly referred to as intensity, *I*, intensity is strictly defined as the radiant power from a point source per unit solid angle, usually measured in watts per steradian [1, 2].

1.1.3.   Frequency, υ

Frequency, *υ*, is the number of oscillations a wave makes per unit time and is typically measured in hertz, Hz, with units of reciprocal seconds, $1/s$ or s^{-1}. To visualize the physical meaning of frequency, imagine sitting on a rock out in the ocean with a stopwatch and counting waves that pass the rock. If you count, say, 120 waves in a minute, the frequency is

$$\frac{120\,\text{waves}}{1\,\text{minute}} \times \frac{1\,\text{minute}}{60\,\text{seconds}} = \frac{120\,\text{waves}}{60\,\text{seconds}} = 2.0\,\text{s}^{-1} \text{ or } 2.0\,\text{Hz}$$

In other words, two waves pass the rock every second. If more waves pass every second, the frequency is higher, and fewer waves per second are associated with a lower frequency.

The speed and wavelength of electromagnetic radiation change as it passes through different media, but the frequency remains the same. As described below, the frequency of EMR is closely related to the energy of the EMR. Therefore, the *frequency* is the characteristic that truly differentiates one wave from another.

1.1.4.   Wavelength, λ

The wavelength, *λ*, is the peak-to-peak distance of the wave, as shown in Figure 1.2. Because it is a distance, wavelengths are typically measured in meters. For example, EMR in the visible portion of the electromagnetic spectrum has wavelengths between 380 and 760 nanometers (nm).

1.1.5.   Energy, *E*

As we will see in all of the subsequent chapters, energy is really the fundamental wave characteristic that matters most in terms of the impact electromagnetic radiation has on matter. We often talk about EMR in terms of wavelengths, frequencies, and wavenumbers, but these ultimately relate to energy. In order to understand the energy of radiation, we must consider the wave/particle duality of light. When EMR propagates through space, it is convenient to focus on its wave properties (frequency, wavelength, amplitude). However, when it interacts with matter, it is useful to think of EMR as a discrete particle that contains a fixed amount of energy that can be transferred to an atom or molecule. That discrete particle is called a photon.

The energy that a photon contains is directly related to its frequency and wavelength, as shown in the two relationships in Eqs. (1.2) and (1.3):

$$E = h\upsilon \tag{1.2}$$

$$E = \frac{hc}{\lambda} \tag{1.3}$$

where h is Planck's constant (6.626×10^{-34} Js) and c is the speed of light (2.998×10^8 m/s). These equations are fundamental to the study of spectroscopy. They also clearly show that photons with higher frequencies (i.e., faster oscillations) and shorter wavelengths are higher in energy than those with slower oscillations and longer wavelengths (see Figure 1.3). To help make a mental association between these relationships, consider the two waves shown in Figure 1.3. Imagine trying to draw a wave of each frequency across the entire length of a chalkboard or white board, and that you only have 15 seconds to go from one end of the board to the other. You have the same amount of time in each case because light propagates through space at the same velocity regardless of wavelength. Clearly, the higher frequency wave – the one with little space between peaks and therefore

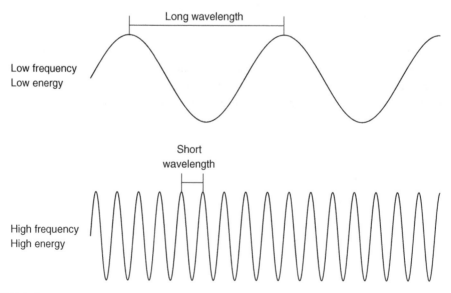

FIGURE 1.3 Depiction of the reciprocal relationships between wavelength and frequency. Longer wavelengths are associated with lower frequencies and lower energies, while shorter wavelengths are associated with higher frequencies and higher energies.

shorter wavelength – will require you to expend a greater amount of energy to fill the board, whereas you can take a leisurely stroll (i.e., exert low energy) along the board when drawing the low-frequency, long-wavelength wave.

From the fact that Eqs. (1.2) and (1.3) are equal to each other, we can see that

$$\upsilon = \frac{c}{\lambda} \tag{1.4}$$

This relationship has implications for the speed and wavelength of EMR as it travels through different media such as air, water, benzene, etc. As radiation passes through matter, its electric field interacts with the electrons in the matter, slowing its propagation, meaning that the speed of light is different in different media. In a vacuum, the speed is 2.998×10^8 m/s, but in everything else, including air, it is slower. However, as the EMR propagates through matter, its frequency is unaffected, so υ remains the same. In order to maintain the equality in Eq. (1.4), when the speed, c, decreases, then the wavelength must also decrease in order to maintain a constant frequency, υ.

The change in the speed of EMR in matter is measured by the refractive index of a substance, η, where

$$\eta = \frac{c_v}{c_i} \tag{1.5}$$

in which c_v is the speed in a vacuum and c_i is the speed in the substance of interest. Because EMR is slower in matter than it is in a vacuum, such that $c_i < c_v$, refractive indices are greater than 1.00. The velocity of radiation in air is within 1% of the velocity in a vacuum such that using $c = 2.998 \times 10^8$ m/s does not generally lead to significant bias in calculations for many of the situations in which we are interested.

It should be noted that the measurement of η depends on the frequency of the EMR used and the temperature. In order to standardize the measurement, the frequency of 5.09×10^{14} Hz (equivalent to 589 nm wavelength light), which is a frequency emitted by excited sodium atoms, is typically used, and values are often measured at 20 °C [3]. The symbol η_D^{20} (or η_D^{25} if values are measure at 25 °C) is often used to denote refractive indices, with the superscript specifying the temperature at which the measurement was made, and the subscript indicating that the sodium D line was used. Some values for common materials are given in Table 1.1 [4–7]. Notice that substances with π-electrons such as toluene, and those with highly polarizable atoms like diiodomethane, have higher refractive indices, meaning that they slow the propagation of EMR more than other substances. Ultimately, the decrease in speed is due to polarization of the atoms and molecules in the material caused by the incoming electric field. The result is a temporary deformation of the electron clouds associated with the atoms or molecules. The energy of the light is then re-emitted by the atoms or molecules with the same frequency, but its progress (i.e., speed) has been slowed by the temporary retention by the material. It is interesting to note that scientists have slowed the speed of light down to a mere 38 miles-per-hour – a typical speed for a car – using Bose–Einstein condensates [8].

TABLE 1.1   Refractive Indices of Some Common Materials

Substance	Refractive index
Hydrogen	1.000 132
Air	1.000 293
Methane	1.000 444
Water (liquid)	1.333
Heptane	1.380
Benzene	1.501
Toluene	1.497
Chlorobenzene	1.525
Quartz	1.544
Diiodomethane	1.737
Diamond	2.417
Gallium (III) arsenide	3.927
Metamaterial[a]	38.6

[a] A polymer film made with a pattern of thin aluminum shapes (see Ref. [4]); measured at 0.3 THz.

EXAMPLE 1.1

Calculate the speed of EMR in diamond.

Answer:

$$\eta = \frac{c_v}{c_i}$$

$$2.417 = \frac{2.998 \times 10^8 \text{ m/s}}{c_{diamond}}$$

$$c_{diamond} = \frac{2.998 \times 10^8 \text{ m/s}}{2.417} = 1.240 \times 10^8 \text{ m/s}$$

Another question:
Calculate the speed of EMR in the metamaterial listed in the table.

Answer:
7.77×10^6 m/s.

1.1.6. Wavenumber, $\bar{v}$

While frequency and wavelength are commonly used to describe EMR, the wavenumber, $\bar{v}$, is also used, particularly when dealing with infrared spectra. The wavenumber is related to energy as shown in the first relationship in Eq. (1.6):

$$E = hc\bar{v} = hv = \frac{hc}{\lambda} \tag{1.6}$$

Wavenumbers are typically measured in units of reciprocal centimeters, cm^{-1}, meaning that they express the number of oscillations that occur per centimeter. Equation (1.7) shows how wavenumbers are related to frequency and wavelength:

$$\bar{\nu} = \frac{\upsilon}{c} = \frac{1}{\lambda} \tag{1.7}$$

While it may seem unnecessary to have three different parameters to describe EMR – frequency, wavelength, and wavenumber – different fields of spectroscopy, for reasons of convenience and historical legacy, use different measures. For example, in UV-visible spectroscopy we usually deal in wavelengths, whereas wavenumbers are commonly used in IR spectroscopy. Of course they are all interchangeable, but to work in any of these fields requires fluency in all three parameters.

EXAMPLE *1.2*

Calculate the energy, frequency, and wavenumber of EMR that has a wavelength of 1.234 μm.

Answers:

$$E = \frac{hc}{\lambda} = \frac{\left(6.626 \times 10^{-34}\ \text{Js}\right)\left(2.998 \times 10^{8}\ \text{m/s}\right)}{1.234 \times 10^{-6}\ \text{m}} = 1.610 \times 10^{-19}\ \text{J}$$

$$\upsilon = \frac{c}{\lambda} = \frac{2.998 \times 10^{8}\ \text{m/s}}{1.234 \times 10^{-6}\ \text{m}} = 2.429 \times 10^{14}\ \text{s}^{-1} = 2.429 \times 10^{14}\ \text{Hz}$$

$$\bar{\nu} = \frac{1}{\lambda} = \frac{1}{1.234 \times 10^{-6}\ \text{m}} = 8.104 \times 10^{5}\ \text{m}^{-1}$$

Another question:
Calculate the energy, wavelength, and frequency associated with EMR with a wavenumber of 543.2 cm^{-1}. Be careful with units.

Answers:
$$E = 1.079 \times 10^{-20}\ \text{J}$$
$$\lambda = 1.841 \times 10^{-5}\ \text{m} = 18.41\ \text{μm}$$
$$\upsilon = 1.628 \times 10^{13}\ \text{s}^{-1} = 16.28\ \text{THz}$$

1.2. THE ELECTROMAGNETIC SPECTRUM

In this section, we explore how atoms and molecules interact with incoming EMR. Specifically, we look at the range of energy carried by electromagnetic radiation of different frequencies and the effects that photons with different energy have on matter.

By measuring these various effects, we learn about the properties of the matter we are analyzing. This is critical, as the different regions of the spectrum ultimately relate to the different spectroscopic techniques described in the coming chapters. The different energies also require significantly different types of detectors, optical components, and mathematical treatment of data in order to make the measurements from which we ultimately gain chemical information.

The electromagnetic spectrum is shown in Figure 1.4.

One of the ways atoms and molecules interact with electromagnetic radiation is to absorb it. This occurs because atoms and molecules can exist in different states that have different, but quantized, energy levels (see Figure 1.5). If the energy of incoming EMR matches the energy difference (ΔE) between two quantum states, then the radiation can be absorbed. This process forms the basis for all common forms of spectroscopy, including UV-visible, IR, and NMR spectroscopy, and is discussed in greater detail later in this chapter and in Chapters 2 and 3.

Figure 1.5 depicts a quantized, electronic transition [9]. In this case, an electron is promoted from a bonding to an antibonding orbital, denoted as π and π^*-orbitals, respectively. Many other types of quantized transitions involving different types of orbitals (e.g., σ-, σ^*-, and n-orbitals) can occur. What is common to each is that the ground and excited states are separated by a fixed energy difference, ΔE. An atom or molecule can absorb an incoming photon if the energy of the photon matches ΔE. In the process, the absorbed energy changes the molecule, for example, by causing a nuclear spin flip as occurs in NMR spectroscopy, an increase in the amplitude of bond vibrations as occurs in IR spectroscopy, or a change in the electron distribution as depicted here for UV-visible spectroscopy.

When thinking about the energy range associated with each region, it is helpful to consider the energy required to break a typical carbon–carbon (C—C) bond, which is approximately 5.8×10^{-19} J (based on the bond energy of approximately 350 kJ/mol). Comparing the photon energies to this bond strength gives some sense of the energy being carried by different photons. Photons with less energy cannot break bonds but can cause other lower energy phenomenon to occur, while those with higher energy can break bonds, which causes significant molecular rearrangements or ionizes atoms and molecules.

1.2.1. Radio-Frequency Radiation (10^{-27} to 10^{-21} J/photon)

The low-frequency, long-wavelength end of the spectrum is known as the radio-frequency range. The waves here are on the order of 0.1 mm to 100 meters in length (i.e., a tiny fraction of an inch to approximately 1 football field) or longer according to the ISO (International Organization for Standardization) definitions of the regions of the electromagnetic spectrum (although we note that the ranges specified seem to allow for some overlap between regions) [10–13]. Recalling the reciprocal relationship between wavelength and energy means that this long-wavelength radiation has exceptionally low energy, which is fortunate for us as radio waves fill the environment and are constantly passing through and around us. There is nothing in our bodies that absorbs this radiation, so it does not have any physiological effect, and we can therefore use it to broadcast radio and television signals.

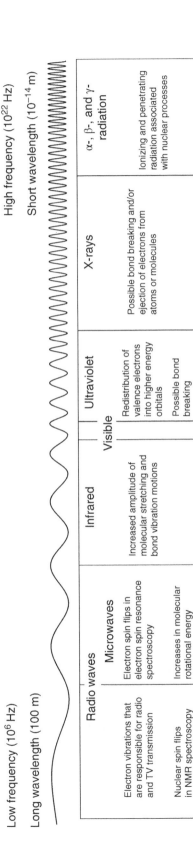

FIGURE 1.4 The regions of the electromagnetic spectrum and the effect that radiation in each region has on matter. Note the labels on the top left and right indicating that the diagram goes from low energy electromagnetic radiation (EMR) on the left to high energy electromagnetic radiation on the right. Also note the reciprocal relationship between energy and wavelength. Long (i.e., high value) wavelength photons on the far left of the diagram have very low energy, and short wavelengths photons (far right on the diagram) have very high energy. The high energy associated with short wavelength photons makes them capable of significant damage to molecules and thus significant negative health effects.

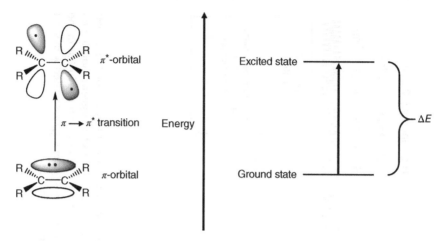

FIGURE 1.5   Depiction of a π–π* transition (left) and the associated energy change, ΔE (right). Note that in the ground state, the electrons in the π-orbital connect both carbon atoms above and below the plane of the bond. The electrons in the π-bond are thus delocalized and shared between both carbon atoms. In the excited state, the π*-orbitals, and thus the electrons in them, are localized on a specific carbon atom. Shared electrons like those in the ground state lower the energy of a molecule compared to localized electron density like that seen in the excited state. In order to transition from the ground state to the excited state, the molecule must absorb a photon that has an energy that exactly matches the energy difference (ΔE) between the ground and excited states.

The EMR used in nuclear magnetic resonance (NMR) spectroscopy is in the radio-frequency range, toward the shorter wavelengths (~0.3 to 0.8 m). Molecules are made of atoms, and atoms contain nuclei, some of which behave as if they are spinning. When placed in a magnetic field, the spins orient with or against the field, known as "spin-up" or "spin-down" orientations. The spin states have slightly different energies. Therefore, when EMR in the low energy, radio-frequency range interacts with nuclei, the nuclei can absorb it. The energy of the EMR is used to change the spin of a nucleus from spin-up (lower energy state) to spin-down (higher energy state). The energy required to do this is almost trivial: approximately 2×10^{-25} Joules per nucleus, six orders of magnitude (one million times) less than that of a C—C bond. Yet, as you may already know, NMR spectroscopy provides valuable information about molecular structures. It is hard to imagine modern chemistry without this technique. So while the energy of the radiation is quite low and the effect it has on matter – a mere nuclear spin flip – is almost imperceptible, this is an important region of the spectrum for chemists.

1.2.2.   Microwave Radiation (10^{-23} to 10^{-22} J/photon)

The microwave region overlaps with the higher frequency portion of the radio wave region, but the two regions are generally distinguished by the technologies used to measure them. The microwave region is associated with two types of spectroscopy: (1) electron spin resonance (ESR) and (2) microwave spectroscopy. The basis of ESR spectroscopy is that electrons, like nuclei, have the property of spin-up or spin-down, which have slightly different energies when samples are placed in a magnetic field. By absorbing EMR in this region, the electron spin is flipped. Just as in NMR, ESR spectra provide structural information by providing insight into the environment of electrons within molecules.

In microwave spectroscopy, the frequency of EMR is such that molecules, which are rotating at their natural rotational frequency, can absorb the radiation and transition to a higher rotational state. This is particularly noticeable for molecules in the gas phase where rotation is not hindered.

1.2.3. Infrared Radiation (10^{-22} to 10^{-19} J/photon)

Progressing through the EMR spectrum toward higher frequency and energy leads to the infrared (IR) region of the spectrum, which can be divided into the far-IR (lowest energy infrared), mid-IR, and near-IR (highest energy infrared) regions.

The bonds in molecules vibrate and stretch with characteristic frequencies. EMR in the IR range of the spectrum coincides with the frequency of bond stretching and bending, so molecules can absorb radiation in this region. IR spectra inform chemists about the bonds that are present in molecules, which couples nicely with NMR information when determining the structure of molecules. IR spectra can also be used to measure concentrations of molecules present in a sample by measuring the amount of EMR absorbed.

The portion of the IR region of the spectrum that is exploited to yield valuable chemical information has grown in the past few decades. Traditionally, most IR spectroscopy was concerned with the range from 4000 to 650 cm^{-1} (2.5–15.38 μm), known as the mid-IR region, which you may have already studied when discussing organic structural analysis. Compared to our reference point of the energy required to break a C—C bond, photons in this region of the spectrum have about one-tenth the energy required to break the C—C bond.

EMR that is slightly higher in energy than mid-IR is known as near-IR radiation. The near-IR (NIR) region, from about 12500 to 4000 cm^{-1} (0.8–2.5 μm), provides chemical information resulting from the overlap of the fundamental frequencies present in the mid-IR. Scientists use NIR spectroscopy to measure the water content of samples, which is important in agricultural studies, as well as to determine protein, fat, sugar, and oil content, which are important to the food and grain industries [14, 15].

Electromagnetic radiation in the far-IR region of the spectrum is lower than that in the mid-IR. The far-IR region therefore falls between the mid-IR and microwave regions. In this region, radiation with wavelengths from 1 cm to 1 mm is often referred to as millimeter wave (MMW) radiation, and from 1 to 0.3 mm is called submillimeter (sub-MMW or sub-mm) radiation. Radiation with shorter wavelengths and therefore higher energy is called terahertz (THz) radiation because the frequency of these waves is on the order of 10^{12} Hz (extending from 30 THz down to about 200 GHz). Spectroscopies that deal in these ranges of the spectrum are used to study biomolecules, semiconductors, polymers, and nanomaterials [16–18]. Most notably, some airport body scanners are based on MMW radiation because such radiation passes through clothing and can detect material – both metals and nonmetals – hidden beneath it. Terahertz techniques are also being explored in the realm of homeland security to detect explosives and chemical agents from long distances [19]. IR radiation is also the radiation used by remote controls, detected by night vision goggles, and used to keep fries warm at fast food restaurants. Again, the energy of the radiation in the infrared region generally causes bonds to vibrate. Much like nuclear spin flips, bond vibrations are generally low energy and do not cause serious disruptions to molecules.

1.2.4. Ultraviolet and Visible Radiation (10^{-19} to 10^{-18} J/photon)

The next highest energy regions of the spectrum are the visible (760–380 nm, red to blue, respectively) and ultraviolet (400–100 nm) regions. The energy of radiation in these regions is approximately 2 000 000 times greater in magnitude than that used in NMR spectroscopy! In that region, the only perturbation to matter is a mere spin flip of the nucleus. In the IR region, the radiation simply causes bonds to vibrate with greater amplitude. A more dramatic effect is observed in the UV and visible regions: The electron distribution within the molecule is altered. In atoms and molecules, electrons are distributed in orbitals, with the highest energy orbitals (i.e., valence orbitals) being populated by π-electrons, bonding (σ) electrons, and lone pair electrons (n-orbitals), all of which are far away from the nuclei relative to the localized core electrons. While valence electrons are those that are furthest away from the nuclei and in the highest occupied energy orbitals, higher, unoccupied orbitals also exist in atoms and molecules. When an atom or molecule absorbs visible or ultraviolet radiation, the energy from the radiation causes a redistribution of electrons from the lowest energy state to a higher energy arrangement of electrons. In other words, the atoms and molecules are temporarily destabilized by the energy they absorb. This does not necessarily mean that the molecule falls apart, but absorption of UV-visible radiation is a higher energy process than those considered thus far and, in some cases, can break bonds, particularly when the radiation absorbed by a molecule is in the UV region of the spectrum.

We have treated the UV and visible portions of the spectrum together because radiation in both regions generally has the same effect on matter – namely, causing electron promotion to higher energy orbitals. Furthermore, UV-visible spectrophotometers typically cover both regions of the spectrum. The two regions are differentiated, however, by the fact that humans can perceive electromagnetic radiation in the visible region, but not in the UV region (or any other region for that matter). They are also differentiated by the fact that in the UV region, the energy of photons is equivalent to or greater than that required to break a C—C bond (see Example 1.3)

Radiation within the UV region is subdivided into the UVA (315–400 nm), UVB (280–315 nm), and UVC (100–280 nm) ranges [10, 11]. These regions are important because of their associations with skin cancer. UVA penetrates the most deeply into the skin, reaching the dermis. UVB does not penetrate as deeply, but affects the epidermis and is responsible for sunburns and blistering. More importantly, UVB also causes the skin cancers that result from UV radiation by causing the dimerization of adjacent base pairs in DNA, making it unreadable by replication enzymes [20]. Sunscreen, which absorbs the UV radiation before it can reach the skin, helps prevent such damage. Even though UVC radiation is more energetic than UVB, it does not cause significant health effects because it is absorbed efficiently by the atmosphere and therefore does not reach us.

1.2.5. X-Ray Radiation (10^{-15} to 10^{-13} J/photon)

UV-visible spectroscopy focuses on exciting the *valence* or highest energy electrons. X-rays have enough energy that they entirely eject *core* electrons from the nuclei, leaving the atom in an ionized state. For this reason, X-ray radiation and all higher energy radiation are referred to as "ionizing radiation." The energy required to eject electrons is different for each atom and thus X-ray spectroscopy can be used to determine the elemental makeup of materials. This is quite commonly used, for instance, to study the materials and pigments used in sculptures, vases, paintings, and other works of art.

X-rays cover the region from approximately 0.10 nm to 1 pm in wavelength. These wavelengths are roughly 10–12 orders of magnitude smaller than the radiation used in NMR spectroscopy, and therefore 10–12 orders of magnitude greater in energy (at least 10 000 000 000–1 000 000 000 000 times greater!). When you get an X-ray at the hospital, the technician typically leaves the room and stands behind a lead-lined wall that shields them from the radiation. This is because continuous or prolonged exposure to X-rays creates reactive, ionized species that can lead to cancer. However, at the doses (or photon fluxes) used in medical X-ray instrumentation, the risks of damage are quite low, and the benefits of revealing health problems, like broken bones, lung tumors, and heart problems, outweigh the risks.

1.2.6. Alpha, Beta, and Gamma Radiation (10^{-13} to 10^{-11} J/photon and Higher)

Alpha (α)-, beta (β)-, and gamma (γ)-radiation are high energy, ionizing radiation created by nuclear processes such as radioactive decay and nuclear fission, with γ-radiation acting much like X-ray radiation but of higher energy. Beta radiation, unlike the other radiation described so far, is not electromagnetic radiation, but rather is composed of high energy electrons that can penetrate more deeply into matter than other types of radiation (e.g., it can penetrate a few millimeters into a block of aluminum) [21–23]. This can be beneficial when trying to analyze materials to different depths. It can also be biologically quite harmful as it can affect the inner layer of skin where new cells form [24]. Alpha radiation, like beta radiation, is also composed of high energy particles. In this case, the particles are composed of two protons and two neutrons (no electrons) and thus are really a helium nuclei bearing a +2 charge.

Alpha, beta, gamma, and X-ray radiation are responsible for the effects that were observed after the nuclear bombing of Hiroshima and Nagasaki, as well as after nuclear accidents such as those at Chernobyl, Three Mile Island, and Fukushima (although some reports assert that no adverse health effects resulted from the accident at Three Mile Island and only very low potential for long-term health effects from the Fukushima accident) [25–27].

Table 1.2 summarizes the approximate energies, wavelengths, frequencies, and atomic and molecular phenomena associated with each region of the electromagnetic spectrum. Figure 1.6 is a graphical representation of the regions of the electromagnetic spectrum and also shows some of the subcategories within each region.

TABLE 1.2   Summary of the Approximate Energy, Wavelength, and Frequency Regions of the Electromagnetic Spectrum

Region	Energy (J/photon)	Wavelength (m)	Frequency (Hz)	Phenomena
Radio	10^{-27} to 10^{-21}	0.0001 to 100 m	10^6 to 10^{12}	Spin flips
Microwave	10^{-23} to 10^{-22}	0.001 to 0.01	10^{10} to 10^{11}	Molecular rotation
Infrared	10^{-22} to 10^{-19}	1.0×10^{-3} to 7.6×10^{-7}	10^{11} to 10^{14}	Bond stretching and bending
UV-visible	10^{-19} to 10^{-18}	7.6×10^{-7} to 1.0×10^{-7}	10^{14} to 10^{15}	Valence electron excitation and bond breaking at high energy end
X-ray	10^{-15} to 10^{-13}	1.0×10^{-10} to 1.0×10^{-12}	10^{18} to 10^{20}	Core electron excitation or ejection and bond breaking
Gamma	10^{-13} to 10^{-11}	1.0×10^{-12} to 1.0×10^{-14}	10^{20} to 10^{22}	Electron ejection, bond breaking

Sources: From Refs. [10] and [11].

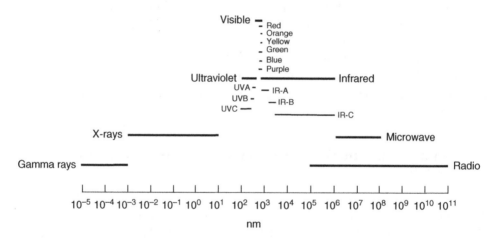

FIGURE 1.6   Graphical depiction of approximate ranges within the electromagnetic spectrum. Wavelengths increase and energy decreases from left to right. Vertical displacement has no meaning other than to allow for easier reading. Source: Reproduced with permission of ANSI (American National Standards Institute).

EXAMPLE 1.3

1. Calculate the wavelength of a photon with an energy equivalent to that of a C—C bond (i.e., approximately 5.8×10^{-19} J).
2. In what region of the electromagnetic spectrum does this wavelength fall?
3. In each pair, identify which process has a higher energy associated with it: (a) a nuclear spin flip or ionizing a molecule? (b) Rearranging the electron distribution in a molecule or bond vibrations?
4. What region of the spectrum is associated with each of the four processes identified in question 3?

Answers:

1. $E = \dfrac{hc}{\lambda} = 5.8 \times 10^{-19} \, J = \dfrac{\left(6.626 \times 10^{-34} \, Js\right)\left(2.998 \times 10^8 \, m/s\right)}{\lambda}$

$\lambda = \dfrac{\left(6.626 \times 10^{-34} \, Js\right)\left(2.998 \times 10^8 \, m/s\right)}{5.8 \times 10^{-19} \, J} = 3.4 \times 10^{-7} \, m = 340 \, nm$

2. The UV region.
3. First pair: ionizing a molecule. Second pair: rearranging the electron distribution in a molecule.
4. Nuclear spin flip, radio; ionizing a molecule, X-ray and gamma; rearranging the electron distribution in a molecule, UV and visible; and bond vibration, infrared.

Additional questions:
   (a) Calculate the wavelength of a photon with an energy equivalent to that of an O=O bond (120 kcal/mol) – note that this energy is given in kcal/mol, not J/molecule as was done above).
   (b) If you wanted photons with energy greater than the O=O bond, would you want longer or shorter wavelengths than that which you just calculated?

Answers:
   (a) 238 nm.
   (b) Shorter (energy and wavelength are reciprocally related).

1.3.   THE PERRIN–JABLONSKI DIAGRAM

The most common types of absorption and emission spectroscopy focus on the UV-visible and IR regions of the spectrum. A Perrin–Jablonski diagram (see Figure 1.7) conveniently summarizes all of the absorption and emission processes that occur in these regions [29–34].

In this diagram, S_0 represents the lowest energy electron distribution of a molecule, which is called the ground state. S_1 and S_2 represent higher energy electronic states. S_0 corresponds to the electron distribution that we typically think of when we draw Lewis structures for molecules, and S_1 and S_2 correspond to the electronic distribution when electrons are promoted into higher energy orbitals such as π^* and σ^* antibonding orbitals. The lines within each of these electronic states represent the vibrational levels of the molecules (i.e., bond stretching and vibrations). Though not shown, within each of the vibrational levels are different quantized rotational levels. In any collection of molecules, the vast majority exist in the lowest electronic and vibrational states (S_0, v_0). This is often demonstrated using the Boltzmann distribution as discussed in a separate section.

This diagram shows the different energy changes associated with microwave (rotational), IR (vibrational), and UV-visible (electronic) spectroscopy and why different frequencies of EMR are needed for each.

Molecules can absorb the energy of an incoming photon if that energy exactly matches the energy difference between two energy levels. In the case of electronic transitions, the energy that is absorbed causes a redistribution of electrons from the lowest energy orbitals into higher energy orbitals (see Figure 1.8).

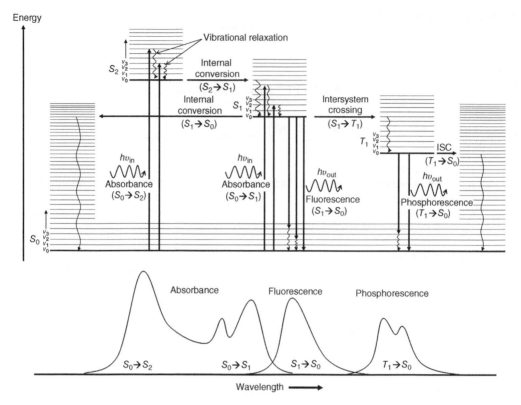

FIGURE 1.7 The Perrin–Jablonski diagram showing various processes that occur when light is absorbed by a molecule. There is a lot of information contained in this figure, and the reader is encouraged to spend some time to fully understand it as it underpins much of the material in this and subsequent chapters. The arrow at the far left is important to note. It indicates that energy is increasing from bottom to top. Horizontal lines represent the energy of different electronic and vibrational levels of molecules. It should be understood that when a molecule undergoes an electronic transition, while the energy of the molecule increases, the electron that is promoted to a higher energy state does not literally go "up", but rather is redistributed into a different molecular orbital. Key processes in this diagram include absorption, vibrational relaxation, internal conversion, fluorescence, intersystem crossing (ISC), and phosphorescence, all of which are described in greater detail in the text. The curves shown under the Perrin–Jablonski diagram indicate the spectral features that could be observed for each type of transition. Source: Adapted in part from Valeur and Berberan-Santos [28]. Reproduced with permission of Wiley-VCH Verlag BmbH & Co. KGaA.

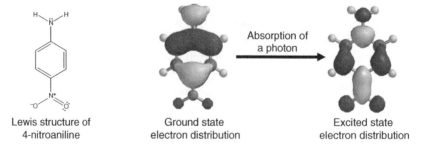

Lewis structure of
4-nitroaniline

Ground state
electron distribution

Excited state
electron distribution

FIGURE 1.8 A visualization of an electronic rearrangement caused by the absorption of a photon. The Lewis structure of 4-nitroaniline is shown on the far left. The electron clouds in the middle and right image depict regions of electron density. When 4-nitroaniline absorbs the energy of a photon, the electrons get redistributed into different orbitals, with the resulting electron distribution being higher in energy than the ground state distribution. Notice that in this case, the electron density shifts away from the amine group toward the nitro group as a result of the electronic transition.

For our discussion, we consider what happens when a molecule transitions from the ground state (S_0/v_0) into the fourth vibrational level of the second excited state (S_2/v_4). In order to go from S_0/v_0 to S_2/v_4, the molecule has to absorb the energy of a photon. So immediately upon absorption, the energy of the molecule increases by exactly the amount of energy the photon had. After being promoted to this new level, the molecule rapidly begins to shed the energy it absorbed. This process is called vibrational relaxation. Vibrational energy is lost through collisions between the excited state molecule and surrounding solvent molecules, with the energy being converted to a miniscule amount of increased kinetic energy of the solvent [35].

After relaxing to the lowest vibrational level within S_2, most molecules undergo a process called internal conversion, which occurs when the lowest vibrational level of S_2 overlaps with higher vibrational levels in S_1. The energy within the molecule is redistributed into a new electronic distribution and different vibrational modes. After this conversion from one excited state to another, additional vibrational relaxation within S_1 occurs, and the molecule quickly relaxes into the ground vibrational level (v_0) of S_1 [35]. Several different processes can take place once an excited molecule is in this state: fluorescence, external conversion, internal conversion, or intersystem crossing.

Fluorescence is the emission of a photon from a singlet excited state molecule, returning the molecule to the singlet ground state. "Singlet state" simply means that all electron spins are paired (one spin-up for every one spin-down). The emitted photon carries away the specific amount of energy required to return the molecule from the ground vibrational state in S_1 back to any of the vibrational levels within S_0. A few molecules also exhibit fluorescence directly from S_2 to S_0, but fluorescence from the S_1 state is much more common [35].

External conversion is a radiationless relaxation process, meaning that no photon is emitted as the molecule loses energy to return to the ground state. It occurs through collisions of the excited state with surrounding solvent molecules, during which the solvent gains energy, while the excited molecule loses it, resulting in a very slight increase in the temperature of the solvent. Nonradiative decay also occurs via internal conversion from the ground vibrational state in S_1 to a high vibrational state in S_0 followed by vibrational relation.

Intersystem crossing involves the redistribution of the energy of the singlet S_1 state into a triplet state, symbolized as T_1 (see Figure 1.7). A triplet state is one in which all of the electrons in a molecule are paired except for two, which both have the same spin. The process of going from a singlet to a triplet state is known as a "spin-forbidden transition." While we use the term "forbidden," in reality some molecules can undergo this process, but the probability of its occurrence is low. Once in the triplet state, molecules undergo vibrational relaxation to the lowest vibrational level. Molecules can then either relax to S_0 via external conversion (i.e., radiationless collisional deactivation) or emit a photon with an energy that matches the energy difference between the lowest vibrational state in T_1 and any of the vibrational states in S_0. Again, this is a quantized transition in that only photons with specific energy matching the energy level gaps are emitted. Emission of this type is referred to as phosphorescence (see Figure 1.7). Like the $S_1 \rightarrow T_1$ transition, the $T_1 \rightarrow S_0$ transition is also spin forbidden and of low probability. This has implications about the timescale on which it occurs as discussed in the next section.

You have observed phosphorescence if you have seen glow-in-the-dark stickers or bracelets. To activate them, they are exposed to light and the molecules absorb that light, promoting the molecules into excited states that ultimately transition into the excited triplet state. When the lights are turned off, you can see the light they emit as they return from the excited triplet state to the ground singlet state. Because this transition is also of low probability, it can take minutes or hours for all of the molecules to eventually return to the ground state, which is why the stickers and bracelets continue to glow even after the light is turned off. The intensity of the glow fades over time as more of the molecules return to the ground state, and fewer are left in the excited state.

1.3.1. Timescales of Events

Besides depicting the different energy scales involved in the various types of spectroscopy and the quantum transitions involved in each, the Perrin–Jablonski diagram also allows for a discussion of lifetimes of the molecules in the various states associated with the different processes (see Table 1.3 for the approximate timescales) [28].

Absorption occurs on a femtosecond (10^{-15} s) timescale. In this short period of time, the electrons are rearranged within the molecule, but the much more massive nuclei do not have time to move. The fact that the nuclei can be viewed as essentially motionless during absorption is known as the Franck–Condon principle and also relates to the Born–Oppenheimer approximation [36, 37].

Vibrational relaxation takes picoseconds to fractions of a nanosecond (10^{-12} to 10^{-10} s), during which time solvent molecules begin to reorient themselves to better interact with the new electronic distribution of the excited state molecule. Internal conversion also occurs on the picosecond timescale.

Fluorescing molecules typically remain in the lowest excited singlet state from 10^{-10} to 10^{-7} seconds before emitting a photon. If nonradiative processes occur faster, then little or no fluorescence is observed. Structural effects and thermal effects play a big role here. Most molecules are not fluorescent because they can lose energy through vibrations and rotations that transfer energy to other molecules. These nonradiative decay mechanisms decrease the likelihood of the molecule relaxing to the ground state via fluorescence. Structurally rigid molecules, however, like polyaromatic hydrocarbons with electron donating and accepting groups on them, tend to be fluorescent. This is simply because relaxation via fluorescence competes on the same timescale as nonradiative decay for

TABLE 1.3 Approximate Timescales for Processes Associated with Spectroscopy

Transition	Timescale
Absorption	10^{-15} s (femtoseconds)
Vibrational relaxation	10^{-12} to 10^{-10} s (picoseconds to subnanoseconds)
Internal conversion	10^{-11} to 10^{-9} s (tens of picoseconds to nanoseconds)
Fluorescence	10^{-10} to 10^{-7} s (subnano- to submicroseconds)
Intersystem crossing	10^{-10} to 10^{-8} s (around a nanosecond)
Phosphorescence	10^{-6} to 10 s (milliseconds to seconds)

Source: From Valeur and Berberan-Santos [28]. Reproduced with permission of John Wiley & Sons.

these molecules. Similarly, fluorescence tends to increase as the temperature of a sample decreases because the molecules are moving more slowly, meaning they bump into one another less frequently and with less energy, thereby increasing their likelihood of fluorescing rather than undergoing nonradiative decay.

As noted above, phosphorescence is spin forbidden. It is therefore an unlikely process and takes a long time, with triplet state lifetimes ranging from milliseconds to minutes. Thus, phosphorescence, when it occurs, is a long-lived phenomenon, as discussed above in the context of glow-in-the-dark stickers and bracelets. As with fluorescence, colder temperatures reduce the nonradiative mechanism of relaxation, increasing the probability that an excited state molecule phosphoresces. It is for this reason that glow-in-the-dark bracelets last longer if stored in a refrigerator or freezer after being exposed to the sun.

1.3.2. Summary of Radiative and Nonradiative Processes

The section above covered a lot of ground. It is an important section as it encapsulates much of modern spectroscopy. Absorption has to occur in order for fluorescence or phosphorescence to occur. Therefore, only molecules that tend to absorb a lot of radiation in the UV-visible region are fluorescent or phosphorescent. There are many pathways by which molecules that have absorbed energy can relax back down to the ground state, with phosphorescence being the slowest and therefore least likely, as other relaxation events usually occur before phosphorescence does. Structure, solvent, and temperature all affect the likelihood of absorption, fluorescence, and phosphorescence. Temperature effects are discussed below, and the effects of the other factors are described in greater detail in subsequent chapters.

1.4. TEMPERATURE EFFECTS ON GROUND AND EXCITED STATE POPULATIONS

Virtually all spectroscopic techniques that are based on absorbance rely on the fact that the population of molecules in the ground state exceeds that in the excited state. The extent of this population difference depends on the energy gap between the ground and excited states and on the temperature. High temperatures mean that molecules have more energy, and this energy can be used to populate higher energy states. The ratio of molecules in a low energy state to that in a high energy state is given by the Boltzmann distribution:

$$\frac{N^*}{N^\circ} = \frac{P^*}{P^\circ} e^{\frac{-\Delta E}{kT}} \tag{1.8}$$

where N^* and N° represent the number of atoms in a sample in the higher and lower states, respectively; P^* and P° are statistical parameters related to the number of states of equivalent energy for the higher and lower states, respectively; ΔE is the difference in energy between the two states in Joules; T is the sample temperature in Kelvin; and k is the Boltzmann constant, 1.3806×10^{-23} J/K. For atomic transitions such as for an s-orbital → p-orbital excitation, P° is 2, as there are two ways an electron can occupy a single s-orbital (either spin-up or spin-down), and P^* is 6 because there are three p-orbitals, each with spin-up/spin-down possibilities, all of which are equivalent in energy.

Consider the case of a transition centered at 500 nm for molecules at 25 °C, which represents an electronic transition in the visible portion of the spectrum. Using $\Delta E = hc/\lambda$ to convert the wavelength to energy and substituting into Eq. (1.8) yields $N^*/N^\circ = 1.12 \times 10^{-42}$ (assuming $P^* = P^\circ$). Taking the reciprocal of this number (i.e., calculating N°/N^*) shows that in a collection of 8.93×10^{41} molecules, only one is in the higher energy state! In other words, virtually all of the molecules exist in the lower energy state for this transition at this temperature. Raising the temperature to 200 °C yields $N^*/N^\circ = 3.71 \times 10^{-27}$. This is still quite small but does represent a 10^{15}-fold increase in the ratio. So while the probability of a molecule existing in the excited state is still miniscule at 200 °C, the odds are much greater at the higher temperature. Equation (1.8) shows that as the energy gap between the states decreases and as the temperature increases, more molecules have a chance of being in the higher energy state. It is critical to note that ΔE and T are both in an exponential term. Therefore, the ratio N^*/N° is quite sensitive to changes in both parameters. The example problem below and those in the end-of-chapter exercises are designed to help you explore both the temperature and energy dependence of the Boltzmann distribution by comparing the ratios for three common types of spectroscopy: UV-visible, IR, and NMR. To maximize the benefit of these problems, pay extra attention to the magnitude of the ratios you calculate and compare them to one another.

EXAMPLE 1.4

1. Calculate the energy change (ΔE) associated with the electronic transition in benzene at 254 nm.
2. Calculate the ratio of excited state to ground state molecules for the transition in part 1 at 25 °C. Let $P^* = P^\circ$. What does this answer tell you about the state of most molecules at 25 °C?
3. If you want to increase the population difference, should you increase or decrease the temperature of the sample?

Answer:

1. $\Delta E = \dfrac{hc}{\lambda} = \dfrac{\left(6.626 \times 10^{-34} \text{ Js}\right)\left(2.998 \times 10^8 \text{ m/s}\right)}{254 \times 10^{-9} \text{ m}} = 7.82 \times 10^{-19} \text{ J}$

2. $\dfrac{N^*}{N^\circ} = \dfrac{P^*}{P^\circ} e^{\left(\frac{-\Delta E}{kT}\right)} = 1\left(2.718^{\frac{-7.82\times10^{-19} \text{ J}}{\left(1.38\times10^{-23} \text{ J/K}\right)\left(298 \text{ K}\right)}}\right) = 2.61 \times 10^{-83}$

3. Interpretation: The vast majority of molecules are in the ground electronic state (approximately 4×10^{82} in the ground state for every 1 in the excited state – the reciprocal of 2.61×10^{-83}).

4. To increase the population difference, cool the sample so that even fewer molecules gain enough random thermal energy to get into the excited state.

Another question:
Repeat the calculation for the aliphatic ketone IR stretch at 1715 cm^{-1} at 25 °C. Hint: Remember to watch units when converting wavenumbers to Joules.

Answer:

$$\frac{N^*}{N^\circ} = 2.53 \times 10^{-4}.$$

Interpretation: About 1 out of every 3950 molecules is in the excited vibrational state at this temperature, so most are still in the ground state, but this ratio is not nearly as skewed as that calculated above for a UV-visible electronic transition. As above, cooling the sample further increases this difference.

1.5. MORE WAVE CHARACTERISTICS

In the previous section we saw a range of effects that EMR has on atoms and molecules. To take advantage of these effects to measure chemical properties, we need to examine additional characteristics and processes associated with EMR.

1.5.1. Adding Waves Together

An electromagnetic wave propagating through space and time can be described using a sine function as shown in Eq. (1.9):

$$y = A \sin(2\pi \upsilon t + \phi) \tag{1.9}$$

where y is the magnitude of the electric field, A is the maximum amplitude of the electric field, t is time, υ is the frequency of the EMR, and ϕ is the phase angle.

Two or more waves traveling through the same space add together to create one resulting wave as shown in Figure 1.9. When the individual waves reinforce one another and create a wave of greater amplitude, the waves are said to constructively interfere. In contrast, destructive interference occurs when two waves cancel each other. For perfect constructive interference, the two waves must have the same frequency and the same phase as one another. Perfect destructive interference occurs when the two waves have the same frequency and are 180° out of phase, as shown in Figure 1.10.

Adding together any number of waves results in a new wave that has its own unique periodicity or beat pattern as shown in Figure 1.9d. What is critical about this is that the reverse process is also true, meaning that any regularly repeating pattern can be deconvolved into the original component waves. This process, depicted in Figure 1.11, is known as Fourier transformation (FT) and lies at the heart of FTIR and FT-NMR spectroscopy. In these techniques, a single signal is recorded in time, and the Fourier transformation is used to determine the frequencies and intensities of the EMR making up the signal. This information is translated into structural and quantitative information about the molecules in the sample.

1.5.2. Diffraction

Diffraction is a process that occurs when EMR propagating through space encounters obstacles, slits, or small holes that are roughly comparable in size to the wavelength of the radiation. It ultimately results from the interference phenomena described above.

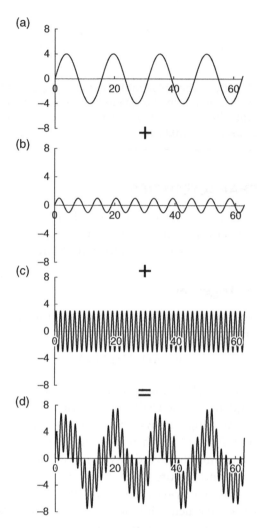

FIGURE 1.9 Three waves (a, b, and c) added together to create a fourth wave (d). Note that each of the waves has a different frequency and amplitude. The amplitudes (y-axis values) at each x-axis value of the first three waves are added to yield the wave in (d). The sum (d) itself has a characteristic pattern. When the process is run in reverse, the single wave in (d) can be deconvolved to yield the frequencies and amplitudes of the three waves that constitute it. This is the basis of Fourier transformation and is depicted in a subsequent figure.

Beams of light that travel different distances can constructively or destructively interfere depending on the exact difference in the path lengths and the wavelengths of the EMR. Consider the example in Figure 1.12 [38, 39] in which two beams of X-rays are reflected off of two atoms separated by a distance, d, in a crystal. Assuming that the X-rays are emitted by the same source, beam 2 travels an extra distance, ABC, compared to beam 1 before striking the detector surface. If the extra distance is a multiple of the wavelength of the light, then constructive interference occurs and the two beams reinforce one another. The requirement for the extra length traveled to be a multiple of the wavelength of the EMR can be likened to two people who start out marching in step with one another, each

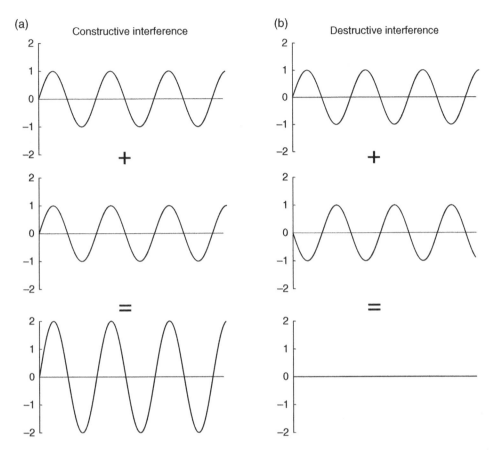

FIGURE 1.10   An illustration of constructive and destructive interference. In both (a) and (b), the top two graphs are added together and create the bottom graph. So the graph on the bottom is the result of adding the y-value of the top two graphs at every point along the x-axis. So in (a), when both curves have a y-value of 1.00 at the same x-value, the result is a y-value of 2.0 at the corresponding x-value. In (b), which represents perfect destructive interference, the value of y in the top curve is always offset by the same value in the negative direction in the second curve. The result is zero amplitude at all x-values.

covering 6 feet of ground for every left/right cycle. If one of the two walks an extra 6 feet, then they are still in step, and the same is true for 12, 18, 24 feet, etc. But if one of them only travels, say, 4 extra feet, then they would no longer be in stride with one another.

Referring back to Figure 1.12, it can be shown using rules of trigonometry that constructive interference occurs when

$$n\lambda = 2d\sin\theta \tag{1.10}$$

where n is any integer (1, 2, 3, etc.), λ is the wavelength of the electromagnetic radiation, d is the spacing between layers of atoms, and θ is the angle of incidence and reflection.

You have likely already observed the effects of diffraction if you have looked at the back of a CD or DVD and seen the rainbow. The tiny pits on the back of CDs and DVDs that encode the information onto the discs diffract light. So when you look at the disc, you

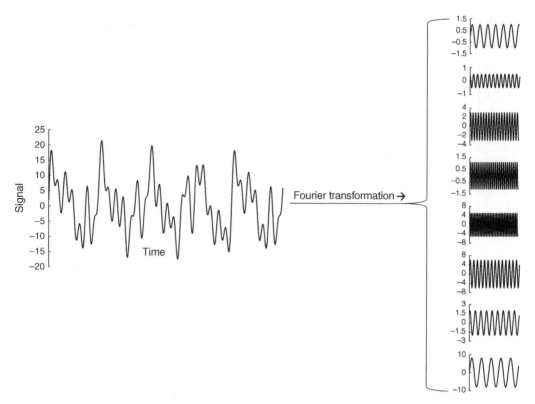

FIGURE 1.11   Depiction of the Fourier transformation during which a single complex signal versus time input (left) gets deconvolved into the constituent frequencies and their associated amplitudes that make up the signal (right). Note that the *y*-axis scales (i.e., amplitudes) are different for the individual waves on the right. The frequency information contained in spectra provides clues about molecular structure (i.e., qualitative information), and the amplitude (i.e., intensity) provides quantitative information related to concentration.

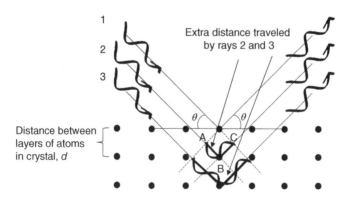

FIGURE 1.12   Schematic of diffraction of electromagnetic radiation from atoms in a crystal. Constructive interference occurs when the extra distance traveled by rays 2 and 3 is a multiple of the wavelength of incoming radiation. Source: Reproduced with kind permission of A. Vantomme, KU Leuven.

(a)   (b)

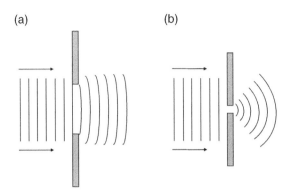

FIGURE 1.13   Bird's-eye view of diffraction. Solid lines represent the crests of waves moving from left to right as they approach and pass through stationary slits. (a) Slits that are much larger than the wavelength of the light cause very little diffraction. (b) Slits that are close to the size of the wavelength passing through them cause significant diffraction.

are seeing the white light that is broken into its constituent wavelengths. Different wavelengths of light (λ) are constructively reinforced at different angles (θ). This same phenomenon of diffraction is also the basis for components in spectrophotometers called diffraction gratings, which we discuss in more detail in Chapter 2.

Diffraction also occurs when EMR passes through a narrow slit or hole that is of the same order of size as the wavelength of the radiation. This causes the radiation to bend and propagate in arcs rather than continuing on unperturbed, as occurs with much wider slits (see Figure 1.13).

Diffraction can be helpful, as in the case of diffraction gratings used in UV-visible spectroscopy and X-ray diffraction techniques. Diffraction can also be deleterious because it sets a lower limit as to how small slits and optical components can be in instruments. It also limits the resolution of images that can be obtained with electromagnetic radiation, a condition called "the Abbe diffraction limit."

We note that while the Abbe diffraction limit seems to dictate the resolution possible using microscopy and thus limit the size of structures than can be resolved, the 2014 Nobel Prize in Chemistry was shared by Eric Betzig, Stefan W. Hell, and William E. Moerner for their work on "the development of super-resolved fluorescence microscopy" [40]. Using techniques they pioneered, it is possible to achieve spatial resolution of structures well below that which the Abbe diffraction limit predicts. A description of their techniques is beyond the scope of this chapter, but explanations of how they overcame the diffraction limit can be found in the citations [41–48].

1.5.3.   Reflection

Reflection occurs when EMR strikes an interface of two materials that have different refractive indices – an air/glass interface, for example (see Figure 1.14). The angle of reflection (θ_{ref}) equals the angle of incidence (θ_{inc}) for a ray that strikes a surface at a given angle relative to the surface normal, as shown in Figure 1.14.

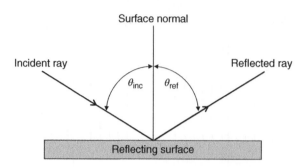

FIGURE 1.14 Reflection of radiation off of a surface. The angle of incidence (θ_{inc}) equals the angle of reflection (θ_{ref}).

For light that strikes an interface at 90° (see Figure 1.15), the fraction of EMR reflected is given by

$$\frac{I_r}{I_o} = \frac{(\eta_1 - \eta_2)^2}{(\eta_1 + \eta_2)^2} \tag{1.11}$$

where I_r is the intensity of the reflected beam, I_o is the intensity of the beam striking the interface, and η_1 and η_2 are the refractive indices of the two materials. The ratio I_r/I_o is also called the reflectivity, R [49]. The fraction of transmitted light (I_t) is given by $1 - I_r/I_o$ assuming no other losses.

The condition of a 90° incident angle is often the situation of interest in spectroscopy – for example, when EMR in a UV-visible spectrophotometer passes through a glass cuvette filled with a sample. In such cases, the loss of EMR due to reflection and hence a loss of signal can be more significant than may first be appreciated. For example,

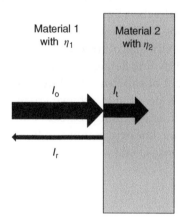

FIGURE 1.15 Diagram of reflection at 90°, which is frequently encountered in spectroscopy, for example, when electromagnetic radiation strikes the surface of a cuvette. I_o is the original intensity of the beam, I_r is the intensity of the reflected beam, and I_t is the intensity of the transmitted beam.

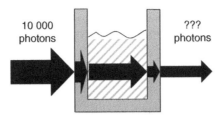

10 000
photons

???
photons

FIGURE 1.16 Diagram of light from a source passing through a quartz cell filled with an aqueous sample.

as shown in Figure 1.16, EMR passing through a filled cuvette encounters four interfaces, with EMR being reflected at each one, such that the intensity being passed on to interface 2 is less than that which strikes interface 1, and that striking interface 4 is less than that at 3, 2, and 1. As you will see when performing the calculation in Example 1.5, the beam that eventually emerges from the cuvette has been reduced in strength by nearly 10%. Keep in mind that in optical systems, many such interfaces can exist with the concomitant reduction in the intensity of the beam. For this reason, many instrument designs try to minimize the number of interfaces in order to maximize the throughput of electromagnetic radiation.

Equation (1.11) shows that the percent of reflected radiation increases as the difference in the refractive indices of the two materials creating the interface increases. It also shows that when the two materials have identical refractive indices (i.e., $\eta_1 = \eta_2$), no EMR is reflected and it all passes through the interface. For this reason, in some applications optical components are bathed in liquids that have refractive indices that match that of glass or the sample to reduce reflections [50, 51]. This is known as refractive index matching and is used in applications where minimizing the loss of EMR, and hence the loss of signal, is required.

EXAMPLE 1.5

1. In a typical UV-visible experiment, an aqueous sample is contained in a quartz cuvette. Assuming 10 000 photons strike the surface of the cuvette, calculate how many actually make it through the cuvette and the sample and emerge out the other side.
2. What percentage is this? Assume that the sample is aqueous, that the light is striking at a 90° angle at each interface, and that the refractive index of air is 1.00. Use the table presented earlier in this chapter to find refractive indices, assuming that the refractive index of an aqueous solution is equal to that of water.

Answer:
The diagram in Figure 1.16 shows that there are four interfaces to consider when answering the question: air/quartz, quartz/water, water/quartz, and quartz/air.

We need to calculate how many photons are reflected, and consequently how many pass through, at each interface, starting at the left air/quartz interface:

$$\frac{I_r}{I_o} = \frac{\left(\eta_{air} - \eta_{quartz}\right)^2}{\left(\eta_{air} + \eta_{quartz}\right)^2} = \frac{\left(1.00 - 1.544\right)^2}{\left(1.00 + 1.544\right)^2} = 0.04573$$

so $\dfrac{I_r}{10000} = 0.04573$, yielding $I_r = 457.3$

So 457 photons are reflected (about 4.5%!), meaning that 9543 are transmitted. At the next interface (quartz/water),

$$I_r = 9543 \times \frac{\left(1.544 - 1.333\right)^2}{\left(1.544 + 1.333\right)^2} \approx 51,$$ so another 51 photons are lost due to reflection at this

interface, meaning that only 9490 photons proceed to the third interface.

Performing similar calculations shows that 51 and 431 photons are reflected at the two remaining interfaces, for a total of 990 photons reflected out of the original 10000 or 9.90%. This shows that even without absorption, the intensity of the light can be significantly diminished due to reflection losses. This becomes particularly detrimental with low intensity sources, potentially causing diminished signal-to-noise ratios.

1.5.4. Refraction

Refraction is the apparent bending of a beam of EMR as it propagates through an interface at an angle. The extent of refraction depends on the angle at which the radiation enters the interface and on the refractive indices of the materials that comprise it. The effect is depicted in Figure 1.17, and the angle to which the beam is bent is given by Snell's law, shown in Eq. (1.12):

$$\frac{\sin\theta_1}{\sin\theta_2} = \frac{\eta_2}{\eta_1} \tag{1.12}$$

where θ_1 and θ_2 are the angles at which the radiation enters and leaves the interface, respectively, both relative to the normal of the interface as shown in the figure and where η_1 and η_2 are the refractive indices of the two materials.

The effect ultimately arises from the fact that the velocity of the EMR is slower in the material with the higher refractive index. The bending of the beam mimics a marching band pivoting to turn a corner [52, 53]. The person on one end must slow down or essentially march in place, while the person on the other end must continue to march rapidly in order to maintain formation while turning the corner. Similarly, the first rays of a beam that enter the higher refractive index material slow down, while the part of the beam that has yet to encounter the interface continues to travel more rapidly until it too encounters the interface and slows down, at which point all parts of the beam are proceeding with the same velocity through the second medium. We commonly observe this effect when objects such as straws or plant stems appear to be bent when placed in a glass or vase filled with water.

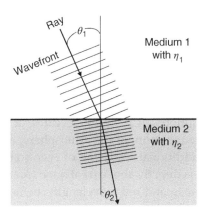

FIGURE 1.17   Refraction of a beam of light as it transitions from one medium to another, such as the transition from air to water. The degree to which the beam is bent is given by Snell's law and depends on the refractive indices of the two media. In this diagram, $\eta_1 < \eta_2$.

1.5.5. Scattering

In addition to the above phenomena of diffraction, reflection, and refraction, when EMR interacts with matter, it can also be scattered. This occurs in several ways, including Rayleigh and Raman scattering and through the Tyndall effect.

1.5.5.1. Rayleigh Scattering. Rayleigh scattering is scattering of EMR due to molecules that are much smaller than the wavelength of radiation being scattered. As EMR interacts with a molecule, like N_2 or O_2, the oscillating electric field of the radiation distorts the electron cloud of the molecule and causes the electrons to oscillate at the same frequency as the radiation. The oscillating electrons then radiate electromagnetic radiation [54]. The wavelength of this scattered radiation is the same as that which induced the electron oscillation. In other words, Rayleigh scattering is an elastic process, meaning that the energy of the scattered photons is the same as that of the incoming photons. The scattering occurs in all directions, but the intensity depends on the angle at which the scattering is being observed.

The intensity of Rayleigh scattering is given by [55]

$$I = \frac{I_o 8\pi^4 \alpha^2 N \left(1 + \cos^2 \theta\right)}{\lambda^4 r^2} \tag{1.13}$$

where I_o is the intensity of the light striking the scatterers, α is the polarizability of the scatterers, N is the number of scatterers per cm^3, θ is the angle of the scattered radiation, r is the distance between the scatterers and the observer, and λ is the wavelength of the scattered radiation. Notice that the intensity of scattering depends on $1/\lambda^4$. Thus, there is a strong dependence on the wavelength of light. This, in fact, is why the sky is blue. The sun essentially emits all wavelengths of visible light. The beams encounter particles that scatter the light as they propagate through space. Blue light, having the smallest wavelengths, is scattered the most, and it is scattered at all angles, so when we look at areas of the sky away from the sun, they appear blue. When we look directly at the sun, it appears yellow because the red, orange, and yellow light is scattered less and therefore reaches our eyes, whereas the blues, greens, and violets are removed by the scattering. One might think that the sun should look red because these wavelengths are least scattered, but there are visual perception and intensity issues regarding the sun's output at different wavelengths that ultimately make it appear yellow during the day. The sun does appear to be orange or red at sunrise and sunset, especially when there are a lot of particles in the sky from forest fires, volcanic eruptions, or pollution, because at sunrise and sunset the light from the sun must travel through more of the atmosphere, and thus it encounters more scattering particles, causing additional scattering of even some of the orange and yellow light relative to when the sun is directly above us at noon (see Figure 1.18). This additional scattering leaves only the orange and red wavelengths behind when we look directly at the sun (i.e., $\theta = 0$).

Rayleigh scattering is also responsible for the line in *America the Beautiful* about "purple mountain majesty." When we look at distant objects, like mountains, our eyes see the blue scattered radiation from the sky, which veils the object. The further away the object is, the more blue radiation is between us and the object, so distant mountains appear bluish or purplish. Artists are aware of this fact and paint objects that they want to seem far away in their scenes with a blue or purple hue – a technique known as "atmospheric perspective" or "aerial perspective." Paintings such as Leonardo da Vinci's *Mona Lisa* and Albert Bierstadt's *Looking up the Yosemite Valley* are examples of works where atmospheric perspective is used.

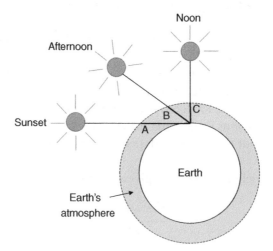

FIGURE 1.18 Rayleigh scattering creates red sunsets because the light from the sun has to travel further through the atmosphere at sunset than at noon, as seen by comparing the distance labeled "A" (sunset) to the one labeled "C" (noon). At noon, the light passes through the least amount of atmosphere and thus encounters the fewest number of scattering particles. In this case, it is mostly the short wavelength blue light that is scattered. Light with intermediate wavelengths associated with orange, yellow, and green are not scattered much and the result is that the sun looks yellow when we look directly at it, while the sky looks blue due to it having been scattered at all angles by the atmosphere. The light at sunset travels through more atmosphere and thus encounters more scattering particles. Because of the extended path, yellows and greens also get scattered to some extent, creating the reddish orange colors in the sky around the sun that we are accustomed to seeing at sunset.

1.5.5.2. Tyndall Effect/Mie Scattering.

The Tyndall effect refers to scattering that occurs off of larger particles (approximately 10% of the wavelength of light striking the particle). A common instance of this is a red laser beam propagating through a beaker of water. If we try to observe the beam from above the beaker (i.e., bird's-eye view) and the water is completely free of particles, we do not see it because none of the light is reflected toward our eyes. When colloidal particles are added to the water (imagine very dilute milk that has globules of protein and fat in it), the beam can be seen going through the solution. We are really seeing the part of the beam that has been scattered by the particles. The amount of radiation scattered varies with the size of the scatterers. Thus, this kind of scattering, which is also known as Mie scattering [56], is used to measure the size of polymers, large biomolecules, and colloids. It is also responsible for the fact that clouds appear white due to the scattering of radiation off of water droplets. As in Rayleigh scattering, the frequency of the scattered radiation is the same as that which originally interacted with the particles. Only its direction of propagation has changed.

1.5.5.3. Raman Scattering.

In Raman scattering, unlike Rayleigh and Mie scattering, the frequency of the scattered radiation is different than that of the incoming radiation. This effect arises when a molecule in a given vibrational state (say, v_0) temporarily absorbs a photon and subsequently reemits it at a different frequency as it relaxes to a different vibrational level than where it started. Two possibilities are depicted in Figure 1.19. In one, the molecule ultimately ends up in a higher vibrational level by emitting a photon of lower energy (longer wavelength) than the original photon. This is called a Stokes shift. In the other, the molecule ends up in a lower vibrational level by emitting a higher energy (shorter wavelength) photon than the original. This is called an anti-Stokes shift. The first situation is much more probable as more molecules are in the lowest vibrational level when the radiation encounters them as was shown in the example problems related to the Boltzmann distribution. Raman spectroscopy has grown to be an important analytical technique and is discussed in much more detail in Chapter 5.

1.5.6. Polarized Radiation

Thus far we have considered nonpolarized electromagnetic radiation, or radiation in which the electric fields of the waves are tilted along any plane centered on the beam path. Figure 1.20 [57] shows four such beams, though more waves at other angles are certainly possible. If we take the perspective of viewing the electric field vectors of the beam coming directly at us, it can be depicted as shown in Figure 1.21a, where again only a limited number of all the possible angles are shown. For any individual wave, however, the electric field vector at any moment can be deconvolved into x- and y-components as shown in Figure 1.21b.

To create polarized radiation, a beam of nonpolarized radiation is passed through a crystal or polarizing filter that selectively removes one component of the radiation. If the x-component is selectively removed, then the beam that emerges from the material is polarized along the y-axis. In the case of complete removal of one component over the other, the radiation is referred to as linearly polarized or plane polarized radiation because the resulting electric component oscillates in a single plane.

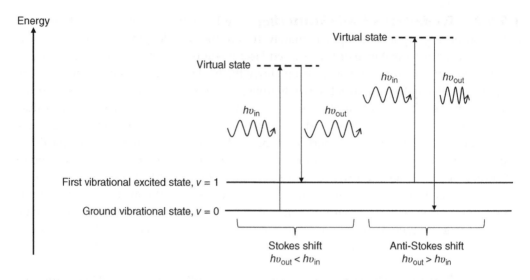

FIGURE 1.19 Depiction of Raman scattering. As with the Perrin–Jablonski diagram, the increasing energy scale from bottom to top is important. Raman scattering involves the temporary uptake of a photon with some energy, and the subsequent reemission of a photon or either lower (Stokes) or higher (anti-Stokes) energy. The former leaves the molecule in a higher vibrational state than it was in prior to absorbing the incoming radiation, and the latter results in the molecule being in a lower vibrational state after emission. In Raman scattering, unlike Rayleigh scattering, the energy of the emitted radiation is different than that of the incoming radiation, and therefore also has a different frequency and wavelength because $E = h\upsilon = hc/\lambda$.

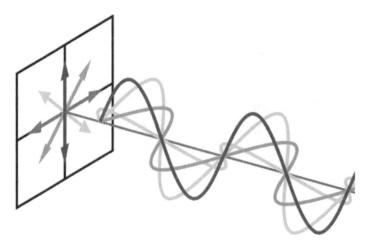

FIGURE 1.20 An illustration of multiple electromagnetic waves propagating through space, each oriented at a different angle relative to one another. Unpolarized light such as this has electric and magnetic components in planes at all angles. Source: Adapted from https://en.wikipedia.org/wiki/File:Wire-grid-polarizer.svg under GNU Free Documentation License. Created by B. Mellish for the English Wikipedia.

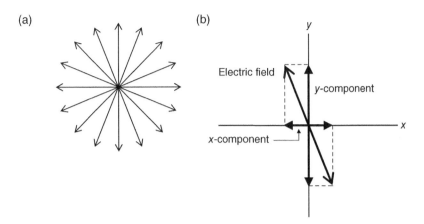

FIGURE 1.21 (a) A view of a few of the electric field components of unpolarized radiation propagating directly at the viewer. Note that the distribution would be random in all directions around the axis of propagation and varying in amplitude over time for each individual wave. (b) An individual electric field vector broken into x- and y-components. The same deconstruction into x- and y-components can be done for all of the waves in part (a).

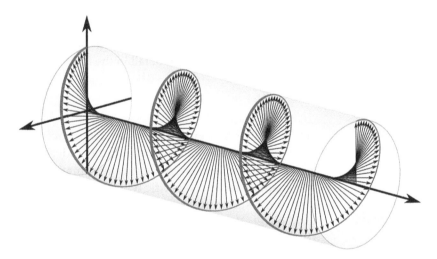

FIGURE 1.22 Depiction of circularly polarized light. The electric field vector (arrows) rotates around the axis as it propagates through space in the direction indicated by the arrow on the right. Source: https://en.wikipedia.org/wiki/Circular_polarization. Licenced under CC BY SA 3.0.

Circularly polarized radiation can also be created. In this case, the resultant electric field vector continuously rotates clockwise or counterclockwise as depicted in Figure 1.22 such that when viewed end on as if the beam were propagating straight at you, the resultant electric field vector appears to rotate in a circle as it comes toward you [58]. Some very helpful diagrams, animations, and videos can be found by searching the terms "linearly polarized light" and "circularly polarized light" on the Internet.

Plane polarized and circularly polarized EMR are used in many areas of spectroscopy. For example, the r- and s-forms of stereoisomers absorb left and right circularly polarized radiation differently from one another, so circularly polarized light can be used to measure the enantiomeric excess of one form of a molecule over another using a technique known as circular dichroism spectroscopy. Secondary structures like alpha helices and beta sheets in biomolecules like proteins and nucleic acids also impart a differential absorption of left- and right-handed circularly polarized radiation. Therefore, circular dichroism spectroscopy is also used to study the structures of biomolecules and how they change with temperature, pH, and other environmental variables.

1.6. SPECTROSCOPY APPLICATIONS

Spectroscopy is one of the most universal and important analytical methods, and as we have indicated throughout this chapter, a wide range of applications exist. For example, NMR and IR spectroscopy are used to determine the structures of molecules. They can also be used in a quantitative manner to determine the concentration of specific substances in mixtures. UV-visible spectroscopy is also commonly used for quantitative analyses. Techniques based on X-ray spectroscopy are used to determine crystal structures and the elemental composition of a wide range of substances. In all of these instrumental methods of analysis, the properties of electromagnetic radiation, how it propagates through space, how it interacts with matter, and the effects it has on matter are important. Specific examples of spectroscopic applications and how they take advantage of the interaction of electromagnetic radiation with matter are described in detail in the following chapters.

1.7. SUMMARY

In this chapter, we focused on the fact that molecules can absorb, reflect, refract, polarize, and scatter electromagnetic radiation. Measurements of these effects underlie a variety of spectroscopic techniques that yield qualitative and quantitative information about atoms and molecules. The next several chapters focus on specific types of spectroscopy, because each different wavelength range of the spectrum dictates instrument design, data collection methods, and the analysis of results that are particular to each mode.

PROBLEMS

1.1 (a) When is it most helpful to think of electromagnetic radiation as a wave?
  (b) As a particle?

1.2 (a) Calculate the energy of a single photon with a frequency of 1.73×10^{17} Hz.
  (b) Calculate the wavelength and wavenumber of the photon in part (a).
  (c) What physical effects could such a photon cause in a molecule or atom?
  (d) Calculate the energy of a mole of these photons.

1.3 Calculate the wavelength of photons having 100.0 kcal/mol of energy (the approximate energy of a mole of C—H bonds). Watch units carefully. What region of the electromagnetic spectrum is this in?

1.4 Consider two photons, one in the visible region of the spectrum with a 700.0 nm wavelength corresponding to red light and the other in the microwave region with a wavelength of 2.000×10^{-2} m. How many times more energetic is the red photon compared to the microwave photon?

1.5 The speed of 516.6 nm light (green light) in germanium is 6.503×10^7 m/s (http://www.filmetrics.com/refractive-index-database/Ge/Germanium). What is the refractive index of germanium at this wavelength?

1.6 Calculate the refractive index of a Bose–Einstein condensate in which the speed of light was found to be 38 miles per hour.

1.7 Why do molecules typically fluoresce from the lowest energy vibrational state of the lowest energy excited state and not from higher energy states?

1.8 What is the difference between fluorescence and phosphorescence?

1.9 (a) NMR spectroscopy probes the difference between spin-up (lower energy) and spin-down states (higher energy) of nuclei in a magnetic field. Common NMR spectrometers operate around 400.0 MHz. Convert this frequency into an energy and then calculate the ratio of spin-up to spin-down states using the Boltzmann distribution equation at $25\,^\circ$C. Let $P^* = P^\circ$.

    (b) What does your answer tell you about the population difference of these two states?

1.10 Complete the table below. For simplicity, assume $P^* = P^\circ$.

Type	Typical radiation	Temperature (°C)	$\dfrac{N^*}{N^\circ}$	$\dfrac{\left(N^*/N^\circ\right)_{300}}{\left(N^*/N^\circ\right)_{25}}$
UV-visible	400 nm	25		
		300		
IR	1600 cm^{-1}	25		
		300		
NMR	400 MHz	25		
		300		

1.11 This problem is designed to use the Boltzmann distribution to provide another way to compare the relative magnitudes of energy in the various regions of the spectrum. This will be done through a critical comparison of the UV-vis, IR, and NMR regions by calculating Boltzmann distribution ratios: (a) Calculate $(N^*/N^\circ)_{NMR}/(N^*/N^\circ)_{UV\text{-}vis}$ (i.e., calculate the ratio of the ratios) at $25\,^\circ$C using values you calculated in Problem 1.10, (b) Calculate $(N^*/N^\circ)_{NMR}/(N^*/N^\circ)_{IR}$ at $25\,^\circ$C using values you calculated in Problem 1.10. (c) Why is the ratio of N^*/N° so much larger for NMR than it is for UV-vis and IR spectroscopy?

1.12 (a) What wavelength of light produces constructive interference at a 30° angle when impinging on atoms separated by 1.5 Å (a typical C—C bond distance) assuming first order diffraction?

    (b) At a 45° angle?

    (c) What region of the spectrum are these in?

1.13 What percent of electromagnetic radiation is reflected at a benzene/diamond interface assuming that the EMR is striking the interface at a 90° angle?

1.14   Which interface will produce more reflected light, water/heptane or water/benzene?

1.15   Consider radiation entering a water/diiodomethane interface at a 10° angle normal to the interface. At what angle will the radiation leave the interface?

1.16   What type of scattering occurs off of molecules that are much smaller than the wavelength of radiation being scattered?

1.17   What kind of scattering is used to measure the size of polymers and larger molecular clusters?

REFERENCES

1. Willard, H.H., Merritt, L.L. Jr., Dean, J.A., and Settle, F.A. Jr. (1988). *Instrumental Methods of Analysis*, 7e, 99. Belmont: Wadsworth Publishing Co.

2. Williamson, S.J. and Cummins, H.Z. (1983). *Light and Color in Nature and Art*, 180. New York: Wiley.

3. Pavia, D.L., Lampman, G.M., Kriz, G.S., and Engle, R.G. (2005). *Introduction to Organic Laboratory Techniques: A Small Scale Approach*, 2e. Salt Lake City: Brooks Cole Publishing Co.

4. Choi, M., Lee, S.H., Kim, Y. et al. (2011). *Nature 470*: 369–373.

5. http://www.engineeringtoolbox.com/refractive-index-d_1264.html (accessed 2 August 2017).

6. Haynes, W.M. (ed.-in-chief); Lide, D.R. and Bruno, T.J. (assoc. eds.) (2014). *CRC Handbook of Chemistry and Physics*, 95e. Boca Raton: CRC Press.

7. Riddick, J.A., Bunger, W.B., and Sakano, T.K. (1986). *Organic Solvents: Physical Properties and Methods of Purification*, Techniques of Chemistry, 4e, vol. *II*. New York: Wiley.

8. http://news.harvard.edu/gazette/1999/02.18/light.html (accessed 2 August 2017).

9. Pavia, D.L., Lampman, G.M., and Kriz, G.S. (2001). *Introduction to Spectroscopy*, 3e, 360. Pacific Grove: Brooks/Cole.

10. International Standard (2007). Space environment (natural and artificial) – process for determining solar irradiances. ISO 21348.

11. http://www.spacewx.com/pdf/SET_21348_2004.pdf (accessed 3 August 2017).

12. https://science.nasa.gov/ems/05_radiowaves (accessed 3 August 2017).

13. http://www.livescience.com/50399-radio-waves.html (accessed 3 August 2017).

14. Williams, P. and Norris, K. (eds.) (2001). *Near-infrared Technology: In the Agricultural and Food Industries*, 2e. St. Paul: American Association of Cereal Chemists.

15. Nielsen, S.S. (2010). *Food Analysis*, 4e, 308. New York: Springer.

16. http://www.npl.co.uk/upload/pdf/091217_terahertz_naftaly.pdf (accessed 3 August 2017).

17. Dexheimer, S.L. (2008). *Terahertz Spectroscopy: Principles and Applications*. Boca Raton: CRC Press.

18. Schmuttenmaer, C.A. http://www.tstnetwork.org/March2008/1aaaSchmuttenmaer_THz_Nano,.pdf (accessed 3 August 2017).

19. Baker, C., Lo, T., Tribe, W.R. et al. (2007). *Proc. IEEE 95*: 1559–1565.

20. http://www.clinuvel.com/en/skin-science/skin-sun/skin-cancer/uv-damage-and-carcinogenesis (accessed 3 August 2017).

21. http://www.oasisllc.com/abgx/radioactivity.htm (accessed 3 August 2017).

22. https://sciencedemonstrations.fas.harvard.edu/presentations/%CE%B1-%CE%B2-%CE%B3-penetration-and-shielding (accessed 3 August 2017).

23. https://www.bcm.edu/bodycomplab/Radprimer/radpenetration.htm (accessed 3 August 2017).

24. Klinkermann, J. (2013). *Conscientious Science*. Bloomington: AuthorHouse, LLC.

25. http://www.world-nuclear.org/information-library/safety-and-security/safety-of-plants/three-mile-island-accident.aspx (accessed 3 August 2017).

26. https://www.nei.org/Master-Document-Folder/Backgrounders/Fact-Sheets/The-TMI-2-Accident-Its-Impact-Its-Lessons (accessed 3 August 2017).

27. http://www.who.int/mediacentre/news/releases/2013/fukushima_report_20130228/en (accessed 3 August 2017).

28. Valeur, B. and Berberan-Santos, M.N. (2013). *Molecular Fluorescence: Principles and Applications*, 2e, 54. Weinheim: Wiley-VCH.

29. Valeur, B. (2002). *Molecular Fluorescence, Principles and Applications*, 35. Weinheim: Wiley-VCH.

30. http://photobiology.info/HistJablonski.html (accessed 4 August 2017).

31. Perrin, F. (1929). *Ann. Phys. (Paris) 12*: 169–275.

32. Perrin, J. (1922). *Trans. Faraday Soc. 17*: 546–572.

33. Jablonski, A. (1933). *Nature 131*: 839–840.

34. Jablonski, A.Z. (1935). *Physik 94*: 38–46.

35. Hercules, D. (ed.) (1966). *Fluorescence and Phosphorescence Analysis: Principles and Applications*, 18. New York: Wiley.

36. IUPAC Gold Book. https://goldbook.iupac.org/html/F/F02510.html (accessed 3 August 2017).

37. Born, M. and Oppenheimer, J.R. (1927). *Ann. Phys. 389*: 457–484.

38. http://www.veqter.co.uk/residual-stress-measurement/x-ray-diffraction (accessed 4 August 2017).

39. https://fys.kuleuven.be/iks/nvsf/experimental-facilities/x-ray-diffraction-2013-bruker-d8-discover (accessed 4 August 2017).

40. https://www.nobelprize.org/nobel_prizes/chemistry/laureates/2014/advanced-chemistryprize2014.pdf (accessed 4 August 2017).

41. Van Noorden, R. (2014). *Nature 514*: 286.

42. Hell, S.W. and Wichman, J. (1994). *Opt. Lett. 19*: 780–782.

43. Hell, S.W. and Kroug, M. (1995). *Appl. Phys. B. 60*: 495–497.

44. Klar, T.A., Jakobs, S., Dyba, M. et al. (2000). *Proc. Natl. Acad. Sci. 97*: 8206–8210.

45. Betzig, E. (1995). *Opt. Lett. 20*: 237–239.

46. Betzig, E., Patterson, G.H., Sougrat, R. et al. (2006). *Science 313*: 1642–1645.

47. Moerner, W.E. and Kador, L. (1989). *Phys. Rev. Lett. 62*: 2535–2538.

48. Sahl, S.J. and Moerner, W.E. (2013). *Curr. Opin. Struc. Biol. 23*: 778–787.

49. Strobel, H.A. and Heineman, W.R. (1989). *Chemical Instrumentation: A Systematic Approach*, 3e, 205. New York: Wiley.

50. Barcelo, D. (2005). *Comprehensive Analytical Chemistry, volume 45, Analysis and Detection by Capillary Electrophoresis* (ed. M.L. Marina, A. Rios and M. Valcarcel), 343. Amsterdam: Elsevier.

51. Poole, C.F. (2003). *The Essence of Chromatography*, 699. Amsterdam: Elsevier.

52. Donnelly, J. and Massa, N. (2007). *Light: Introduction to Optics and Photonics*. Boston: The New England Board of Higher Education.

53. DeValois, K.K. (2000). *Seeing*. New York: Academic Press.

54. Miles, R.B., Lampert, W.R., and Forkey, J.N. (2001). *Meas. Sci. Tech.* *12*: R33–R51.

55. Perkampus, H.H. (1993). *Encyclopedia of Spectroscopy* (trans. H.C. Grinter and R. Grinter), 508. Weinheim: Wiley-VCH.

56. Parker, S.P. (ed.-in-chief) (1987). *Optics Source Book*, 223–224. New York: McGraw-Hill Book Company.

57. https://en.wikipedia.org/wiki/File:Wire-grid-polarizer.svg (accessed 3 August 2017).

58. https://en.wikipedia.org/wiki/Circular_polarization (accessed 3 August 2017).

FURTHER READING

1. Avram, M. and Mateescu, G.D. (1970). *Infrared Spectroscopy: Applications in Organic Chemistry* (trans. L. Birladeanu). New York: Wiley.

2. Colthrup, N.B., Daly, L.H., and Wiberley, S.E. (1990). *Introduction to Infrared and Raman Spectroscopy*, 3e. San Diego: Academic Press, Inc.

3. Hercules, D. (ed.) (1966). *Fluorescence and Phosphorescence Analysis: Principles and Applications.* New York: Wiley.

4. Hollas, J.M. (2004). *Modern Spectroscopy*, 4e. Chichester: Wiley.

5. Ingle, J.D. Jr. and Crouch, S.R. (1988). *Spectrochemical Analysis.* Englewood Cliffs: Prentice Hall, Inc.

6. Jaffe, H.H. and Orchin, M. (1962). *Theory and Applications of Ultraviolet Spectroscopy.* New York: Wiley.

7. Lakowicz, J.R. (2006). *Principles of Fluorescence Spectroscopy*, 3e. New York: Springer.

8. Parker, S.P. (ed.-in-chief) (1987). *Optics Source Book.* New York: McGraw-Hill Book Company.

9. Pavia, D.L., Lampman, G.M., and Kriz, G.S. (2008). *Introduction to Spectroscopy*, 4e. Pacific Grove: Brooks/Cole.

10. Perkampus, H.H. (1993). *Encyclopedia of Spectroscopy* (trans. H.C. Grinter and R. Grinter). Weinheim: Wiley-VCH.

11. Rao, C.N.R. (1967). *Ultra-Violet and Visible Spectroscopy: Chemical Applications.* New York: Plenum Press.

12. Smith, E. and Dent, G. (2005). *Modern Raman Spectroscopy: A Practical Approach.* Chichester: Wiley.

13. Tasumi, M. and Sakamoto, A. (eds.) (2015). *Introduction to Experimental Infrared Spectroscopy: Fundamentals and Practical Methods.* Chichester: Wiley.

14. Valeur, B. and Berberan-Santos, M.N. (2013). *Molecular Fluorescence: Principles and Applications*, 2e. Weinheim: Wiley-VCH.

15. Williamson, S.J. and Cummins, H.Z. (1983). *Light and Color in Nature and Art.* New York: Wiley.

UV-VISIBLE SPECTROPHOTOMETRY

2

Of all of the different instrumental techniques, ultraviolet–visible (UV-visible) spectroscopy is among the easiest to understand and execute with reasonable reliability. The relative simplicity of the principles and measurements makes it a useful tool in scientific laboratories and throughout society. It can be used to determine the hops content of beer, the nitrate levels in groundwater, and the concentration of pesticides in municipal wastewater.

You have probably already made a number of UV-visible absorbance measurements – perhaps to create a calibration curve to measure the concentration of a specific analyte or to record the absorbance spectrum of a molecule to better understand quantum theory. So you are likely familiar with how to measure absorbance values.

In this chapter we stress that every SINGLE absorbance measurement is actually the result of TWO analytical measurements. To be explicit, absorbance is determined by comparing the intensity of light that passes through a sample that contains the analyte of interest (i.e., the "sample") and one that does not (i.e., a "reference" or "blank"). A consideration of the definition of absorbance and the Beer–Lambert law will show how both of these measurements are combined to yield a single absorbance value. It is thus important to remember throughout this chapter, and certainly whenever making any spectrophotometric measurement, that any measured absorbance value is only as good as each of the two measurements comprising it – *an error in one results in an error in the measured absorbance*.

In this chapter, we also examine a number of ways in which UV-visible spectrophotometers are designed to acquire the two different measurements that contribute to absorbance readings. In most systems, all that is displayed as output data is the overall absorbance value, rather than independent measurements on the reference and the sample. Thus, we must diligently remember that absorbance readings from such instruments are always the result of two measurements.

Spectroscopy: Principles and Instrumentation, First Edition. Mark F. Vitha.
© 2019 John Wiley & Sons, Inc. Published 2019 by John Wiley & Sons, Inc.

2.1.   THEORY

2.1.1.   The Absorption Process

As was shown in Chapter 1, the amount of energy associated with a photon of light corresponds to its wavelength and frequency according to the Planck relationship:

$$E = h\upsilon = \frac{hc}{\lambda} \tag{2.1}$$

in which h is Planck's constant (6.626×10^{-34} Js) and c (2.998×10^8 m/s) is the speed of light in a vacuum. The energy of a photon is calculated in joules, given these units. Molecules absorb UV-visible electromagnetic radiation (light) when the energy of the light matches an energy gap in the molecule. Energy gaps are the result of different electronic arrangements within the molecule. The arrangements are the result of electrons being distributed into different molecular orbitals. The different arrangements and their corresponding energy levels are often referred to as the "ground state" (i.e., the lowest energy electron distribution and the one in which the molecule usually exists) and "excited states." Under typical conditions, essentially all but the smallest fraction of molecules exist in the ground state. Excited state distributions are higher in energy than the ground state distribution and are achieved by imparting energy to molecules, often in the form of electromagnetic radiation. Multiple excited state distributions can exist for one molecule. Differences between the energy of the ground state and that of any of the excited states are referred to as energy gaps. These gaps are quantized (i.e., of a finite and fixed value – Figure 2.1), but the exact energy difference depends on the structure of the molecule being investigated. For many organic molecules, particularly aromatic or highly conjugated species, the energy

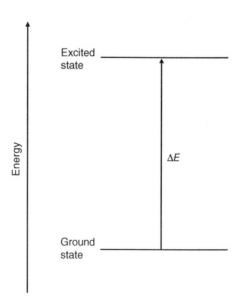

FIGURE 2.1   A line diagram depicting the energy gap, ΔE, between ground and excited state molecular orbitals.

gaps between ground and excited state electron distributions correspond to the energy of electromagnetic radiation in the UV-visible portion of the spectrum (i.e., approximately 200–800 nm). In order for a molecule to absorb light, the energy of the light must match a specific, quantized energy gap. In other words, not all wavelengths of light will be absorbed equally by a specific molecule. Furthermore, because structurally different molecules have different energy gaps dictated by their different electronic arrangements, not all molecules absorb the same wavelengths of light equally well. In this way, the absorption profile, or spectrum, of a molecule can be used to help identify it.

When light of the proper wavelength impinges on a molecule that can absorb it, an electron from the ground state is "promoted" to, or "transitions" to, a higher energy state. This causes the electron distribution to change. In UV-visible spectroscopy, these transitions often involve the promotion of lone pair electrons (also referred to as n-electrons) or electrons found in double bonds (i.e., π-electrons) into antibonding molecular orbitals (e.g., π* orbitals). Such transitions are referred to as n → π* and π → π* transitions, respectively (see Figures 2.2 and 2.3) [1–4]. In saturated molecules containing heteroatoms with lone electron pairs – such as O, N, S, and Br – other transitions such as n → σ* are theoretically possible. But these transitions are generally higher in energy and therefore require wavelengths of light that exceed the energy of 200–800 nm electromagnetic radiation – the range typically associated with UV-visible spectroscopy. For any given molecule, the lowest energy electronic transition occurs when an electron in the *h*ighest *o*ccupied *m*olecular *o*rbital (highest in terms of energy) absorbs a photon and is promoted to the *l*owest *u*noccupied *m*olecular *o*rbital. These transitions are referred to as HOMO/LUMO transitions.

Because these transitions are quantized, in theory they should be exceedingly sharp and correspond to a narrow range of wavelengths. However, within any electronic state, the actual energy can be slightly altered by the numerous vibrational and rotational energy levels that also exist. This results in multiple transitions of slightly different energy layered on top of the one main electronic transition (e.g., π → π*). These transitions all have slightly different energies and thus require slightly different wavelengths of light to be absorbed. The result is that a broad range of wavelengths, rather than just a discrete wavelength or two, are absorbed by molecules. Furthermore, UV-visible spectroscopy is often performed on solutions, meaning that the analyte molecules are surrounded by solvent molecules. Collisions with the solvent molecules and with the container walls cause small variations in the ground and excited state energies of the analytes. This, in turn, means

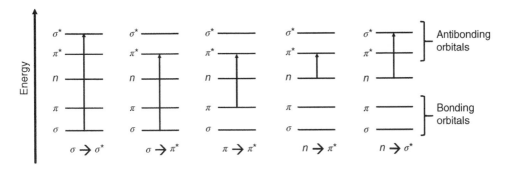

FIGURE 2.2   Energy diagram showing transitions between σ, n, π, π*, and σ* orbitals.

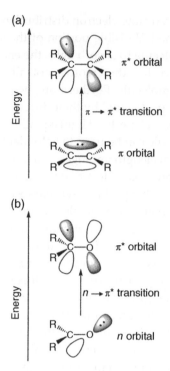

FIGURE 2.3   Representations of the change in electron distribution within orbitals associated with (a) π → π* (bonding to antibonding) and (b) n → π* (nonbonding to antibonding) electronic transitions.

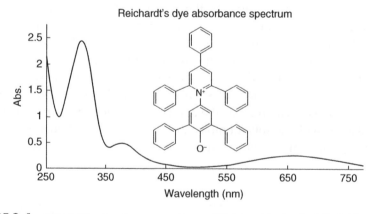

FIGURE 2.4   UV-visible absorbance spectrum of Reichardt's dye in 1,2-dichloroethane.

that a wider range of wavelengths are absorbed in solution compared to the range absorbed by the same analyte in the gas phase [5, 6]. It can also be the case that transitions from the ground state to two different excited states partially overlap in energy, meaning that the resulting spectrum shows a combination of two or more electronic transitions. For all of these reasons, UV-visible spectra tend to have very broad absorbance bands, such as those shown in Figure 2.4. The absorbance bands are still centered on a main electronic transition, but other factors clearly broaden these spectral features.

Figure 2.4 also illustrates that the same molecule absorbs different wavelengths of light to different extents. To be more explicit, the spectrum in Figure 2.4 was recorded with a fixed concentration of the molecule known as Reichardt's dye. So even though the concentration remained the same, the amount of light absorbed at each wavelength was different. Thus, there must be something about the molecular structure that makes it likely that some wavelengths are absorbed more than others.

Absorption can occur when a photon collides with a molecule. Each molecule can be thought of as having a cross-sectional area for photon capture, and the photons must pass within this area in order to be absorbed. The cross-sectional area varies with wavelength and represents, in effect, the probability that photons are captured by any given molecule. The closer the energy (i.e., wavelength) of the light is to the energy of the most probable electronic transition within a molecule, the greater the probability that the photon gets absorbed. If the energy of the light does not match the energy of an electronic transition, the light passes through the solution without being absorbed. This is observed in Figure 2.4 around 500 nm and also in the long-wavelength (low energy) portion of the spectrum around 800 nm. So even for a fixed molecule at a fixed concentration, the *absorbance is not constant, but rather depends on the wavelength of light.* Some wavelengths yield high absorbance values (e.g., λ = 305 nm in Figure 2.4), and others yield low, or zero, absorbance (e.g., λ = 500 and 800 nm). *Therefore, whenever we talk about the absorbance of a solution, we must specify the wavelength at which that absorbance was obtained.* This fact also has important implications for quantitative analyses based on the Beer–Lambert law discussed below. Later in the chapter we discuss how wavelengths are chosen for analytical work and show how different instruments are designed to physically separate UV-visible light into its component wavelengths.

EXAMPLE *2.1*

What is the lowest energy transition possible for a carbonyl group?

Answer:
n $\rightarrow$ π^*; a nonbonding electron from the oxygen lone pairs is promoted to the unoccupied π^* orbital (see Figure 2.3).

2.1.2. The Beer–Lambert Law

In this section, we examine what happens when a beam of light is directed into a solution. Such an examination is important because it:

1. Explains the wavelength dependence of absorption.
2. Shows why it is necessary to make two separate measurements to obtain one absorbance value.
3. Reveals a logarithmic relationship that underpins absorbance measurements, which has critical implications for the analytical reliability of high and low absorbance values.

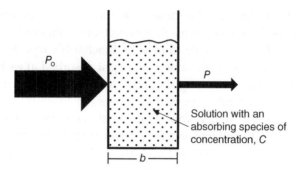

FIGURE 2.5 Schematic diagram of the change in light intensity due to absorption. P_o is the power (intensity) of the light incident on a cell, with path length, b, containing a solution of an absorbing species at a concentration, C. P is the light intensity after absorption and scattering (not depicted in this diagram) occurs.

A simple schematic of the result of light absorption by a sample is shown in Figure 2.5. A light beam with an initial power, P_o, impinges on a sample containing a fixed concentration of an absorbing species. Some of the light is absorbed, such that the power, P, of the beam as it exits the sample is reduced. The Beer–Lambert law shows that the change in the power of the beam is related to the concentration and structure of the absorbing species, as well as the length of the sample through which the beam travels, also called the path length.

The derivation of the Beer–Lambert law begins by considering a beam of *monochromatic* radiation of power, P, (power is energy per unit time striking a given detector area) traveling through an infinitely small distance, dx, of an absorbing medium [7]. The decrease in power due to absorption, $-dP$, is proportional to the thickness of the distance traveled, as shown in the following equation:

$$-dP = kPdx \tag{2.2}$$

where k is a proportionality constant. The proportionality of the decrease in power and thickness assumes that the absorbing species is dispersed equally throughout the distance being considered (i.e., that the concentration of the species is constant).

Rearrangement of the equation yields

$$\frac{-dP}{P} = kdx \tag{2.3}$$

This equation shows that the fraction of light absorbed over a short distance, dx, is proportional to that distance, meaning that if the distance is doubled, the fraction of light that is lost is also doubled.

Integrating both sides over a finite distance, b, and defining the initial power of the beam as P_0 and the power of the beam emerging after traversing the distance b as P, yields

$$-\int_{P_o}^{P} d\ln P = k\int_{0}^{b} dx \tag{2.4}$$

The result is

$$\ln P_0 - \ln P = \ln\left(\frac{P_0}{P}\right) = kb \tag{2.5}$$

This is known as Lambert's law and states that for parallel monochromatic radiation passing through an absorbing medium of constant concentration, the radiant power decreases logarithmically as the path length, b, increases arithmetically [8]. This can be understood in simple terms by considering four consecutive "slices" of a solution of equal thickness. If each slice absorbs half the radiant power that impinges on it, then as the beam exits the first slice, it is reduced to half its original power. The next slice reduces that half in half again, as do the next two slices. In other words,

$$P = \left(P_0 \times \tfrac{1}{2}\right) \times \tfrac{1}{2}\right) \times \tfrac{1}{2}\right) \times \tfrac{1}{2}\right)\ldots$$

First slice
Second slice
Third slice
Fourth slice

This leads to the logarithmic, rather than linear, relationship between P_0/P and the total path length the light travels, b.

The dependence of P_0/P on the concentration of the absorbing species can be determined in a similar manner if the wavelength is again held constant (i.e., monochromatic light), but now also holding the path length fixed. In this case, the number of absorbing species that collide with photons is proportional to the concentration, C. Thus,

$$-dP = k'PdC \tag{2.6}$$

where k' is a proportionality constant. Following similar manipulation as in the equations above and integrating from $C = 0$ to $C = C$ yields

$$\ln\left(\frac{P_0}{P}\right) = k'C \tag{2.7}$$

This is Beer's law [8]. In their most convenient forms, Beer's law and Lambert's law are combined into one equation:

$$\log\left(\frac{P_0}{P}\right) = abC \tag{2.8}$$

where "a" is a proportionality constant that incorporates the constants k and k' and a factor of 2.303 introduced by the conversion from the natural logarithm to the base-10 logarithm. This relationship is the Beer–Lambert law, but is often simply referred to as Beer's law. The equation is fundamentally important in that it shows the logarithmic relationship between the ratio, P_o/P, and the path length, b, and concentration, C. Thus, a plot of $\log(P_o/P)$ versus either path length or concentration is linear, assuming all else is held constant. In chemical analyses, we are typically interested in determining the concentration of a species; thus, it is the relationship between P_o/P and C that is most important, but the dependence on path length has implications in how we conduct spectroscopic measurements.

Because it is the *logarithm* of P_o/P that is proportional to concentration, we define the absorbance, A, as

$$A = \log\left(\frac{P_o}{P}\right) = abC \tag{2.9}$$

This relationship makes it explicit that any single absorbance value is really the result of two measurements. Both the initial P_o and final (P) power of the light being passed through a sample must be measured in order to calculate the absorbance of a sample. While most spectrophotometers ultimately provide a readout of just absorbance values, the instruments have to measure two light power values – P_o and P – associated with the reference and sample solutions, respectively, in order to display a single absorbance value.

2.1.2.1. Transmittance. Transmittance, T, is related to absorbance. Transmittance is simply the fraction of light that passes through the solution without being absorbed:

$$T = \frac{P}{P_o} \tag{2.10}$$

The percent transmittance, % T, is simply the transmittance multiplied by 100 and is a measure of the percent of the initial power of the light beam that reaches the detector:

$$\%T = T \times 100 \tag{2.11}$$

Thus, absorbance, transmittance, and percent transmittance are related via the equations

$$A = \log\frac{P_o}{P} = \log\frac{1}{T} = -\log T = 2 - \log(\%T) \tag{2.12}$$

2.1.2.2. Molar Absorptivity. The final important aspect of the relationship $A = abC$ to consider is the proportionality constant, a. This constant accounts for the fact that molecules do not absorb all wavelengths with equal probability, even when the concentration of the molecule is fixed. Thus, if one changes nothing but the wavelength of the monochromatic light impinging on the absorbing species, the absorbance is still likely to change, as is depicted in Figure 2.4. When both the path length and concentration are fixed, only the proportionality constant, a, can mathematically account for the variation in absorbance with wavelength. Thus, the proportionality constant depends on wavelength, and due to the common use of molar concentration in lab measurements, the Beer–Lambert law is most frequently written as

$$A_\lambda = \varepsilon_\lambda bC \tag{2.13}$$

Here, ε is referred to as the molar absorptivity and the subscript λ makes the wavelength dependence explicit. While molar concentration is most commonly used in the Beer–Lambert law, other concentration scales can be used with an understanding that the proportionality constant changes accordingly. Regardless of which concentration scale is used, because absorbance is a dimensionless quantity (resulting from the cancelation of units in the ratio P_0/P in the definition of absorbance), the units of ε must cancel the units of the quantity, bC. When using molar concentration, the units for ε are Lmol^{-1}cm^{-1}, as path lengths are usually given in cm and molar concentration is measured in mol/L.

EXAMPLE 2.2

What is the transmittance, percent transmittance, and absorbance for a sample with $P_0 = 10\,000$ and $P = 5000$ arbitrary units? If the concentration of the absorbing species is 4.00 µM and the path length is 1.00 cm, what is the molar absorptivity at the wavelength the measurements were recorded?

Answer:

$$T = \frac{P}{P_0} = \frac{5000}{10000} = 0.5000$$

$$\%T = T \times 100 = 0.5000 \times 100 = 50.00\%$$

$$Abs = -\log T = -\log(0.5000) = 0.3010$$

$$Abs = 0.3010 = \varepsilon bc = \varepsilon (1.00 \text{ cm})\left(4.00 \times 10^{-6} \text{ M}\right)$$

$$\varepsilon = \frac{0.3010}{(1.00 \text{ cm})\left(4.00 \times 10^{-6} \text{ M}\right)} = 7.53 \times 10^4 \text{ Lcm}^{-1}\text{mol}^{-1}$$

Another question:
If the reference signal (P_0) in the analysis above is incorrectly measured to be 8000 arbitrary units, instead of 10 000, what is the percent error in the measured transmittance?

Answer:

$$\text{With } P_0 = 8000, T = 0.6250$$

$$\% \text{ error} = \frac{0.625 - 0.5000}{0.5000} \times 100 = 25.00\% \text{ error}$$

EXAMPLE 2.3

What is the concentration of a sample with a percent transmittance of 63.5% using a cell with a 0.100 cm path length, if the molar absorptivity of the sample at the wavelength being used is 13895 Lcm^{-1}mol^{-1}?

Answer:

$$\text{Abs} = -\log T = -\log(0.635) = 0.197$$

$$\text{Abs} = 0.197 = \varepsilon bc = \left(13895 \, \text{Lcm}^{-1}\text{mol}^{-1}\right)(0.100 \, \text{cm})(c)$$

$$c = \frac{0.197}{\left(13895 \, \text{Lcm}^{-1}\text{mol}^{-1}\right)(0.100 \, \text{cm})} = 1.42 \times 10^{-4} \, \text{M}$$

It is important to note that ε is related to molecular properties. Specifically, ε increases as the wavelength of light, and hence its energy, nears the energy of a main electronic transition within a molecule, such as the $\pi \rightarrow \pi^*$ and $n \rightarrow \pi^*$ transitions mentioned earlier. If the energies are closely matched, ε will be high (values of 10000–100000 Lmol^{-1}cm^{-1} are common for conjugated aromatic systems), allowing for the detection and quantitation of concentrations in the milli- to micromolar range. If the energy of the light does not match the energy of an electronic transition, it is unlikely the light will be absorbed, ε will be close to or equal to zero, and as a consequence, the absorbance will also be close to zero even if the concentration is quite high. In such cases, quantitation at the selected wavelength is difficult or impossible and other conditions must be found.

The relationship $A = \varepsilon bC$ is easy to remember when considering the conditions that increase the probability of light being absorbed by a sample. Naturally, increasing the concentration of the absorbing species increases the absorbance because the probability that a photon collides with an absorbing molecule increases. Likewise, as the path length is increased, photons travel further through the solution, thereby improving their chances of colliding with an absorbing molecule and thus of being absorbed. Lastly, the molar absorptivity, as discussed above, reflects the likelihood that photons of different wavelengths get absorbed. So if the molar absorptivity is high, the chance for absorption is higher than when the molar absorptivity is low. Thus, Beer's law can always be reconstructed by considering the main factors that increase the probability of photon absorption:

1. Longer path lengths.
2. Higher concentrations.
3. Closer matches between the energy required for an electronic transition and the energy of the light entering the solution.

The fact that all three are linearly related to the absorbance is no coincidence, but rather is a result of the convenient definition of absorbance as the logarithm of the ratio of P_o and P.

As a final note, while the linear relationship between absorbance and concentration ultimately plays an important role in using UV-visible spectroscopy for analytical purposes, and is therefore the one that often gets emphasized, it is the underlying logarithmic relationship between P_o/P and concentration that dictates the analytical reliability of absorbance measurements and that guides the design and proper use of UV-visible spectrophotometers.

2.1.3. Solvent Effects on Molar Absorptivity and Spectra

In the previous section we showed that the molar absorptivity reflects the likelihood of a molecule absorbing a photon of a specific wavelength and that the likelihood increases as the energy (i.e., wavelength) of the photon nears that of an electronic transition within the molecule. It is important to recognize that the energy of electronic transitions depends on the solvent that the absorbing species is dissolved in. This dependence has potentially important implications when using the Beer–Lambert law to determine concentrations.

Molar absorptivities depend on the solvent that the molecule is dissolved in because the energy of the molecule's ground and excited states, and hence the energy difference between them, is influenced by intermolecular interactions. For example, take p-nitroaniline as the absorbing solute. The lowest energy transition (i.e., longest wavelength) for the molecule occurs when a lone electron on the amine nitrogen is promoted to a π^* antibonding orbital [9]. This transition is shown in Figure 2.6 using Lewis structures and representations generated using computational chemistry to calculate the HOMO and LUMO electron distributions. Note that the electron density around the amine group is lower in the excited state (LUMO) than in the ground state (HOMO). This means that the hydrogen atoms in the excited state have a greater partial positive charge and therefore can form stronger hydrogen bonds with the solvent in the excited state than they do when the molecule is in the ground state.

First consider dissolving p-nitroaniline in a nonpolar, nonhydrogen bond-accepting solvent like cyclohexane. In cyclohexane, the ground and excited states of p-nitroaniline will have some specific energy levels as depicted in Figure 2.7. Next consider dissolving p-nitroaniline in a polar solvent capable of accepting hydrogen bonds, such as methanol. Because of the hydrogen bonding (and other polarity effects), the ground state energy of p-nitroaniline is lower in methanol than in cyclohexane. The excited state energy is also lowered because of the hydrogen bonding. Realize, however, that because the excited state of p-nitroaniline forms *stronger* hydrogen bonds due to the greater partial positive charge on the hydrogen atoms, *the decrease in the excited state energy is greater than the decrease in the ground state energy* when comparing methanol to cyclohexane as the solvent [9, 10]. In other words, there is a *differential stabilization* of the ground and excited states as depicted in Figure 2.7. The energy gap between the HOMO and LUMO for p-nitroaniline shrinks when it is dissolved in methanol compared to cyclohexane. Because the energy gap decreases in methanol, the wavelengths of light that are absorbed and cause excitation increase. Lastly, because the molar absorptivity, ε, reflects the likelihood of a photon of a specific wavelength being absorbed, it must be concluded that the molar absorptivity of the analyte molecule depends on the solvent.

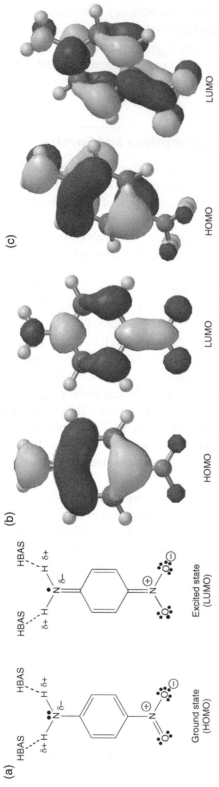

FIGURE 2.6 Representations of the ground and excited electronic states of *p*-nitroaniline. (a) Lewis structures: HBAS stands for "hydrogen bond-accepting solvent." (b) Molecular orbitals calculated using semiempirical computational methods. (c) Same as (b) but with the molecule rotated so that the electron density above and below the ring can be seen. The molecule is oriented the same way in all three figures, that is, with the amine group on the top and the nitro group on the bottom. Note that the electron density around the $-NH_2$ group is much lower in the excited state than in the ground state. This lack of electron density makes the partial positive charge on the hydrogen atoms much greater. This increases their strength of hydrogen bonding with solvents that can accept hydrogen bonds, ultimately lowering the energy of the excited state. The stabilization arising from hydrogen bonding is greater for the excited state than it is for ground state because the partial positive charge on the hydrogen atoms is greater in the excited state than in the ground state. This shrinks the energy difference between the two states in hydrogen bond-accepting solvents relative to when *p*-nitroaniline is dissolve in solvents that do not accept hydrogen bonds as depicted in the next figure.

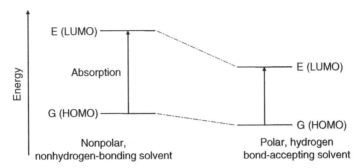

FIGURE 2.7 An energy diagram showing the differential stabilization of the ground and excited states of *p*-nitroaniline in going from a nonpolar, nonhydrogen-bonding solvent (e.g., cyclohexane) to a polar, hydrogen bond-accepting solvent (e.g., methanol) [1, 3, 4].

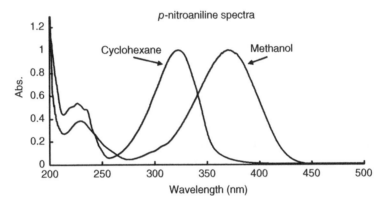

FIGURE 2.8 The dependence of the spectrum of *p*-nitroaniline on solvent. The absorbances in each solvent are normalized to the maximum absorbance in the longest-wavelength band to make the shift in the spectrum easier to observe.

The dependence of the UV-visible spectrum of *p*-nitroaniline on the solvent is shown in Figure 2.8. Clearly, the molar absorptivities are different at given wavelengths depending on the solvent. Specifically, as predicted above, the wavelength of maximum absorbance shifts to longer wavelength (i.e., lower energy) in methanol, the polar, hydrogen-bonding solvent, compared to cyclohexane, the nonpolar, nonhydrogen-bonding solvent.

Such changes in the molar absorptivity of a species as a function of solvent are called "solvatochromic shifts" to reflect both the <u>**solv**</u>ent influence and the fact that they shift the wavelengths (and hence color, or "<u>**chroma**</u>") absorbed by the molecule.

While we focused on *p*-nitroaniline, the same reasoning can be generalized to all absorbing species whose ground and excited states differentially interact with the solvent or sample matrix. Furthermore, while we focused on hydrogen bonding, all forms of intermolecular interactions (e.g., dipole–dipole, dispersion, dipole-induced dipole, and hydrogen bonding) can cause solvatochromic shifts [1–4, 9, 11]. In all cases, the relative interaction abilities of the ground and excited state of the specific analyte being studied must be considered to understand and/or predict solvent effects. It is also important to realize that the energy gap between the HOMO and LUMO can increase or decrease as a result of solute/solvent interactions, and thus solvatochromic shifts can be toward higher or lower wavelengths. The specific direction and magnitude of the shift depends on the electron distributions in the ground and excited states and their effect on the strength of intermolecular interactions between the absorbing molecule and the solvent.

Early in this section we stated that the solvent dependence on molar absorptivities has important implications for the application of the Beer–Lambert law in quantitative analyses. Specifically, to be rigorously valid, the solvent (i.e., matrix) in which the analyte is dissolved must also be used to prepare the calibration standard solutions. If this is not the case, the molar absorptivity of the analyte may be different in the samples than in the standards. This would result in an incorrect determination of the analyte concentration.

2.2. UV-VISIBLE INSTRUMENTATION

In order to make absorbance measurements and to collect spectra, UV-visible spectropho-tometers must perform several tasks. They are therefore comprised of many components, each dedicated to a particular task. The components must:

1. Provide electromagnetic radiation in the UV-visible portion of the spectrum.
2. Separate the light into its component wavelengths (this is necessary because quantitative analyses require monochromatic radiation).
3. Selectively pass light of a specific wavelength through the sample and reference solutions and systematically varying the wavelength if a full spectrum is desired.
4. Measure the intensity of the light that passes through the sample and reference (i.e., measuring P and P_o, respectively).
5. Convert the measured light intensities into absorbance readings.

We first discuss the components that are typically used to accomplish these tasks and then consider the overall design of spectrophotometers. We will see how these compo-nents are arranged in a variety of designs to make single wavelength UV-visible absor-bance measurements or to collect spectra.

In general, spectrometer components are often arranged as depicted in Figure 2.9, illustrating that light from the source is passed through a wavelength selector, directed onto the sample and/or solvent (also referred to as a reference or blank), and

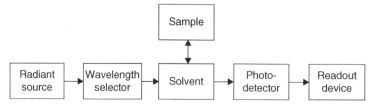

FIGURE 2.9 A simple block diagram of a UV-visible spectrometer for measuring the absorption of electromagnetic radiation. Source: From Willard et al. [12]. Reproduced with permission from Frank A. Settle, Jr.

strikes the detector. The detector converts light energy into an electrical signal, which ultimately gets transformed into a final digital readout that we can interpret.

Because spectrophotometers perform multiple tasks, and because there are a number of components that can be used to perform each task, the following discussion that ultimately leads to the design of spectrophotometers has a number of sections in it. Without guidance, it could be easy to forget the ultimate goal, which is to understand how the components of a spectrophotometer work together to provide the two measurements of light intensity that go into every absorbance measurement. The outline below will guide us through the discussion of each of the individual components. It may be helpful to periodically refer back to it to see where different topics fit along the path toward a complete understanding of the design and performance of spectrophotometers.

Spectrophotometer components
- Sources of visible and ultraviolet light
  - Halogen lamps (visible light)
  - Deuterium and hydrogen lamps (ultraviolet)
  - Xenon flash lamps
- Wavelength selectors
  - Filters
    - Interference filters
    - Absorbance filters
  - Monochromators
    - Components
    - Slits
    - Diffraction gratings
      Echelette
      Echelle
      Holographic
    - Mirrors
  - Configurations
    - Fastie–Ebert
    - Czerny–Turner
    - Littrow
- Sample holders
  - Cuvettes
  - Fiber optics
- Detectors
  - Phototubes
  - Photomultiplier tubes
  - Photodiodes, photodiode arrays, and charge-coupled devices

Spectrophotometer designs
- Single beam
- Double-beam-in-space
- Double-beam-in-time
- Diode array

2.2.1.   Sources of Visible and Ultraviolet Light

Electromagnetic radiation sources in absorption spectrophotometry have two basic requirements. First, they must provide sufficient radiant energy over the wavelength region where absorption is to be measured. Second, they should maintain a constant intensity over the time interval during which measurements are made. If the first criterion is not met, too little light will reach the detector, leading to large measurement errors. If the second criterion is not met, absorbance values will be incorrect due to inconsistencies between reference and sample light levels.

Unfortunately, most radiation sources do not provide sufficient intensity output over the entire range of wavelengths (200–800 nm) in the UV-visible region of the spectrum. For this reason, many spectrophotometers have two radiation sources: a deuterium or hydrogen discharge lamp that provides electromagnetic radiation in the UV range (200–350 nm) and a tungsten filament lamp that provides visible light (350–800 nm).

2.2.1.1.   Deuterium and Hydrogen Discharge Lamps.   Deuterium and hydrogen lamps (see Figure 2.10) function by electrically exciting D_2 or H_2. The excited state D_2 or H_2 subsequently dissociates and emits a photon in the UV range of the spectrum.

The process can be summarized as follows:

$$D_2 + E \;\rightarrow\; D_2{}^* \;\rightarrow\; D' + D'' + h\upsilon$$

where E is the electrical energy imparted to the deuterium, $D_2{}^*$ is an excited state deuterium molecule, and D' and D'' are deuterium atoms (deuterium is an isotope of hydrogen with one proton and one neutron in the nucleus) [13]. Recall that the total energy must be conserved for any reaction and that the energy of excitation of deuterium is quantized.

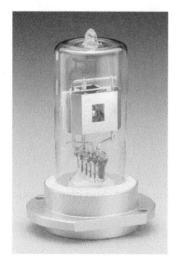

FIGURE 2.10   Picture of a deuterium lamp. Source: Reproduced with permission of Thermo Fisher Scientific.

The kinetic energy of the two atomic deuterium species that are the product of the reaction can vary continuously over a wide range of energies. Therefore, the energy of the electromagnetic radiation emitted as a result of the gas-phase dissociation of D_2^* – which must account for any difference in the energy of the products compared to the reactants – also varies continuously. In the case of deuterium and hydrogen, the emitted energy happens to fall in the UV range of the spectrum, making D_2 and H_2 lamps useful radiation sources for UV-visible spectrophotometers.

Heated electrodes, to which a 40 V potential difference is applied, maintain the radiation emission. A vital feature of these lamps is an aperture between the cathode and anode, which constricts the discharge to a narrow path and produces an intense ball of radiation of about 0.6–1.5 mm in diameter. D_2 produces more intense radiation than H_2, making the use of D_2 lamps in spectrophotometers much more widespread than H_2 lamps. The voltage applied to the lamp must be stable in order to maintain a constant lamp intensity. For this reason, regulated power supplies are often incorporated into the electronics of the spectrometer.

Figure 2.11a shows a plot of the intensity of D_2 emission as a function of wavelength. Notice that the radiation quickly decays in the visible range of spectrum, generally limiting the usefulness of D_2 and H_2 lamps to the 160–380 nm range. Because glass strongly absorbs below 350 nm, the windows of D_2 and H_2 lamps must be made of quartz. In addition to the broad emission in the UV portion of the spectrum, D_2 lamps produce narrow emission lines in the visible portion of the spectrum due to atomic deuterium emissions. These lines, particularly those at 486.0 and 656.1 nm, are useful for calibrating wavelengths in spectrophotometers, but they can also be problematic in some applications.

2.2.1.2. Tungsten Filament Lamps.

Electromagnetic radiation above 350 nm and into the near-infrared (NIR) up to 2500 nm is usually provided by incandescent tungsten filament lamps. In these lamps, a coiled tungsten filament is heated by an electric current, causing it to emit light. The filament is enclosed in a sealed glass bulb filled with an inert gas or in a vacuum. Filament lamps are rugged, low cost, and intense enough for nearly all absorption work in the visible and near-ultraviolet regions. Figure 2.11b shows the spectral profile of a tungsten-halogen lamp in the UV-visible range. Notice that the intensity is different at all wavelengths and falls dramatically toward the ultraviolet region.

Tungsten-halogen lamps are a special class of incandescent lamps in which iodine is added to the gas inside the bulb. The iodine reacts with tungsten that has sublimed from the filament. The resulting volatile WI_2 decomposes when it encounters the hot filament, redepositing the tungsten metal onto the filament and regenerating I_2. Thus, the filament is continuously being cleaned and regenerated by this process. As is the case with D_2 lamps, a constant intensity of the emitted light of tungsten lamps is highly dependent on consistent applied voltage, necessitating the use of regulated power supplies. It should be noted that neither D_2 nor tungsten-halogen lamps emit strongly in the region from about 300 to 600 nm, as is clear from the combined output plot shown in Figure 2.11c. Thus, analytically, the signals in this region are weaker than in other regions and are therefore potentially subject to noise effects.

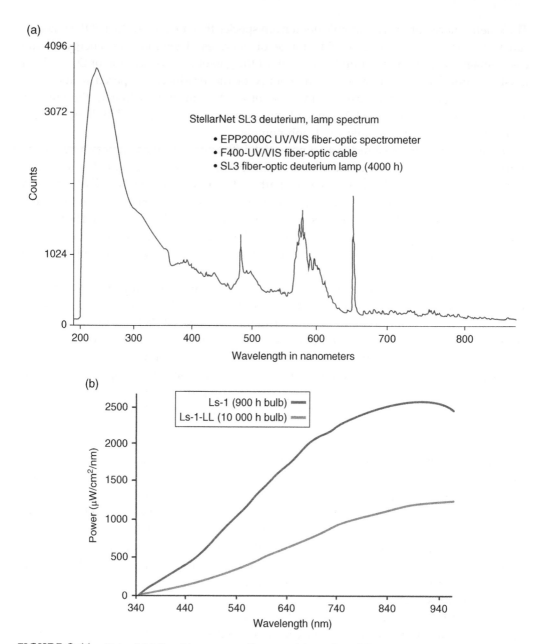

FIGURE 2.11 Plot of (a) D_2, (b) tungsten filament, (c) combined D_2 with tungsten (on the next page), and (d) xenon flash lamp emission (on the next page) as a function of wavelength. Notice in the combined output that some of the distinct spectral features of the deuterium emission, like the sharp emission band at 656.1 and 486.0 nm, are visible above the broad output from the tungsten lamp. Also notice that the intensity between 300 and 650 nm is much lower than at higher and lower wavelengths. Sources: (a) and (c) Reproduced with permission of StellarNet Inc., Tampa, FL. (b) and (d) Reproduced with permission of Ocean Optics, Tampa, FL.

(c)

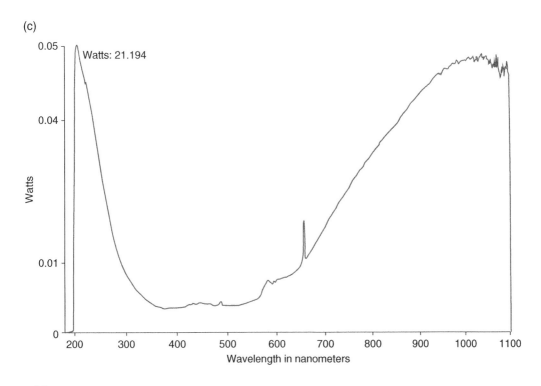

(d)

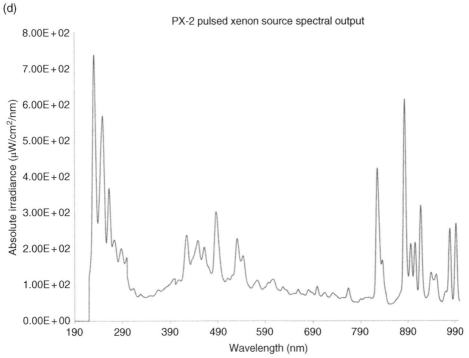

FIGURE 2.11 *(Continued)*

2.2.1.3. Xenon Flash Lamps. Xenon flash lamps (also called pulsed xenon lamps) produce light by applying a high voltage to xenon gas. The xenon atoms are ionized by the voltage, which then allows a plasma to form. The plasma emits electromagnetic radiation. The voltage is rapidly pulsed through the gas at a rate of around 100 Hz, creating short, high intensity bursts of electromagnetic radiation. As can be seen in Figure 2.11d, the xenon produces electromagnetic radiation that ranges from the UV to the near infrared region of the spectrum. While several distinct peaks are present in the output, the intensity at all wavelengths is sufficiently high that xenon lamps can be used to collect absorbance values across the entire UV-visible spectrum. Because of this, some spectrophotometers have a single xenon flash lamp rather than having both a deuterium and tungsten lamp.

2.2.2. Wavelength Selection: Filters

As we have seen, Beer's law is based on the assumption of monochromatic radiation. Thus, spectrophotometric methods usually require the isolation of discrete bands of radiation. There are two common methods for isolating a narrow range of wavelengths: filters and monochromators. Two main kinds of filters – interference and absorption – and two main kinds of monochromators – prism-based and grating-based – are frequently used. These four methods of wavelength selection are discussed in the following sections.

2.2.2.1. Interference Filters. Interference filters, as the name implies, are based on the phenomenon of optical interference (see Chapter 1). A simple two-interface filter, known as a Fabry–Perot filter, consists of a dielectric material (CaF_2, MgF_2, or SiO) sandwiched between two parallel, partially reflecting metal films, as shown in Figure 2.12.

The filters select particular wavelengths out of a broad spectrum of light using internal reflections to produce constructive and destructive interference as depicted in Figure 2.12. Some portion of the incident light (beam 1) passes through the filter (beam 2), while another portion is reflected from surface B back to surface A (beam 3). A portion of this reflected radiation is again reflected from surface A through the dielectric layer and exits as beam 4, parallel to (actually coincident with) beam 2. Thus, the path traveled by beam 4 is longer than that traveled by beam 2 by exactly twice the thickness of the dielectric spacer film. For constructive interference to occur, the wavelength of the light *in the dielectric material* must be a multiple of the additional path length that the light travels. For this reason, the thickness of the filter is controlled to be 1/2, 1, or 3/2 times that of the wavelength that is desired. The equation that predicts which wavelength is passed is

$$\lambda = \frac{2t\eta}{m} \tag{2.14}$$

In this equation, η is the refractive index of the dielectric material, t is the thickness of the dielectric material, and m is an integer (0, 1, 2, ...) and referred to as the order of interference. This equation assumes that the polychromatic light strikes the surface of

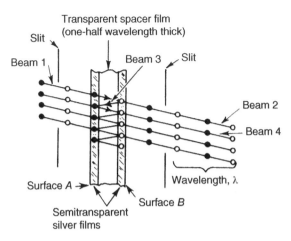

FIGURE 2.12 Schematic of an interference filter and the path of light rays through the filter. Polychromatic radiation strikes surface A from the left, undergoes partial reflection, enters the spacer film made of a dielectric material, and again undergoes partial reflection off of surface B (beam 3). For radiation with a wavelength that is twice the thickness of the spacer film, constructive interference results in the beam being passed and emerging from the right side of the filter (beams 2 and 4). Radiation with wavelengths that are not a multiple of the film thickness experience destructive interference and are therefore not observed on the right side of the filter. Source: From Willard et al. [12]. Reproduced with permission from Frank A. Settle, Jr.

the filter at 90° (the wavelength shifts to shorter wavelengths with increasing angle of incidence). The factor $2t$ accounts for the path length of the light traveling from surface B back to surface A. Thus, filters of different thicknesses isolate different wavelengths. The refractive index appears in the equation to account for the fact that the wavelength of light decreases when it travels through material that has a refractive index different than that of air. The fact that the order, m, is any integer shows that more than one wavelength is transmitted by the filter. For example, if a filter is designed to pass 500 nm light, it also passes 250 nm (500/2), 166.7 nm light (500/3), etc. For this reason, these filters are often surrounded by glass that absorbs all but the main transmission line.

Partial reinforcement occurs for other wavelengths, so filters actually transmit a narrow band of wavelengths, with the maximum intensity occurring at the wavelength that satisfies Eq. (2.14). The bandwidth is usually 10–15 nm wide at half the maximum intensity, more commonly known as full width at half maximum (FWHM). The intensity of the light transmitted through the filter can be considerably less than the original intensity, but this is a tradeoff made to obtain the narrow wavelength band that is necessary for reliable absorbance measurements. Some filters with high transmission (80% or better) are available, but these have larger bandwidths and can therefore cause deviations from Beer's law because the criteria of monochromatic light are less closely approximated (see Section 2.4.3.3 about deviations from Beer's law below).

EXAMPLE 2.4

Calculate the first-order wavelength of light that an interference filter passes if the dielectric material has a thickness of 165 nm and a refractive index of 1.28. What other wavelengths are also passed?

Answer:

$$\lambda = \frac{2t\eta}{m} = \frac{2 \times 165 \text{ nm} \times 1.28}{1} = 422 \text{ nm}$$

Higher-order ($m = 2, 3, 4$, etc.) wavelengths will also be passed (211, 141, 106 nm).

2.2.2.2. Absorption Filters. Absorption filters derive their effects from bulk interactions of radiation with material. Some types rely on scattering, and in others, true absorption occurs. The goal in designing such a filter is to have it absorb all wavelengths of light except that which is of interest. However, due to the nature of the absorbance process, infinitely narrow bands of monochromatic light cannot be transmitted. Rather, a band of wavelengths, centered on a main wavelength, is passed. Absorption filters are produced from a variety of materials, but the most common are made from dyes embedded in glass or gelatin sealed between two glass windows. While these filters are generally cheaper than the interference filters described above, they are characterized by broader transmission bands and poorer transmittances. Typically, the narrower the band, the lower the transmittance.

Another type of filter, known as cutoff filters, generally transmits 100% of the light above or below a specified wavelength and absorbs all other wavelengths. The cutoff, however, is not infinitely sharp, but rather has a slope that can span up to 100 nm (see Figure 2.13). These types of filters are often used to eliminate second- and third-order light from monochromators and interference filters. They can also be coupled together, as shown in Figure 2.13, to pass a limited range of wavelengths. In this case, a short-wavelength filter (i.e., one that transmits short wavelengths and absorbs high wavelengths) is coupled with a long-wavelength filter. The area in which their slopes overlap allows transmittance of a band of wavelengths. Such an arrangement is often referred to as a "notch filter." As depicted in Figure 2.13, the transmitted light is characterized by the maximum transmittance (in this case, only 25%), the spectral bandwidth (FWHM, here about 20 nm), and spectral bandpass (total wavelength range transmitted, here about 50 nm). The central wavelength that is transmitted can be controlled by varying the wavelengths at which the cutoff filters have their effects. Also, smaller bandwidths can be produced by choosing cutoff filters that have more overlap, but this clearly decreases the intensity of the transmitted light. Here again we see a tradeoff between obtaining nearly monochromatic light (i.e., narrow bandwidths) and the intensity of the transmitted light. This is almost always the case, and we must consider it when selecting components for UV-visible instruments.

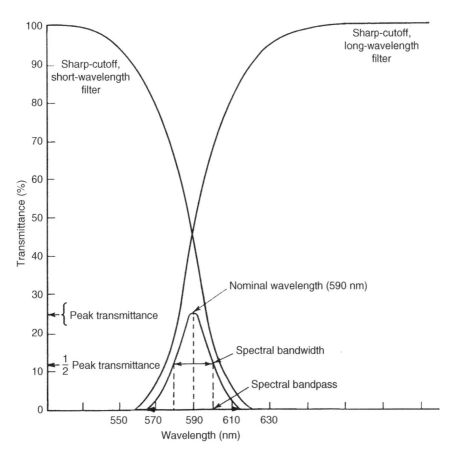

FIGURE 2.13 Spectral transmittance characteristics of two sharp cutoff filters combined to create a notch filter that passes a narrow range of wavelengths. Source: From Willard et al. [12]. Reproduced with permission from Frank A. Settle, Jr.

As suggested earlier, you may find it helpful to pause at this point and return to the outline of spectrophotometer components on page 53 to get a sense of where we are in our overall consideration of the design of spectrophotometers.

2.2.3. Wavelength Selection: Monochromators

Both absorbance and interference filters are useful in certain applications, but they do not allow for much experimental flexibility. For each wavelength of interest, a different filter needs to be inserted into the instrument. Some instruments have filter wheels with which multiple filters can be selected, but even then, analyses are limited to the wavelengths isolated by those filters.

In many cases, it is desirable to record an entire UV-visible spectrum – a continuous plot of absorbance versus wavelength such as that shown in Figure 2.4. For this purpose, instruments with monochromators are more commonly used and much more flexible than filter-based instruments. In contrast to the single wavelength selection of filters, monochromator systems can systematically and continuously scan through the entire wavelength range. A monochromator, which is a separate collection of components housed within spectrophotometers (see Figure 2.14), generally consists of:

1. An entrance slit that provides a narrow optical image of the radiation source.
2. A collimating mirror that focuses the radiation coming through the entrance slit onto a prism or grating.
3. A grating or prism that separates the incident radiation into its component wavelengths.
4. A collimating mirror that reforms images of the entrance slit on the exit slit.
5. An exit slit that controls the wavelength resolution of the instrument by isolating the desired spectral band by blocking all of the dispersed radiation except that within the desired range.

The heart of a monochromator is the grating or prism that separates light into its component wavelengths in space. The wavelength of light that is focused onto the exit slit and subsequently through the sample is selected by rotating the grating or prism. In the

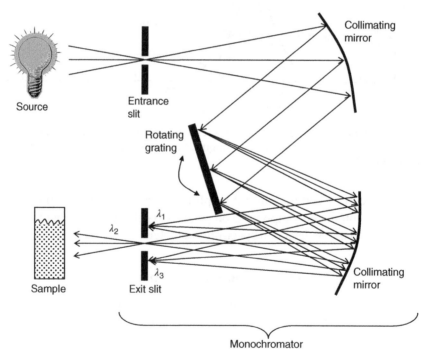

FIGURE 2.14 Monochromator schematic. Polychromatic light from the source enters the entrance slit, is collimated by the mirror, and is reflected toward the grating. The grating disperses the light into its constituent elements – for clarity, only three wavelengths, labeled λ_1, λ_2, and λ_3, are shown here, but for white light, all the wavelengths in the visible region would be present. The dispersed light is collimated by the second mirror and focused on the plane of the exit slit. The exit slit allows only a narrow range of wavelengths to pass through, while all others are physically blocked. The emerging beam of nearly monochromatic light then passes through the sample where it is absorbed or transmitted.

following sections we first discuss the slits because the slits allow light from the radiation sources into and out of the monochromator and ultimately allow the isolation of nearly monochromatic light to pass into the sample. Then we examine how prisms and gratings disperse light into its component wavelengths and how the optics of a monochromator work to focus the light output onto the sample.

2.2.3.1. *Slits.* In spectrophotometers, the electromagnetic radiation emitted by the D_2 (or H_2) and tungsten lamps located outside of the monochromator is focused on the entrance slit of the monochromator. The slits are generally rectangular, with an adjustable width. Inside the monochromator, the rays diverge from the entrance slit and illuminate the collimator mirror, which reflects and focuses them on to a prism or grating. Leaving the collimator, the parallel set of rays is a broadened image of the entrance slit. This rectangle of radiation must be large enough to illuminate the entire side of the prism or length of the grating. The prism or grating separates the incident polychromatic radiation into an array of neighboring monochromatic rectangles, each of which leaves the dispersing device at a slightly different angle, essentially forming a "rainbow" (see Figure 2.14). The dispersed radiation is subsequently imaged onto the second collimating mirror. This mirror reflects and focuses the dispersed radiation onto the focal plane of the monochromator, where the exit slit allows a narrow band to pass. The distance between the second collimating mirror and the exit slit is called the *focal length* of the monochromator and influences both the light gathering power and dispersion of monochromators. The focal length of many spectrophotometers is 0.5 m. The width of the entrance and exit slits is usually set equal to one another such that the image of the entrance slit just fills the exit slit.

To understand the spectral region that is isolated by the slits, consider Figure 2.15 [14, 15]. Let ABCD be the image of a monochromator band at the exit plane, and let EFGH be the dimensions of the exit slit. The numbers 1, 2, 3, … correspond to positions of the monochromator wavelength setting. Suppose that purely monochromatic radiation is being passed through the monochromator. The plot next to the slit images in Figure 2.15 shows the fraction of light transmitted by the system as a function of the position of the image of the entrance slit as it moves along the focal plane and passes over the exit slit. The position on the focal plane depends on the angle of the grating relative to the collimating mirrors. Up to the point where the leading edge of the image, BD, reaches position 1, no radiation is transmitted by the exit slit. When BD reaches position 2, half of the total radiation is passed. In position 3, the instrument set-point matches the wavelength of the radiation coming through the monochromator. When the grating is in this position, the exit aperture is filled, and the transmitted intensity is 100% of that available. As the image reaches position 4, the transmittance falls to one-half, and at position 5 it falls to zero.

The case above considered monochromatic radiation. However, prisms and gratings disperse polychromatic radiation into a continuous band of wavelengths (i.e., a "rainbow"). In this case, the image that is projected through the exit slit is not purely monochromatic but rather contains a very narrow range of wavelengths centered on the main wavelength of interest. So when polychromatic radiation is present, the x-axis in the triangular plot represents the range of wavelengths passed at the setting of the central wavelength. The central wavelength is dictated by the angle of the grating or prism relative to the collimating mirror. The range that is passed is characterized by the *spectral bandwidth*, which is the difference in the wavelengths between the points where the transmittance is half the maximum. For example, if the central wavelength being transmitted at point 3 in Figure 2.15 is 600 nm, and 598 and 602 nm light is transmitted at positions 2 and 4, respectively, then the bandwidth is 4 nm.

The spectral bandwidth is determined by the dispersing ability of the grating or prism and the physical width of the slits, as shown in the equation

$$\text{Bandwidth} = WD^{-1} \tag{2.15}$$

Here, W is the actual, physical slit width, often on the order of millimeters, and D^{-1} is the reciprocal of the linear dispersion of the monochromator in units of nm/mm. Linear dispersion, discussed in more detail below, is a measure of how far apart different wavelengths of light are separated on the focal plane. For example, if 300 and 800 nm

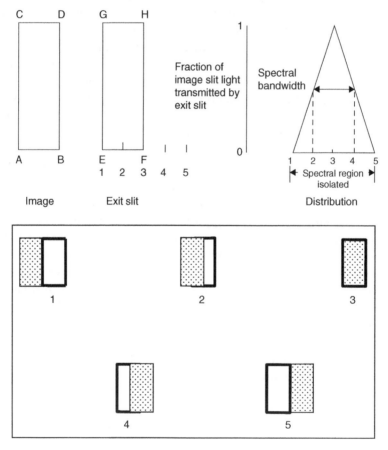

FIGURE 2.15   (Top) Slit distribution function when the image size and the exit slit are identical. The cone on the right plots the intensity of the light (vertical axis) versus the position of the image from the entrance slit on the exit slit. (Bottom) Another way of visualizing the image from the entrance slit of a narrow band of wavelengths (represented as the hashed rectangle) as it moves across the fixed exit slit. Source: From Willard et al. [12]. Reproduced with permission from Frank A. Settle, Jr.

light are 250 mm apart in the focal plane (see Figure 2.16), then the linear dispersion is 250 mm/500 nm, or 0.500 mm/nm. So the reciprocal linear dispersion is 1 nm/0.500 mm, or 2.0 nm/mm.

The bandpass of a monochromator, $\Delta\lambda$, is given by

$$\Delta\lambda = 2WD^{-1} \tag{2.16}$$

and is the sum of the image width and the exit slit width in wavelength units. The bandpass specifies the spectral region isolated for a given slit width (see Figure 2.15) and impacts resolution. Specifically, for baseline resolution of two spectral features, the bandpass must be just slightly smaller than the separation of the spectral features ($d\lambda$ in wavelength units).

As these equations show, the wider the slits, the greater the range of wavelengths passed. Wider slits allow more light to reach the detector, making it easier to detect the

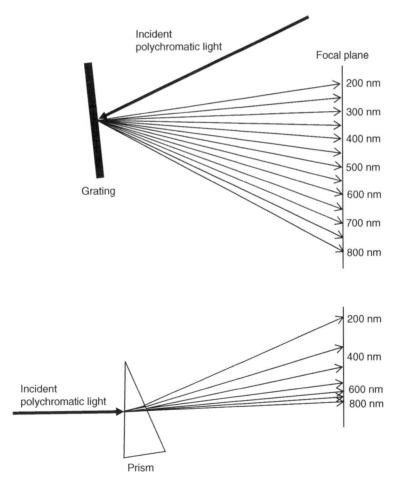

FIGURE 2.16 Depiction of the linear dispersion provided by a grating as opposed to the nonlinear dispersion typical of a prism. Source: From Willard et al. [12]. Reproduced with permission from Frank A. Settle, Jr.

light and improving the signal-to-noise ratio (S/N) of absorbance measurements. However, Beer's law assumes monochromatic radiation, and polychromatic radiation causes deviations from linearity (the reason for this is discussed in Section 2.4.3.3 below). Furthermore, wider slits cause a decrease in wavelength resolution and tend to "smear out" sharp spectral features as shown in Figure 2.17. For example, the spectral features around 515 and 550 nm in spectrum 1 in Figure 2.17, which was recorded with the largest slits, are considerably broader and flatter than those in spectra 4 and 5 (recorded with narrow slits). This broadening and flattening is a distortion of the true spectrum, which is better represented by spectra 4 and 5.

Narrow slits, in contrast, improve wavelength resolution and reduce the chances that deviations from linearity will arise. But narrow slits allow less light to reach the detector and, in the extreme, can cause serious signal-to-noise degradation because the noise inherent in the detectors becomes significant compared to the signal produced when light strikes it. This degradation in S/N is apparent in the noise around 480–500 nm and 530 nm in plot 5 of Figure 2.17, which was recorded with the smallest slit widths. Thus, in all applications, there is a tradeoff between wavelength resolution and absorbance precision. Slit widths are therefore selected depending on which is more important for a particular application.

One final note about slits is that when actually using a spectrophotometer, the slit widths are often listed on a nanometer scale, with typical slit width settings being anywhere from 0.1 to 20 nm. Recognize *that this is not the actual physical width of the slit* that is being set, but rather the bandwidth that emerges from the monochromator. The actual physical slit is much wider than this, being in the millimeter range. If the slits were actually closed down to a physical width in the nanometer range, the light would be diffracted by the slits because the slits would be near the same size as the wavelength of light.

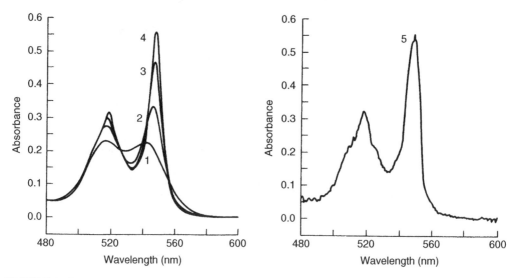

FIGURE 2.17   Effect of spectral bandwidth on the observed absorption band shapes of cytochrome *c*. Spectral bandwidths are (1) 20 nm, (2) 10 nm, (3) 5 nm, (4) 1 nm, and (5) 0.08 nm (shown in a separate plot for clarity). Source: From Willard et al. [12]. Reproduced with permission from Frank A. Settle, Jr.

2.2.3.2. Prisms and Diffraction Gratings. The main purpose of a monochromator is to separate light into component wavelengths such that individual wavelengths may be isolated and, if desired, to also allow for a sequential scan through a range of wavelengths. Prisms and gratings are capable of dispersing rays of different wavelengths at different angles and therefore can spatially spread out polychromatic light according to its constituent wavelengths.

2.2.3.3. Prisms. Prisms are wedges of crystalline quartz. Because quartz has a different refractive index than air, when light passes through a prism, the rays are bent at an angle according to Snell's law (see Chapter 1). Shorter wavelengths are bent to a greater extent than longer wavelengths. The result when polychromatic radiation strikes a prism is that different wavelengths are dispersed at different angles in space (this is depicted on Pink Floyd's "Dark Side of the Moon" album cover). The dispersion, however, is nonlinear, and the result is that the longer wavelengths are more cramped together in space and the shorter wavelengths are more spread out (see Figure 2.16). This has two important instrumental consequences. Firstly, it is harder to resolve the longer wavelengths than the shorter. Secondly, if one wants to scan through the wavelengths to record an entire spectrum, the prism must be rotated at varying speeds to compensate for the nonlinear behavior. This complicates the design of the instrument. Despite these complications, early scanning spectrophotometers used prisms as the dispersive device. Grating monochromators, which are described in the next section, have since superseded prisms, because gratings are more easily fabricated than prisms and because they produce linear dispersion of light rather than the nonlinear dispersion produced by prisms.

2.2.3.4. Diffraction Gratings. Diffraction gratings are finely grooved surfaces that, as their name implies, cause diffraction of light such that different wavelengths of light constructively interfere at different angles. The grooves must be rigorously straight, equally spaced, and parallel to one another for high wavelength resolution. A cross-sectional profile of a typical diffraction grating known as an echelette-type grating is shown in Figure 2.18. A typical grating is 10–25 cm wide and usually has between 1000 and 3000 grooves per mm. Thus, each groove may be only 500 nm wide. The grooves are also commonly referred to as lines or blazes.

Each groove of the grating has a broad face exposed to the incident polychromatic radiation. The radiation strikes the face of the groove at an angle, i, relative to the grating normal (i.e., 90°). Because the grooves are approximately the same size as the wavelengths of light striking them, each groove causes the polychromatic light striking it to be diffracted. This causes the dispersion of the light in all directions, and this occurs at every groove. So each groove essentially acts as a separate radial source of radiation separated in space from one another [16]. The waves reflected from each source then constructively or destructively interfere with one another as they propagate away from the grooves in all directions. To understand how different wavelengths of polychromatic light appear at different angles in space, consider two parallel beams of polychromatic light incident on the surface of the grating, as shown in Figure 2.19. Beam 2 travels further than beam 1, creating a difference in the distance traveled by each beam. When the difference in the distance is a multiple of the wavelength striking the detector, constructive interference

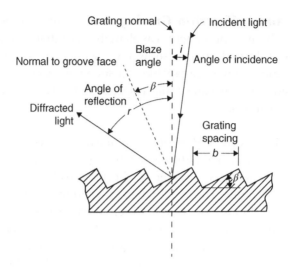

FIGURE 2.18 Cross section of a diffraction grating showing the angles of a single groove, which are microscopic on an actual grating. Source: From Willard et al. [12]. Reproduced with permission from Frank A. Settle, Jr.

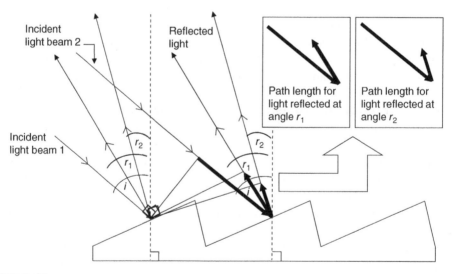

FIGURE 2.19 The figure depicts the effect of different angles of reflection on the path length traveled by the light. The same angle of incidence, i, relative to the grating normal (vertical dashed line) is incident on the blazes. Two different angles of reflection of the diffracted light, r_1 and r_2, are also depicted. The inset shows that there is clearly a difference in the path lengths traveled by light diffracted at the two different angles – the path length for each is highlighted by the bold lines. This difference in path length means that different wavelengths of light are constructively reinforced at different angles, effectively dispersing the incident polychromatic light into its component wavelengths. Source: From Willard et al. [12]. Reproduced with permission from Frank A. Settle, Jr.

occurs; otherwise, destructive interference occurs. It is evident in Figure 2.19 that the difference in the distance clearly depends on the angle of reflection, r, and the angle of incidence (both relative to the grating normal). Thus, constructive interference occurs for different wavelengths at different angles of incidence and reflection. By varying the angles, one can vary the wavelength that is observed. Or, put another way, the grating disperses polychromatic radiation into its constituent wavelengths, separating the wavelengths in space according to the angles at which constructive interference occurs. Reference [16] presents an excellent article about gratings that also has instructions for a 3D printable kit that demonstrates the constructive overlap of wavelengths quite well.

The mathematical relationship that predicts which wavelengths are constructively reinforced as a function of the incident (i) and reflection (r) angles is

$$b(\sin i \pm \sin r) = m\lambda \tag{2.17}$$

where b is the distance between adjacent blazes and m is a whole number designating the diffraction order. A positive sign applies in the grating formula when incident and diffracted beams are on the same side of the grating normal as depicted in Figure 2.19, and a negative sign applies when they are on opposite sides, as shown in Figure 2.18 [12, 17].

The grating formula shows that the incident radiation is diffracted into several orders, which are whole number multiples of the primary constructive wavelength, as shown in Figure 2.20. Interference filters also exhibit this phenomenon. This can be problematic in that the light that is passed through the sample is not truly monochromatic, and polychromatic light can cause nonlinearities in Beer's law. It can also create spectral artifacts when scanning over a broad range of wavelengths. We must therefore keep it in mind when using gratings. Modern gratings, however, are designed to concentrate most of the light energy from the source into the wavelength region of interest.

EXAMPLE 2.5

What first-order wavelength is constructively reinforced if the angle of incidence, i, in Figure 2.19 is +50° and the angle of reflection, r_2, is +15° with a blaze density of 2000 blazes/mm?

Answer:

$$b(\sin i \pm \sin r) = m\lambda$$

$$b = \frac{1\,mm}{2000\,blazes} = 5.000 \times 10^{-4}\,mm/blaze$$

$$\lambda = \frac{b(\sin i \pm \sin r)}{1} = \frac{5.000 \times 10^{-4}\,mm/blaze(\sin 50 + \sin 15)}{1}$$

$$\lambda = \frac{5.000 \times 10^{-4}\,mm/blaze(0.766 + 0.259)}{1}$$

$$\lambda = \frac{5.125 \times 10^{-4}\,mm}{blaze} \times \frac{10^{6}\,nm}{mm} = 512.5\,nm$$

Another question:
   (a) What second-order wavelength is constructively reinforced if the angle of inci-
       dence, i, in Figure 2.19 is +65° and the angle of reflection, r_1, is +45° with a blaze
       density of 1500 blazes/mm?
   (b) What first- and third-order light are also passed under these conditions?

Answers:
   (a) 538 nm
   (b) 1076 nm (first order), 359 nm (third order)

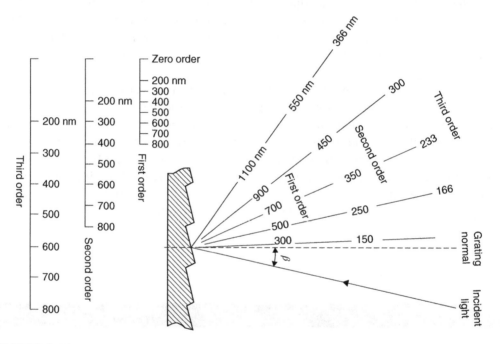

FIGURE 2.20 Overlapping orders of spectra from a reflection grating. The overlap results from
the fact that light is diffracted according to the equation $m\lambda = b(\sin i + \sin r)$, where the order, m, is
a whole number. As a consequence, the angle at which 700 nm first-order light appears also has
350 and 233 nm light due to the second-order ($m = 2$) and third-order ($m = 3$) diffraction. Also
notice the scales on the left – they show the overlap in the orders and also clearly demonstrate that
higher-order diffraction has a higher linear dispersion, meaning that the wavelengths are spread
out over a greater distance. Source: From Willard et al. [12]. Reproduced with permission from
Frank A. Settle, Jr.

2.2.3.4.1. **Angular and Linear Dispersion.** Gratings are characterized by their
ability to separate different wavelengths of light in space. This is measured by both
angular and linear dispersion. *Angular dispersion* is measured as $dr/d\lambda$, where dr is the
range of angles over which a range of wavelengths, $d\lambda$, is dispersed by a grating with
blazes of width, b, when the angle of the diffracted light relative to the grating normal

is r. If the incident angle, i, is kept constant and m is the diffraction order, the angular dispersion is given by

$$\frac{dr}{d\lambda} = \frac{m}{b\cos r} \tag{2.18}$$

Linear dispersion, D, measures the range of wavelengths separated by a distance, dx, along the focal plane. In other words, it is a measure of how spread out the wavelengths are. For example, the top image in Figure 2.16 depicts a grating separating light from 200 to 800 nm over a specific distance on the focal plane in the image. If a second drawing were made with the wavelengths even more spread out from the top of the page to the bottom, then the grating in the second drawing would have a greater linear dispersion than the one shown in the figure.

The linear dispersion is related to the angular dispersion and the focal length of the monochromator, F, by the equation

$$D = \frac{dx}{d\lambda} = \frac{F\,dr}{d\lambda} = \frac{mF}{b\cos r} \tag{2.19}$$

To visualize linear dispersion, imagine a rainbow (i.e., λ = 400 to 800 nm) spread out over a Post-it note, with red light at one edge and blue light at the other. Now compare this to spreading out the rainbow onto a poster board. In the second case, dx, the distance over which the wavelengths are spread out is much greater. Because the range of wavelengths, $d\lambda$, is the same but dx is much larger with the poster board, the ratio $dx/d\lambda$, and hence the linear dispersion, is much greater in the case of the poster board compared to the Post-it note. This is important because large linear dispersions mean that different wavelengths of light are separated by large distances. This makes it easier to pick out a specific wavelength and therefore obtain better wavelength resolution. Equation (2.19) also shows that small blaze distances (i.e., greater blaze density) and longer focal lengths increase linear dispersion. If one considers a linear dispersion of 0.312 mm/nm and that UV-visible radiation has wavelengths in the range of 200–800 nm, then the polychromatic light is spread out into the rainbow (and the nonvisible UV) over a distance of 187 mm (or almost 7.5 inches) in the focal plane.

Because $\cos r \approx 1$ for reflection angles up to 20°,

$$D \approx \frac{mF}{b} \tag{2.20}$$

Equation (2.20) implies that for grating monochromators, unlike prisms, the wavelengths are separated essentially linearly along the focal plane when r is small. Instrumentally, this simplifies scanning through the wavelengths. It also means that the ability to resolve wavelengths is fairly constant across the entire spectrum. In contrast, prisms tend to have the long wavelengths bunched together, making it more challenging to get the same resolution at longer wavelengths than at shorter ones.

The reciprocal linear dispersion, D^{-1}, is often cited when characterizing monochromators. Given the equation above, the reciprocal linear dispersion is simply

$$D^{-1} \approx \frac{b}{mF} \tag{2.21}$$

The units on D^{-1} are typically nm/mm, so the reciprocal linear dispersion measures the range of wavelengths that are dispersed onto one mm of the focal plane. Monochromators with large linear dispersions (and hence small reciprocal linear dispersions) separate the wavelengths of light better than monochromators with smaller linear dispersions. As shown above in Eq. (2.16), a small reciprocal linear dispersion leads to a small bandpass, which is a good thing because it means that a very narrow range of wavelengths (i.e., nearly monochromatic light) is being passed by the monochromator. So monochromators with large linear dispersions, and hence small reciprocal linear dispersions, offer better performance than those with smaller linear dispersions and larger reciprocal linear dispersions.

2.2.3.4.2.   Wavelength Resolution. The resolution of a grating, defined as $\lambda/\Delta\lambda$, is given by the equation

$$R = \frac{\lambda}{\Delta\lambda} = mN \tag{2.22}$$

where m is the diffraction order and N is the number of blazes illuminated by the incident light coming from the entrance slit. Typical resolutions are between 10^3 and 10^4. The equation shows that better resolution is obtained by illuminating large gratings with narrowly spaced blazes (i.e., high N) and also by working at higher orders. However, working at higher orders with standard echelette gratings is problematic because the orders begin to severely overlap in space as shown in Figure 2.20. To work at higher orders, a different approach and a different type of grating are used, as explained in the next section.

EXAMPLE 2.6

(a) What are the linear dispersion in mm/nm and the reciprocal linear dispersion (in nm/mm) of a monochromator if electromagnetic radiation with wavelengths from 400 to 800 nm is dispersed onto a Post-it note that is 3.00 inches wide? Repeat the calculation for a poster board that is 28.0 inches wide. Which has the higher linear dispersion? Which has the smaller reciprocal linear dispersion?

(b) What is the resolution of a grating operating in the first order if the grating is 25.0 mm wide and has 30000 grooves illuminated by a rectangle of polychromatic radiation that is 18.0 mm wide? Is the resolution better or worse if the grating is operated in the second order?

(c) If the linear dispersion of a monochromator is 0.180 mm/nm and the slits are 2.00 mm wide, what are the bandwidth and the bandpass of the monochromator?

Answers:

(a)
$$D = \frac{dx}{d\lambda} = \frac{3.00 \text{ in}}{800 - 400 \text{ nm}} \times \frac{2.54 \text{ cm}}{1 \text{ in}} \times \frac{10 \text{ mm}}{1 \text{ cm}}$$

$= 0.191 \text{ mm/nm}$ for the Post-it note example

$$D^{-1} = \frac{1}{D} = \frac{1}{\dfrac{0.191 \text{ mm}}{nm}}$$

$= 5.24 \text{ nm/mm}$ for the Post-it note example

$$D = \frac{dx}{d\lambda} = \frac{28.0 \text{ in}}{800 - 400 \text{ nm}} \times \frac{2.54 \text{ cm}}{1 \text{ in}} \times \frac{10 \text{ mm}}{1 \text{ cm}}$$

$= 1.78 \text{ mm/nm}$ for the poster board example

$$D^{-1} = \frac{1}{D} = \frac{1}{\dfrac{1.78 \text{ mm}}{nm}}$$

$= 0.562 \text{ nm/mm}$ for the poster board example

These calculations show that the poster board example has a higher linear dispersion and a smaller reciprocal linear dispersion.

(b) $R = mN$

where N is the number of grooves illuminated. The grating in the problem has 30 000 grooves over 25.0 mm, but only 18.0 mm is illuminated. So

$$N = \frac{30000 \text{ grooves}}{25.0 \text{ mm}} \times 18 \text{ mm} = 21600 \text{ grooves illuminated},$$

so

$R = 1 \times 21600 = 21600$ for the first-order $(m = 1)$ resolution

If the grating is operated in the second order, the resolution is $R = 2 \times 21\,600 = 43\,200$, which is better resolution, but the intensity of the second-order light may be lower than that of the first-order diffraction, so the first order is used.

(c) Bandwidth $= WD^{-1}$

$$D^{-1} = \frac{1}{D} = \frac{1}{\dfrac{0.180 \text{ mm}}{nm}} = 5.56 \text{ nm/mm}$$

So the bandwidth $= WD^{-1} = 2.00 \text{ mm} \times \dfrac{5.56 \text{ nm}}{\text{mm}} = 11.1 \text{ nm}$

The bandpass $= 2 \times WD^{-1}$, so the bandpass $= 22.2$ nm. This tells us that under the conditions stated, a physical slit width of 2 mm allows a wavelength range of 22.2 nm through the slit.

Another question:
What focal length is required to achieve a linear dispersion of 2.500 mm/nm if $m = 1$, $b = 555.5$ nm, and $r = 18°$? Would this be practical?

Answer:
$F = 1321$ mm $= 4.33$ ft. This would make the footprint of the instrument quite large and may therefore not be practical.

2.2.3.4.3. Echelle Gratings. Echelle gratings, in contrast to echelette gratings, have only a few grooves per millimeter (typically 300 grooves/mm or less). However, they are used in a very high order – 20–120th orders are common. A grating with so few lines and used in such a high order has severe overlap of orders. Another grating or prism that disperses at right angles to the echelle is used to disperse the radiation before or after it falls on the echelle, thus giving a spectrum in two dimensions. The main echelle grating has a very high linear dispersion and resolving power due to the high order, m (see Eqs. (2.19) and (2.22) for the effect of m on D and R). Typical resolutions and linear dispersions for echelle gratings are an order of magnitude greater than those for echelette gratings [18].

2.2.3.4.4. Holographic Gratings. Historically, the production of high-quality echelette gratings by mechanical means was difficult because of the exacting requirements of parallel grooves of incredibly small dimensions. For this reason, once a high-quality grating was made, it would serve as a "master" grating from which other slightly lower-quality replicas would be made.

Because of the difficulty of making gratings by mechanical means, a method for producing gratings using lasers was developed. These gratings are known as holographic gratings [19]. They are made by using two collimated beams of monochromatic laser radiation to produce interference fringes in a photosensitive material deposited on glass. The portion of the photosensitive material exposed to the laser beams where they constructively interfere is then washed away, creating a grooved, sinusoidal structure. The grating is then coated with an appropriate reflective layer of metal (typically aluminum) and can be used in the same manner as a ruled (echelette) grating. Because holographic gratings are recordings of optical phenomena, they have absolutely no "ghosts," which are aberrations created by imperfections in the position of the grooves. They also have very low stray light levels because stray light is scattered by irregularities in the spacing of the grooves and by imperfections in the planarity of the grooves' reflective surface. However, the efficiency (i.e., light intensity transmission) of holographic gratings is lower than that of ruled gratings. Holographic gratings can be produced with as many as 6000 grooves/mm in sizes up to 600 × 400 mm, theoretically providing better linear dispersion and wavelength resolution than ruled gratings (see previous equations). Different groove spacings are obtained by changing the wavelength of the laser beams. *Blazed holographic gratings* are ones in which one edge of each sinusoidal wave has been truncated to create a grating that has some "sawtooth" character, like echelette gratings. These gratings have higher efficiencies but still offer the main advantages of holographic gratings.

2.2.3.4.5. Mirrors. Mirrors are used to collimate and focus the light coming into and exiting a monochromator. They also reflect the light onto the dispersive device and gather the light that is reflected from the grating. In order to maximize the signal-to-noise ratio, it is desirable that monochromators gather as much of the radiant energy from the lamps as possible and transmit it to the detector. The light gathering power of a monochromator is measured by a quantity called the *f*/number, or *speed*, that is given by

$$f = \frac{F}{d} \tag{2.23}$$

where F is the focal length of the collimating mirror and d is its diameter. Specifically, the f/number is a measure of the ability of the collimating mirror to collect radiation that emerges from the entrance slit. Clearly, as the diameter of a mirror increases, it captures more incoming light, so smaller f/numbers are associated with greater light gathering power. In fact, the light gathering power increases as the inverse square of the f/number. The f/numbers of typical monochromators range from 1 to 10.

As an aside, f/numbers are also used in photography, with larger diameter apertures being associated with smaller f/numbers. Larger apertures allow more light into the camera than smaller ones, assuming the shutter is open for the same amount of time. Because more light gets in, larger apertures are used in combination with fast shutter speeds to "stop" or "freeze" motion when photographing things like sporting events, animals in motion, and waves crashing on rocks.

It is important to keep in mind that in any optical system, interactions of light with lenses and mirrors always cause some partial loss of light – about 0.4% loss per interface. In most applications, these losses are insignificant, but in some cases, they are quite detrimental to the analysis. Thus, designing systems with a minimum number of interfaces is desirable.

2.2.4.   Monochromator Designs: Putting It All Together

There are a number of ways in which the components of a monochromator (i.e., the slits, grating, and mirrors) can be arranged, as indicated in the outline on page 53. We discuss three common systems: the Fastie–Ebert, Czerny–Turner, and Littrow monochromators.

2.2.4.1.   Fastie–Ebert Monochromators. The optical features of a Fastie–Ebert monochromator are shown in Figure 2.21 [20–24]. Entrance and exit slits are on either side of the grating. A single concave spherical mirror is used as a collimating and focusing mirror. Polychromatic radiation from the lamps enters the monochromator, strikes the left side of the mirror, and is collimated and reflected toward the grating. The grating disperses the radiation into its component wavelengths. The radiation is then reflected off the right side of the collimating mirror and focused onto the exit slit. The wavelength that is passed by the exit slit is selected by pivoting the grating about its vertical axis.

2.2.4.2.   Czerny–Turner Monochromators. The Czerny–Turner monochromator (see Figure 2.22) is closely related to the Ebert monochromator, with two separate concave mirrors replacing the single collimating mirror [25]. It is generally used in more expensive systems for better resolution and stray radiation control. It is a quite common configuration for monochromators, so the brevity of this section should not be misinterpreted regarding its importance in spectrophotometer design.

2.2.4.3.   Littrow Monochromators. The Littrow configuration uses a single mirror. The source illuminates the upper portion of the mirror. The beam is collimated and directed to the grating where it is dispersed into its constituent wavelengths. In this

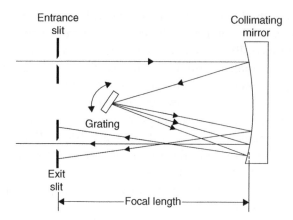

FIGURE 2.21 Bird's-eye view of the Fastie–Ebert mounting. Electromagnetic radiation enters the entrance slit, strikes the collimating mirror, and is reflected to the grating where it is dispersed into its constituent wavelengths. The dispersed light is reflected off of the collimating mirror and onto the plane of the exit slit. The exit slit allows a narrow band of wavelengths to pass through, while all others are physically blocked by the wall of the monochromator. The Fastie–Ebert mount uses a single collimating mirror, whereas the Czerny–Turner monochromator (Figure 2.22) has two separate collimating mirrors. Source: From Willard et al. [12]. Reproduced with permission from Frank A. Settle, Jr.

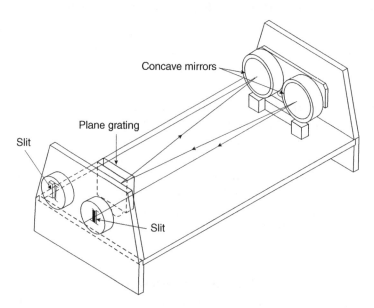

FIGURE 2.22 The Czerny–Turner mounting. This arrangement operates in the same manner as the Fastie–Ebert arrangement with the exception of having two distinct concave collimating mirrors rather than a single mirror. Source: From Willard et al. [12]. Reproduced with permission from Frank A. Settle, Jr.

arrangement, the grating's axis is parallel to and in the plane of the grooves (see Figure 2.23). The diffracted beam is sent back to the lower portion of the same mirror where it is focused on the exit slit. The upper and lower portions of the same slit assembly are used as entrance and exit slits. This configuration is commonly used for small spectrophotometers.

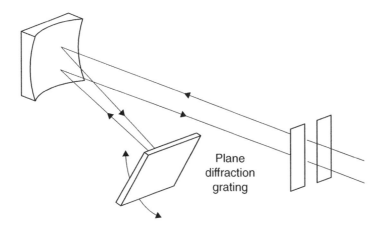

FIGURE 2.23 Littrow mounting with the diffracted beam below the source beam. This arrangement is useful for creating compact spectrophotometers. Source: From Willard et al. [12]. Reproduced with permission from Frank A. Settle, Jr.

2.2.4.4. Additional Notes About Monochromators.

Other monochromator configurations such as the Seya–Namioka, the Wadsworth, the Eagle, and the Rowland circle mounting are also used, each having specific benefits and drawbacks for different applications [12].

2.2.4.5. Double Monochromators.

A double monochromator consists of two monochromators used in series, with the output from the first monochromator serving as the input for the second. Thus, the exit slit of the first monochromator serves as the entrance slit of the second. With two identical monochromators used sequentially, the dispersion and resolution are approximately doubled, and the stray radiation is greatly reduced. We will see below that stray radiation causes serious deviations from Beer's law. For example, if the intensity of the stray radiation in each monochromator is 0.1% of that of the primary beam – the usual situation in the ultraviolet and visible regions – then the double monochromator reduces this to 0.0001% (0.1% of 0.1%). The transmission of a double monochromator is lower than that of either monochromator by itself, so there is a three-way tradeoff between throughput (and therefore S/N) on the one hand and resolution and stray radiation levels on the other.

2.2.4.6. Sample Holders.

Once the desired wavelength of light has been isolated by using either a filter or a scanning monochromator, it must be passed through the sample or reference solution (or gas). The sample and reference are typically held in square cuvettes made of plastic, glass, or quartz. For the most accurate work, square cuvettes should have perfectly parallel windows and be positioned such that the windows are perfectly perpendicular to the beam of radiation. Cylindrical cells are sometimes used in inexpensive instruments, and when they are used, it is important to position them the same way each time.

The cells used in UV-visible spectroscopy are usually 1 cm in path length, but cells are available from 0.1 cm or less up to 10 cm or more. Smaller path length cells are used if the volume of the available sample is small or if its absorbance is quite high. Considering Beer's law ($A = \varepsilon bC$), we can see that replacing a cell with a 1 cm path length with one that has a 0.1 cm path length decreases the measured absorbance by a factor of 10. Larger cells (e.g., 10 cm) are often used for gaseous samples in which the concentration of absorbing species is low. The larger path length thus increases the total number of absorbing species in the path of the light and can increase the absorbance to a point where it is more reliably measured.

Naturally, the cells must be made of a material that itself does not absorb in the wavelength region of interest. Plastic, glass, and quartz cells are common. Plastic cells are inexpensive but can only be used in the visible portion of the spectrum because they strongly absorb UV light. Glass cells are also fairly inexpensive but they cannot be used with wavelengths lower than 350 nm, because they, too, absorb strongly in the UV region. In order to perform measurements in both the UV and visible ranges, it is necessary to use quartz cells. Quartz is highly purified crystalline SiO_2, whereas glass is an amorphous and impure form of silicate. Quartz cuvettes transmit in the UV region of the spectrum (200–400 nm). They also transmit visible light and thus can be used over the entire UV-visible range. They are fragile and expensive (several hundreds of dollars for a matched pair). Quartz cells must therefore be handled carefully and are often reserved for applications that absolutely require them. Because quartz and glass cells are difficult to tell apart based on appearance, most manufacturers have some type of label, such as the letter "Q" or "G," at top of the cells.

The importance of the cell material is often overlooked. Because they are cheaper, plastic and glass cells are commonly found in laboratories and routinely used for many applications. It is therefore easy to fall into a habit of not thinking about the type of cuvette to use. This is problematic, however, if glass and plastic cells are used for UV measurements. Because they strongly absorb UV light, very little light reaches the detector, which leads to high, incorrect, and noisy absorbance readings. Thus, it is important to always consider the region in which the measurement is being made and to use the appropriate cuvettes.

Making good spectroscopic measurements requires additional considerations regarding the seemingly simple cells. For example, when using double-beam spectrophotometers (see Section 2.3), the cells should be matched, meaning that they have identical transmission properties. Briefly, in a double-beam spectrophotometer, the light shines through two cuvettes at once: the reference, which is associated with P_0 in the definition of absorbance, and the sample, which corresponds to P. Imagine if the two cells are not matched – meaning that one absorbs, reflects, or scatters a little more light than the other. Now imagine putting those cells in two independent beams of equal intensity light, one cell in each beam. If one does not transmit as much light as the other, $P_0 \neq P$. Consequently, the absorbance will not equal zero (because abs $= \log P_0/P \neq \log 1.00$ if $P_0 \neq P$). Thus, there will be a contribution to the absorbance that does not derive from the sample itself. This causes inaccurate, biased measurements. For the same reason, it is important to reproducibly position cuvettes that are used in single-beam instruments, which are commonly used for simple single wavelength absorbance measurements. It is also important to keep the windows of cuvettes scrupulously clean. Fingerprints, grease, dirt, and lint left by touching the cuvettes alter their transmission properties and lead to inaccurate absorbance measurements.

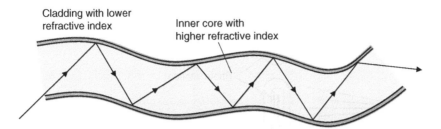

FIGURE 2.24 Depiction of total internal reflection in a fiber optic. Electromagnetic radiation traveling in the core of the fiber optic, which has a higher refractive index than the cladding, is totally internally reflected at the core/cladding interface if the angle of interaction is greater than a critical angle. The total internal reflection allows the radiation to propagate down the fiber optic and be carried to a sample or a detector.

Naturally, if the transmission were affected in EXACTLY the same way for both cells, the effect would cancel, but this is extraordinarily unlikely, and the best practice is to keep the cells clean.

Most cuvettes have two opposing sides that are "frosted," which are the sides that the light beam will not pass through. These sides are intended for handling, leaving the clear window sides, which the beam does pass through, free of fingerprints and dirt. It is also advisable to handle the cuvettes at the very top because this portion is not in the path of the light beam.

Fiber optics are useful for remote or online sensing of samples that cannot be readily contained in a standard cell – for example, when the absorbance of a flowing solution is to be measured. Fiber optics transmit radiation via total internal reflection. Total internal reflection occurs when electromagnetic radiation traveling in a transparent medium strikes an interface with a medium of lower refractive index at an angle of incidence that is larger than a certain value, known as the critical angle (see Figure 2.24). Optical fibers therefore consist of a high refractive index core surrounded by a lower refractive index material. For absorbance measurements, the incident beam is carried to the solution by the fiber optics, and the transmitted beam can then be brought back to the instrument by another fiber-optic bundle.

2.2.5. Detectors

Once the desired wavelength has been isolated from the polychromatic sources and passed through the sample, it must be detected in order to yield absorbance values. Detectors function by converting electromagnetic radiation into an electric current or voltage, which is then converted into an absorbance readout. There are three main types of detectors used in spectrophotometers: phototubes, photomultiplier tubes, and photo-diode detectors.

2.2.5.1. Phototubes.
A phototube contains an anode and a radiation-sensitive cathode sealed in a glass tube under vacuum (see Figure 2.25). The anode is held at a positive potential relative to the cathode. The cathode is in the form of a half-cylinder and

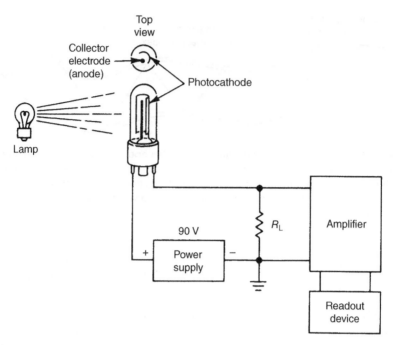

FIGURE 2.25   Phototube and its associated circuitry. Source: From Willard et al. [12]. Reproduced with permission from Frank A. Settle, Jr.

is coated with a photoemissive material. Electrons are emitted from the photoemissive material when photons strike the surface. The emitted electrons, being negative in charge, accelerate across the intervening vacuum toward the positive anode. This flow of electrons creates a current in the anode or a voltage if the current runs through a resistor. Importantly, the number of electrons ejected is directly proportional to the number of photons striking the surface. Thus, the current or voltage that is generated is also proportional to the intensity of the light reaching the detector. The signal is then amplified (along with any noise from the detector) using electronic amplifier circuits.

Phototubes are referred to as "single-stage detectors" because there is just one step involved in converting the incoming light into the electronic signal that gets processed. In contrast, photomultiplier tubes, which are discussed below, are called "multi-stage detectors" because after the light strikes the photoemissive surface, the ejected electrons undergo several stages of multiplication that dramatically increase the final electronic current that serves as the analytical signal.

The photoemissive material coating the cathode must have a high absorption coefficient for the incident radiation. It must also have a low work function (the minimum energy of a photon that will cause the ejection of an electron) in order to extend its spectral range to longer wavelengths (lower energies). Ideally, the response of the cathode would be constant for photons of all wavelengths; however, certain wavelengths of light cause certain materials to eject more electrons per photon than others. The response of different photoemissive surfaces is shown in Figure 2.26. Cathode quantum efficiency, the parameter in the illustration, is the average number of electrons emitted from the photocathode

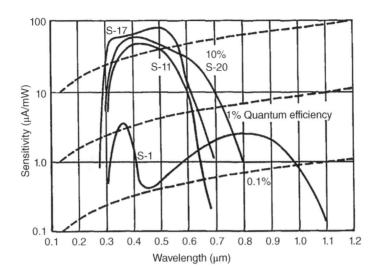

FIGURE 2.26 Sensitivity of photoemissive surfaces. S-1 is Ag—O—Cs, S-11 is Cs—Sb, S-17 is also Cs—Sb, S-20 is (Cs)—Na—K—Sb. Source: Designations from Turner [26]; From Anderson and McMurtry [27]. Reproduced with permission of The Optical Society.

divided by the number of incident photons. It is apparent from the graph that S-20, which is made of Cs–Na–K–Sb, has a broad response over the UV-visible portion of the spectrum. Ga–As (not pictured) has an even flatter response that extends to about 900 nm. Other materials such as Sb—Cs and Ag—O—Cs offer higher sensitivities in the UV and near-infrared (800–1100 nm) regions, respectively, but with varying response in these regions.

The primary source of noise in photoemissive detectors at room temperature is generally shot noise, which is due to the fundamental quantized nature of electromagnetic radiation and electrical currents. Photons arrive at the cathode randomly, even though the overall intensity of the radiation is constant. Thus, photoelectrons are emitted from the cathode and arrive at the anode randomly. Yet, the long-term rate of photoelectron pulses at the anode is constant and proportional to the radiant power. So the random nature of the process creates minor fluctuations on top of a nearly continuous signal.

Phototubes are generally less expensive than the photomultiplier tubes discussed below and thus are frequently used in less expensive spectrophotometers. They are also less sensitive, so they are used mainly with high intensity sources.

2.2.5.2. *Photomultiplier Tubes (PMTs).*

Photomultiplier tubes, as their name implies, act to multiply the original signal generated when light strikes a photoemissive surface [28]. They are composed of (1) a photocathode similar to that described above for a phototube, (2) a series of dynodes that eject multiple electrons when one electron strikes their surface, and (3) an anode that ultimately collects the electrons and generates a current in response to incident light striking the detector. A bird's-eye view of a PMT is shown in Figure 2.27. Light enters through a quartz window and strikes the photosensitive cathode. The emitted electron is accelerated toward the first dynode by holding that dynode at a positive voltage relative to the cathode (+75–100 V). The high-speed electrons cause

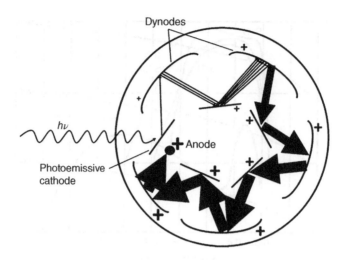

FIGURE 2.27 Bird's-eye view of a photomultiplier (PMT). A photon strikes the photoemissive cathode and ejects an electron. The electron is accelerated toward first dynode, which is held at a positive voltage relative to the photocathode. Multiple electrons are ejected from dynode 1 for every electron that strikes it. This cycle of acceleration with subsequent ejection of electrons continues at all of the dynodes until the electrons are collected at the anode. Source: Courtesy of Radio Corporation of America.

multiple electrons to be ejected from the first dynode surface (BeO, GaP, and CsSb are common). These ejected electrons are then focused and accelerated toward the second dynode, which is held at 75–100 V higher than the first dynode. Repeating this electron-multiplying process produces a current avalanche that finally impinges on the anode. Internal current amplification, or gain, is achieved by this process. The number of electrons ejected from a dynode is dependent on the energy of the incoming electrons. Thus, the number of ejected electrons depends on the velocity of the incoming electrons, which in turn depends on the potential difference between each successive dynode. By increasing the potential difference, electrons are accelerated to higher speeds, causing more electrons to be ejected from the next dynode, thereby ultimately increasing the gain of the PMT. Phototubes, in contrast, have no internal amplification. However, the signal from both phototubes and photomultiplier tubes can be amplified using external circuitry.

PMTs are much more sensitive than phototubes because of the internal multiplication that takes place. This is advantageous in UV-visible spectroscopy because detecting low analyte concentrations requires differentiating between two large light signals – the reference or blank, which ideally lets through nearly all of the light, and the very dilute sample, which absorbs only a bit more light than the reference. Being able to amplify these small differences thus contributes to lower detection limits when using PMTs compared to phototubes.

It is important to recognize that because PMTs are incredibly sensitive, intense light causes irreversible damage to them. Thus, when powered up, PMTs should never be opened to light from the surroundings or other bright sources. PMTs are also subject to the same sources of noise as phototubes, such as stray radiation from space, ohmic resistance, spontaneous thermal emission of electrons, radioactive decay, etc. In PMTs, however, this noise is partially amplified by the same process that multiplies the radiation of interest.

2.2.5.2.1. Photon Counting. Because of the multiplicative nature of the process, PMTs are much more sensitive than simple phototubes. In fact, PMTs are sensitive enough to detect pulses arising from individual photons that fall on the photocathode at low light levels. This allows for photon counting. The number of photons counted per unit time represents the intensity of the light striking the detector and thus serves as the analytical signal. Photon counting generally results in higher S/N ratios than measurements of the average output current at low power levels. To count individual photons, a wideband amplifier is used in series with a pulse-height discriminator to eliminate low-amplitude spurious pulses that originate from thermal emission of electrons from the dynodes. These pulses are amplified to a lesser extent than pulses that originate at the photocathode because they have not gone through all of the multiplication steps. The equipment and signal processing techniques for photon counting are more complex and expensive than for regular analog signal processing, but some low-light-level applications – such as fluorescence, chemiluminescence, and Raman spectroscopy – benefit from the improved signal-to-noise ratios. Photon counting is not often required for UV-visible spectroscopy because the light levels are typically strong enough that analog signal processing suffices.

2.2.5.3. Semiconductor Photodiodes.

Photodiodes operate on a completely different principle than phototubes and PMTs. They are constructed from a semiconducting silicon p–n junction as shown in Figure 2.28. P-type material ("p" for positive) is created by doping an element such as boron, which has only three valence electrons, into the crystal structure of silicon, which has four valence electrons. The boron thus looks like it lacks an electron and creates a "hole" in the material. Conversely, n-type ("n" for negative) is created by doping silicon with an element that has five valence electrons, such as phosphorus. By applying a reverse-bias voltage using an external voltage source, the "holes" (lack of electrons) in the p-material and the electrons in the n-material are separated from one another as depicted in Figure 2.28. In this state, the conductance of the junction between the p- and n-materials is reduced to such an extent that no current flows through the diode. However, when photons interact with the intrinsic region of the diode, which is

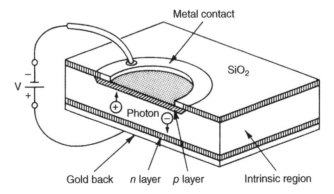

FIGURE 2.28 Construction of a planar diffused p–n junction photodiode. P-type material has "holes" and n-type material has "excess" electrons. When a voltage is applied across the p–n junction, holes and electrons separate as depicted in the figure. Source: From Willard et al. [12]. Reproduced with permission from Frank A. Settle, Jr.

made of a high-resistivity silicon material, electrons are promoted to the conduction band where they can act as charge carriers. This allows current to flow through the diode. The magnitude of the current is proportional to the power of the incident radiation and is also dependent on the wavelength of light striking the detector. In this way, semiconductors are used to detect and quantify electromagnetic radiation.

2.2.5.4. *Photodiode Array Detectors (PDA).*

2.2.5.4. Photodiode Array Detectors (PDA). Many tiny photodiodes can be assembled in a linear or two-dimensional array in which each diode gathers a signal simultaneously. A tiny capacitor is coupled to each diode and charged to a given level before illumination of the diode. When photons strike the diode, charge leaks off the capacitor. After the signals are obtained, each diode in the array is scanned and independently recharged to its initial charge state. The time-integrated current required to recharge each individual capacitor is recorded and is proportional to the number of photons that struck each separate diode. Through this process, the power of light (e.g., P or P_0 in Beer's law) striking each diode is measured.

With a photodiode array detector (PDA), different light levels at different positions in space can be recorded. This differs from phototubes and PMTs, where light must strike only one area of space (the photocathode) to be recorded. To foreshadow parts of the discussion below regarding instrument design, consider the effect of these different approaches to recording a spectrum (i.e., absorbance as a function of wavelength). For phototubes and PMTs, the absorbance at different wavelengths must be recorded over *time*, with each separate wavelength being passed through the sample and on to the detector *sequentially*. When using a PDA, a grating can be used to spread out polychromatic light into its constituent wavelengths and simultaneously illuminate hundreds of tiny diodes that are separated in *space* and thus collect an entire spectrum *at one time*. There are benefits and drawbacks to both approaches and these are discussed below.

Linear photodiode arrays are commercially available with 128 to as many as 4096 diodes per array. The diodes themselves can be 25 µm wide or smaller and are about 2.5 mm high. Thus, an entire array containing several hundred or thousands of diodes is several centimeters long. Photodiode arrays make it possible to rapidly record an entire spectrum. They significantly simplify the design of spectrophotometers and have been used in many areas of analytical chemistry [29–32]. However, they do not provide the sensitivity or the resolution that is possible with monochromators combined with PMTs.

2.2.5.5. *Charge-Coupled Devices (CCDs).*

2.2.5.5. Charge-Coupled Devices (CCDs). A variation of photodiode array detection is the charge-coupled device (CCD). These are two-dimensional arrays made up of diodes in both rows and columns. In these devices, the readout of each diode in the array is triggered by a transfer pulse, which essentially acts to "push" the charges that accumulated in the diodes toward one end of the array, where they are rapidly and sequentially recorded. CCDs are thus used to measure light intensity in a two-dimensional field, like a small section of a Petri dish, for example. They are relatively expensive compared to simple phototubes, PMTs, and PDAs and create a large amount of spatially resolved information that demands extensive computer memory, but because the costs of the detectors and computer memory have fallen, CCDs are common components in laboratory instrumentation.

2.3. SPECTROPHOTOMETER DESIGNS

As indicated in the outline on page 53, having described the various components of a spectrophotometer and how they function, we can now discuss how the components are arranged within an instrument to measure absorbance. Recall that there are two main things that can be learned about a sample by making absorbance measurements. First, Beer's law tells us that the absorbance at a specific wavelength is proportional to the concentration of absorbing species in solution ($A = \varepsilon bC$). Thus, quantitative information can be obtained. Second, different molecules absorb more or less strongly at different wavelengths because of their different electronic ground and excited state distributions. Therefore, the UV-visible spectrum of a molecule serves as a structure-dependent fingerprint. In this section, we discuss how various instrument designs are used for single wavelength absorbance measurements and to collect entire UV-visible spectra (i.e., a plot of absorbance versus wavelength).

Also recall that early in the chapter we emphasized the fact that each absorbance measurement results from two independent measurements of light intensity, P and P_o, and that these comprise the definition of absorbance: $A = \log P_o/P$. Furthermore, because the absorbance of a sample varies with wavelength, the quantities P_o and P must both be measured at every wavelength of interest. Therefore, in this section we discuss how the different spectrophotometer designs make these two intensity measurements for each sample analyzed.

2.3.1. Single-Beam Spectrophotometers

The simplest instruments for UV-visible absorbance measurements are single-beam spectrophotometers [33]. The light path and components of a single-beam instrument are shown in Figure 2.29. In this instrument, light from the tungsten lamp is focused through a lens and passes through the entrance slit. It strikes the grating, which disperses it into its constituent

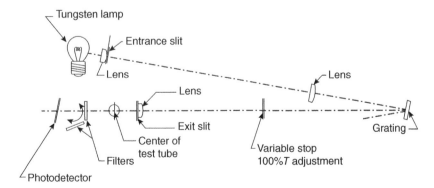

FIGURE 2.29 Top view of the optical arrangement of the single-beam Thermo Scientific™ SPECTRONIC™ 20 visible spectrophotometer. Light from the tungsten lamp is directed through a slit and onto the grating where it is diffracted into its constituent wavelengths. A narrow range of wavelengths selected by the exit slit passes through the test tube that contains the sample where some light may be absorbed, and is ultimately detected by the photodetector. Source: Reproduced with permission of Thermo Scientific™.

wavelengths. The lens focuses the dispersed light onto the focal plane of the exit slit. The slit allows only a narrow band of wavelengths to pass through it, while all other wavelengths are physically blocked. Light of the selected wavelength then passes through the test tube that contains the sample. The sample absorbs some of the light if the energy of the light matches an electronic transition of the molecules in the sample. The light that is not absorbed (i.e., the light that is transmitted through the sample) then goes through filters that remove the higher-order wavelengths. The light then strikes the detector, which in single-beam instruments can be a phototube, photomultiplier tube, or silicon diode detector, depending on the particular instrument.

Typically, the reference solution (usually the solvent for the analyte) is inserted into the instrument first. At this point, two readings are usually taken – one that sets the 100% transmittance and one that sets 0% transmittance. To set the 0% transmittance, a shutter that entirely blocks the light beam is put in place. Any current (i.e., the dark current) generated by the detector in the absence of light is recorded and essentially set equal to zero in the "mind" of the instrument. Sometimes you can even hear a "click" when the shutter is moved into and out of place. Note, though, that the current is not really zero but rather the dark current is subtracted from all future readings to define 0% transmittance. The 100% transmittance is then set by opening the shutter and allowing light of the selected wavelength to pass through the reference solution. The full strength of the beam does not, in reality, reach the detector even if the reference sample does not absorb at the selected wavelength. There are reflection and scattering losses that occur at the surfaces of the cuvette, plus potential scattering from dust and other particles in the solution. The intensity of the light that does reach the detector generates a current, which, by designating the solution as a reference, is associated with 100% transmittance. In other words, by designating a solution as a reference, we are telling the instrument that this is the maximum amount of light expected. This, then, serves as the current that is associated with P_0 in the definition of absorbance ($A = \log P_0/P$). It is one of the two measurements required to determine an absorbance.

To determine the absorbance of a sample, you then remove the reference solution and replace it with the sample. Because the sample contains the absorbing analyte(s), some of the light is absorbed, decreasing the intensity of the transmitted beam and consequently decreasing the current generated when the light strikes the detector. This current is P – the second measurement required to determine an absorbance value. For the absorbance value to reflect ONLY the absorbance due the analyte, all other possible reasons for changes in the transmitted light intensity must be eliminated. For example, imagine two cuvettes, each filled with the exact same solvent, but one cuvette has multiple fingerprints and/or scratches on the sides. If the clean, unscratched cuvette is used to set the reference signal (P_0), and the second cuvette is used as the "sample," because the second cuvette will prevent more light from reaching the detector than the first, a nonzero absorbance will be observed even if both cuvettes contain the exact same solvent and should, therefore, produce an absorbance of 0.00. Thus, in a single-beam instrument, the same cuvette should be used for the reference and the sample because each cuvette reflects and scatters light in its own particular way. Furthermore, cuvettes should be wiped free of external dirt and fingerprints and positioned in the same way each time because different parts of the same cuvette may interact with light in slightly different ways.

There are several important consequences of the fact that two independent measurements (three if the 0% transmittance setting is included) comprise each absorbance value. Practically, it means that any absorbance value calculated from the signals arising from the transmitted light *is only as good as the reference measurement*. This solution and measurement that we often treat cavalierly affects every subsequent measurement. Thus, it is critical and should be prepared accordingly. Furthermore, the reference solution should be analyzed frequently to check for any changes or errors that might relate to it. It should produce a zero absorbance reading every time – if not, something has changed and the data that has been collected since the last time the reference solution was checked should not be trusted. Lastly, single-beam instruments are prone to "drift" – variations in intensity readings due to changes in the performance of the source, detector, and other electronics. Thus, the 0% and 100% transmittance settings should be rechecked often over the course of data collection. This issue is discussed in more detail below.

You may have already gotten the sense that one drawback to single-beam instruments is that they require a lot of manual manipulation – first, putting in the reference, blanking the instrument, removing the reference, putting in the sample, and then recording the measured absorbance. If you want to collect an entire spectrum, this needs to be repeated at every wavelength of interest because the lamp intensity, transmission properties of grating and lenses, and detector sensitivity all vary with wavelength. So to record a spectrum from 200 to 800 nm at 1 nm intervals requires 1200 changes of the reference and sample solutions to measure P and P_o at each successive wavelength if the same cuvette is used for all measurements. Thus, single-beam instruments are not often used for recording entire spectra. Double-beam or diode array instruments, which are described below, are much more commonly used to record spectra.

Single-beam instruments are generally inexpensive (less than $5000), rugged, and easy to maintain. Some instruments do have electronics that allow for scanning by first recording and storing the entire blank spectrum with a reference solution in place (i.e., P_o as a function of wavelength), then recording P as a function of wavelength with the sample in place, and lastly converting the stored data into the final spectrum.

In some single-beam systems, the monochromator is replaced by filters. Monochromators allow for any wavelength to be chosen, but filter systems are restricted to isolating wavelengths of the filters that are available.

In most single-beam instruments the slits are generally not adjustable, so the bandwidth is fixed, typically between 2 and 20 nm. Some instruments do not include a deuterium lamp and thus do not cover the UV portion of the spectrum, but many modern instruments are capable of covering the range from 190 to 1100 nm, which includes the UV, visible, and a portion of the near infrared (NIR) regions of the spectrum. Those that cover the NIR often have a diode detector in addition to or in place of a phototube or PMT in order to detect the low energy NIR radiation. Thus, single-beam instruments can range from simple filter-based systems that are limited to the visible and NIR regions and have just a simple phototube or photodiode detector, up to systems that scan the full range from 190 to 1100 nm using a holographic grating, with two light sources and two detectors. Naturally, the price tends to scale with the complexity and flexibility of the instrument. Despite all of the potential limitations, single-beam instruments are quite appropriate for single wavelength absorbance determinations and are thus applicable to studies based on Beer's law (i.e., concentration determinations). Because a vast number of spectrophotometric studies rely on Beer's law, single-beam spectrometers are important staples of most analytical laboratories.

2.3.1.1. Checking for True Zero Absorbance.

Because of the electronic ease of setting zero and 100% transmittance – simply pressing a button on most instruments – it is easy to just insert the reference, press the button, and let the instrument perform calculations accordingly. But consider the following scenario: Imagine an exceptionally dilute sample of p-nitroaniline (e.g., 10^{-5} M) dissolved in toluene. p-Nitroaniline absorbs strongly at 200 nm, so it would be tempting to conduct an analysis at this wavelength. It might make sense, then, to insert a cuvette of pure toluene into the spectrometer as a reference, set the wavelength to 200 nm, hit the "blank" or "background" button, and allow the instrument to automatically set the 0 and 100% transmittance. This would be followed by inserting the sample with the p-nitroaniline, for which the instrument will return some absorbance value (that is what they are made to do – no questions asked). Procedurally, everything seems to be in order: A wavelength at which the analyte absorbs was selected, a proper solution was used to "zero" the instrument, and then the sample was analyzed. The problem, however, is that the solvent itself absorbs strongly at 200 nm (all aromatic compounds absorb strongly in the UV region). Thus, the solvent absorbs a significant amount of the light so that very little is actually reaching the detector. So, what the instrument is really being asked to do is to discriminate between two potentially very weak signals – not a lot of light is being passed by either the reference or the sample solution. Analytically, this situation is prone to inaccurate and irreproducible results. For this reason, if you are not completely familiar with the absorbance properties of the solvent and the sample you are investigating, it is highly recommended that you first "zero" the instrument with water at the analytical wavelength of interest because water does not absorb anywhere in the UV-visible region. In this way, water is acting as a true "blank" – a sample that allows the maximum amount of light through to the detector. Then the "reference" is inserted *as if it were a sample*. If it has a significant absorbance (0.5 or above), it should be noted that the light intensity that ultimately serves to create the signal is being diminished by the reference itself, thus making the absorbance measurement more difficult. If an absorbance above 0.5 for the "reference" relative to water is observed, then another wavelength of analysis should be sought. To understand why, consider that an absorbance of 2 means that only one photon out of 100 reaches the detector and that an absorbance of 3 means only one out of 1000 reaches the detector. These numbers are the result of the definition $A = \log P_0 / P$. Thus, if the reference sample has a high absorbance, it will be difficult to measure any additional absorbance caused by an analyte. Analytically, *it is very difficult to accurately measure a small difference when very little light is getting through to the detector in the first place.* Even though the instrument will "zero" with apparent ease and continue to provide absorbance readings, we must recognize that it is merely electronically and blindly following what it is manufactured to do. We, as users, must do the proper checks to make sure that the numbers instruments produce are meaningful and analytically reliable. This requires that each absorbance measurement is based on a technically sound establishment of 0 and 100% transmittance (i.e., a good "reference" measurement) and on an equally sound measurement of the absorbing sample. Further recommendations for conducting proper UV-visible analyses are discussed in subsequent sections.

2.3.2. Scanning Double-Beam Instruments

In contrast to the single-beam instruments described above, a double-beam instrument splits or chops the light from the sources into two equally intense beams and sends them down two independent paths – one beam passing through the sample solution and the other beam passing through the reference solution [34]. Thus, the measurements of P and P_0 are made simultaneously without having to take out the reference and replace it with the sample solution as needs to be done with a single-beam instrument. Because this comparison between reference and sample is continuous, it is convenient to use double-beam instruments to scan through a range of wavelengths and record the absorbance as a function of wavelength – in other words, to record a UV-visible spectrum. Double-beam instruments are more expensive than single-beam instruments, ranging from $10 000 to $40 000 or more. In the following section we look at two different types of double-beam instruments – "double-beam-in-time" and "double-beam-in-space."

2.3.2.1. Double-Beam-in-Time Instruments. A diagram of a double-beam-in-time instrument is shown in Figure 2.30. In this arrangement, radiation from either the visible or ultraviolet source (A) reflects off a mirror (M_1) and enters the grating monochromator in the Czerny–Turner configuration through the entrance slit (S_1). The beam is reflected off the first collimating mirror (M_2) and onto the grating (G). The diffracted light is then reflected off the second collimating mirror (M_3) onto the plane of the exit slit (S_2). A narrow range of wavelengths exits the slit and is reflected off of a series of mirrors (M_4–M_6). The beam then encounters the rotating chopper wheel (CW_1), which has alternating open spaces and mirrored surfaces. When the beam encounters the open portion of the chopper wheel, it proceeds through the opening to a mirror (M_7), which directs it through the near cell in the diagram (C_1). When the light passes through a cell, it can be absorbed by the components in the cell. The two cells contain either the sample or the reference solution.

When the beam encounters the mirrored portion of the rotating chopper wheel, the beam reflects off the mirrored surface onto another mirror (M_8) and through the far cell (C_2). After passing through the cells, both beams reflect off mirrors (either M_9 or M_{10}) and then encounter another chopper wheel (CW_2). Note that the chopper wheels are in sync, such that when the open portion is present at the first chopper wheel (CW_1), the mirrored portion is present at the second (CW_2). In this configuration, light that has passed through the cell labeled C_1 strikes a mirror (M_9), reflects off the second chopper wheel (CW_2), reflects off another mirror (M_{11}), and then strikes the detector (D). If the reference is in cell C_1, the light intensity striking the detector is P_0. When the mirrored portion of the first chopper wheel is in place, light is directed through the other cell (C_2), onto a mirror (M_{10}), through the open portion of the second chopper wheel, onto the final mirror (M_{11}), and then to the detector (D). In this case, if the second cell contains the sample, the signal is a measure of P. Taking the ratio of P to P_0 yields the transmittance, which can be converted into an absorbance.

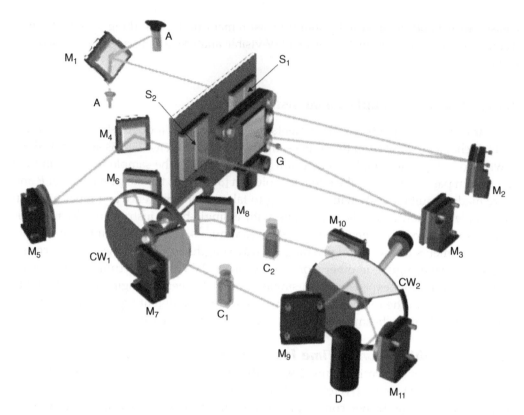

FIGURE 2.30 Diagram of a double-beam-in-time spectrophotometer (Cary 100 – courtesy of Varian Associates, Inc.). Polychromatic electromagnetic radiation from the UV and visible sources (A) is directed off a mirror (M_1), through the monochromator entrance slit (S_1), onto a collimating mirror (M_2) and onto the grating (G) that disperses it into its constituent wavelengths. The dispersed electromagnetic radiation reflects off the second collimating mirror (M_3) and is focused onto the plane of the exit slit (S_2). A narrow band of wavelengths exits the slit and is directed by mirrors onto the chopper wheel (CW_1). As explained in more detail in the text, depending on whether the open or mirrored portion of the chopper wheel is in place, the light goes through either the reference or the sample cell where it can be absorbed. Whatever light is not absorbed passes through the solution and is ultimately directed onto the detector by mirrors and the second chopper wheel. In this way, independent measurements of P to P_0 are obtained. Taking the ratio yields the transmittance of the sample, which is electronically converted into an absorbance value. The chopper wheel spins continuously so that P and P_0 (and hence absorbance) are constantly updated. While the chopper wheel is spinning, the grating is rotated, which allows different wavelengths to be passed through the reference and sample solutions. In this way, the absorbance is measured at different wavelengths to obtain an entire spectrum without having to manipulate the solutions once the scan is started. Source: Reproduced with permission of Agilent Technologies, Inc.

The optical chopper wheels spin at a high speed. While that is happening, the grating is rotated to change the wavelength that passes through the exit slit of the monochromator. As the different wavelengths exit the slit, the detector alternately senses the intensity of light coming from the sample and reference paths – reference, sample, reference, sample, etc. This is how the instrument records the P_0 and P signals as the instrument scans through the wavelengths. By rapidly sensing the power of both beams, and by knowing how fast the grating is being rotated (and thus how fast the wavelength is changing), an entire spectrum is recorded without any manual manipulations of the solutions once the scan is started.

An advantage of this design is that if the intensity of the source changes, all but the highest frequency fluctuations are passed through both the reference and sample solutions because of the rapid rotation of the chopper. When the ratio P_0/P is then calculated electronically, the increase or decrease in source intensity cancels, leaving the absorbance value unaffected. This automatic compensation for drift is a significant advantage over single-beam instruments, which despite some attempts to compensate for such effects, are more prone to drift over time. Measuring the ratio in this way also automatically compensates for variations in the lamp intensity, transmission properties of gratings and lenses, and the sensitivity of detectors as a function of wavelength that cause P_0 to vary with wavelength.

2.3.2.2. Double-Beam-in-Space Instruments.
Another method for creating a double-beam instrument, although less popular than the double-beam-in-time design, is the double-beam-in-space arrangement. Here, a double beam is created by literally splitting the beam in half, focusing one-half on the reference cell and the other half on the sample cell. Figure 2.31 shows a schematic of such an arrangement.

Many of the components of a double-beam-in-space instrument are the same as a double-beam-in-time instrument. The main difference is that a half-mirror or beam splitter replaces the chopper as the device that separates the beam into two. Of course, a chopper does not literally separate the beam but rather redirects the entire beam alternately along two different paths. In contrast, the half-mirror or beam splitter in a double-beam-in-space

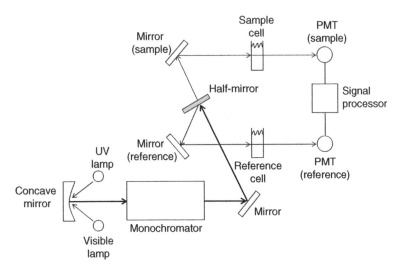

FIGURE 2.31 Diagram of a double-beam-in-space spectrophotometer. Polychromatic light from UV and visible lamps is directed through the monochromator, where it is dispersed into its constituent wavelengths. The beam splitter (half-mirror) reflects half the light and transmits the other half. The two beams that are created pass through the sample and reference cells simultaneously and proceed to the sample and reference PMTs. The ratio of the two signals is taken and converted into the absorbance.

instrument does split the beam into two, with half the intensity sent along two different paths simultaneously. The other significant difference is that the double-beam-in-space instrument requires two separate detectors, whereas the double-beam-in-time design requires only one. The signal generated by the detector associated with the reference path serves as P_o, whereas P is associated with the sample path signal.

The same advantage that was described above for double-beam-in-time instruments (i.e., compensation for source fluctuations and wavelength-dependent properties of the sources, optical components, and detectors) is also true for double-beam-in-space designs. The main disadvantage of the double-beam-in-space design, however, is that the two detectors must have equal sensitivities to one another. Imagine if this is not the case. Then, even if the same solution is put in both cells (i.e., the true abs = 0.00), the measured absorbance would not be zero. More specifically, if the sample detector gives a greater response to the same intensity of light than the reference detector, P would be greater than P_o and result in a negative absorbance value. If the converse is true, a positive absorbance value would result. In either case, the reading would be erroneous. For this reason, great care is taken by manufacturers to use matched detectors (i.e., ones that produce the same response at all wavelengths).

Another disadvantage of the double-beam-in-space design is that the intensity of the beam that goes through both the reference and sample cells is at most half of the total intensity from the sources. In contrast, in the double-beam-in-time design, all of the output from the sources (at the selected wavelength) is being passed through the sample or reference cells. Thus, the double-beam-in-time instrument produces stronger signals with the concomitant advantage of improved signal-to-noise ratios.

2.3.2.3. Double-Beam Measurement Considerations and Performance Characteristics. In double-beam instruments, the use of matched cells is critical. Commercial suppliers sell matched sets of cells so that both cells reflect and/or scatter the same amount of light and thus do not themselves contribute to the absorbance measurements. Quartz cells are commonly used in double-beam instruments because double-beam instruments are commonly used to record an entire spectrum, which often includes wavelengths in the UV region where glass and plastic cells absorb. In contrast to single-beam instruments that have fixed slit widths and thus fixed bandwidths, scanning instruments have variable slits that are typically controlled electronically. Remember that while the instrument setting may say 1 nm slits, the slit itself is not that narrow, but rather the setting reflects the range of wavelengths that are being passed through the slit. Narrow slits provide better wavelength discrimination but allow less light to reach the detector.

Another common feature of all scanning spectrophotometers is the ability to scan at a variety of rates. Typical scan rates vary anywhere from 1 nm/min up to several thousand nm/min. Scanning at a high speed allows for the rapid collection of an entire spectrum, but can also distort spectral features, particularly for sharp, narrow spectral features. The effect is somewhat like the effect of having slits that are too wide, as shown in Figure 2.17. Scanning at slow rates (e.g., 1 nm/min) avoids distortions, but will obviously take longer to cover the same wavelength range. For example, scanning from 200 to 800 nm at a scan rate of 10 nm/min means that it will take one hour to collect the spectrum. If accurate wavelength and absorbance values are needed, this compromise must be made, but if the instrument is being used just to get an idea of the spectral features of a new compound and only qualitative information is desired, faster scan rates will suffice. So again, the user must think about the kind of information that is desired and adjust the operating conditions appropriately. In practice, the easiest thing to do is collect a spectrum at a fast scan rate and then collect another spectrum at a slower rate. If no difference is noted between the two, then the faster scan rate is acceptable and will save time for repetitive measurements. If differences are observed, then continue to record spectra at successively slower scan rates until no distortions to the spectra are noted.

The technical specifications for the PerkinElmer LAMBDA 1050 UV/Vis/NIR spectrophotometer are shown in Table 2.1. This is a research-grade instrument and thus provides an example of the kinds of high performance characteristics commercially available. Other instrument manufactures offer comparable systems. First note that the wavelength range covers 175–3300 nm and thus extends well into the near infrared region of the spectrum. Also notice the low stray light levels and that the wavelength resolution of the grating is ≤0.05 nm in the UV region. Perhaps the most important thing to note, however, is the photometric accuracy. This is a measure of the difference between the absorbance that is read using the instrument and the true absorbance for various standards. For the filters from NIST, the accuracy is only 0.003 or 0.002 absorbance units. In general, most people believe absorbance readings only to this level of accuracy. Thus, even though the instrument may read out in more significant figures, the additional digits beyond the thousandths of an absorbance unit should be viewed with skepticism. Note, however, that the reproducibility is much greater. So even though the reproducibility is quite high, the accuracy of the results may not warrant reporting more than three or four significant digits.

2.3.3. Photodiode Array Instruments

In both the single-beam and double-beam instruments described above, light is passed through the sample one wavelength at a time. In other words, the continuous light provided by the deuterium and tungsten-halogen lamps is first dispersed into its constituent wavelengths, and then individual wavelengths are sequentially passed through the sample and/or reference. Diode array instruments reverse this [35]. All of the light coming from the sources is passed through the sample at the same time and then broken up into its constituent wavelengths, which are dispersed in space and impinge upon a linear diode array detector. A schematic of such an instrument is shown in Figure 2.32. The grating does not rotate, so each wavelength of light gets dispersed to the same diode each time a measurement is made. Thus, each diode is, in a sense, "dedicated" to measuring a particular narrow range of the spectrum. The signal from each diode is recorded to create of plot of absorbance versus wavelength. The beauty of the diode array is that the intensity of the light reaching the diode array detector *for every wavelength* is measured in essentially a matter of seconds. This is a considerable advantage over scanning instruments that can take anywhere from several minutes to an hour to record a high-quality spectrum.

Note that diode array instruments are like single-beam instruments in that they have only one sample holder and one main beam of light. Thus, recording a spectrum requires

TABLE 2.1 Technical Specifications for the PerkinElmer LAMBDA 1050 UV/Vis/NIR Spectrophotometer

Technical description and specifications	LAMBDA 1050
Principle	Double-beam, double-monochromator, ratio-recording UV/Vis/NIR spectrophotometer controlled with microcomputer electronics by DELL PC or compatible personal computer
Optical system	All reflecting optical system (SiO_2 coated) with holographic grating monochromator 1440 lines/mm UV/Vis blazed at 240 nm and 360 lines/mm NIR blazed at 1100 nm, Littrow mounting, sample thickness compensated detector optics
Beam splitting system	Chopper (46+ Hz, Cycle: Dark/Sample/Dark/Reference, Chopper Segment Signal Correction Reference, CSSC)
Detector	Photomultiplier R6872 for high energy in the entire UV/Vis wavelength range. Combination of high-performance Peltier-cooled InGaAs detector, 2 options: narrow band covering 860–2500 nm and Peltier-cooled PbS detector for1800/2500–3300 nm in the NIR wavelength range
Source	Prealigned tungsten halogen and deuterium. Utilizes a source doubling mirror for improved UV/Vis/NIR energy
Wavelength range	175–3300 nm (N_2 purge required below 185 nm)
UV/Vis resolution	≤0.05 nm
NIR resolution	≤0.20 nm
Stray light	
At 200 nm (12 g/L KCl USP/DAP method)	>2 A
At 220 nm (10 g/L NaI ASTM method)	≤0.00007 %T
At 340 nm (50 mg/L $NaNO_2$ ASTM method)	≤0.00007 %T
At 370 nm (50 mg/L $NaNO_2$ ASTM method)	≤0.00007 %T
At 1420 nm (H_2O 1 cm path length)	≤0.00040 %T
At 2365 nm ($CHCl_3$ 1 cm path length)	≤0.00050 %T
Wavelength accuracy	
UV/Vis	±0.08 nm
NIR	±0.300 nm
Wavelength reproducibility	
UV/Vis (deuterium lamp lines)	≤0.010 nm
NIR (deuterium lamp lines)	≤0.040 nm
Standard deviation of 10 measurements UV/Vis	≤0.005 nm
Standard deviation of 10 measurements NIR	≤0.020 nm
Photometric accuracy	
Double Aperture Method 1 A	±0.0003 A
Double Aperture Method 0.5 A	±0.0003 A
NIST 1930D Filters 2 A	±0.0030 A
NIST 930D Filters 1 A	±0.0030 A
NIST 930D Filters 0.5 A	±0.0020 A
$K_2Cr_2O_7^-$ Solution USP/DAP method	±0.0080 A
Photometric linearity	
(Addition of filters UV/Vis at 546.1, 2 nm slit 1 s integration time)	
At 1.0 A	±0.0060 A
At 2.0 A	±0.0160 A

TABLE 2.1 (Continued)

Technical description and specifications	LAMBDA 1050
At 3.0 A	±0.0050 A
NIR at 1.0 A (1200 nm)	±0.0005 A
NIR at 2.0 A (1200 nm)	±0.0010 A
Photometric reproducibility	
Standard deviation for 10 measurements, 2 nm slit, 1 s integration time	
1 A with NIST 930D Filter at 546.1 nm	≤0.00016 A
0.5 A with NIST 930D Filter at 546.1 nm	≤0.00008 A
0.3 A with NIST 930D Filter at 546.1 nm	≤0.00008 A
Photometric range	
UV/Vis	8 A
NIR	8 A
Photometric display	Unlimited
Bandpass	0.05–5.00 nm in 0.01 nm increments UV/Vis range
	0.20–20.00 nm in 0.04 nm increments NIR range
	Fixed resolution, constant energy, or slit programming
Photometric stability	≤0.0002 A/h
After warm-up at 500 nm, 0 A, 2 nm slit, 2 s integration time, peak to peak	
Baseline flatness	±0.0008 A
190–3100 nm, 2 nm slit	
0.20 s UV/Vis, no smoothing applied	
0.24 s NIR integration time, no smoothing applied	
Photometric noise RMS	
1 s integration time, UV/Vis, PMT, 2 nm slit	
0 A and 190 nm	≤0.00010 A
0 A and 500 nm	≤0.00005 A
4 A and 500 nm	≤0.00100 A
6 A and 500 nm	≤0.00500 A
2 A and 1500 nm (NIR, PbS series slit)	≤0.00010 A
3 A and 1500 nm, (NIR, PbS series slit)	≤0.00250 A
3 A and 1500 nm (NIR, wideband InGaAs servo slit)	≤0.00010 A
3 A and 1500 nm (NIR, narrowband InGaAs servo slit)	≤0.000025 A
Primary sample compartment dimensions	$(W \times D \times H)$ 200 × 300 × 220 mm^3
Secondary sample compartment dimensions	$(W \times D \times H)$ 480 × 300 × 220 mm^3
Purging	
Optics	Yes
Sample compartment	Yes
Instrument dimension $(W \times D \times H)$	1020 × 630 × 300 mm^3
Instrument weight	~77 kg
Digital I/O	RS 232 C
Light beam	90 mm above the base plate
	120 mm beam separation
	3–12 mm beam height
Instrument requirements	Power 90 VAC–250 VAC, 50/60 Hz; 250 VA
	Temperature 10–35 °C
	10–70% relative humidity, noncondensing

Source: Reproduced with permission of PerkinElmer, Inc., ©2007–2018.

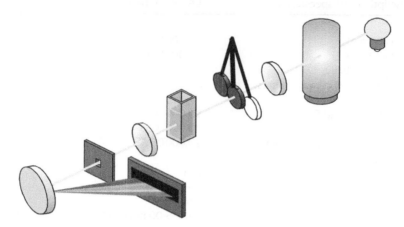

FIGURE 2.32 Schematic of an Agilent Cary 8454 diode array spectrometer. Polychromatic light from tungsten (visible) and deuterium (UV) lamps passes through the sample and is then dispersed into its constituent wavelengths by the grating. The dispersed light is directed onto a diode array, where individual diodes detect specific wavelengths. An entire spectrum is recorded by first measuring the light intensity striking the diodes when a reference solution is in the cuvette (i.e., measure P_0 at each wavelength) and then repeating the measurement with the sample solution in the cuvette (i.e., measure P). Source: Reproduced with permission of Agilent Technologies, Inc.

the same manual manipulation of first inserting the reference, recording P_0 (simultaneously at every wavelength), then replacing the reference with the sample, and recording the signals again. The instrument stores the P_0 readings and takes the ratio to the P readings at every diode to generate the absorbance value as a function of wavelength. A schematic of the general process is shown in Figure 2.33.

Because the diodes work by discharging the voltage across them as light strikes them, an integration time must be set; otherwise the voltage would quickly be entirely dissipated, and all diodes would yield the same saturated value. The term "integration time" can be thought of as being equivalent to the summing up of the photons striking each diode over a short period of time. Most integration times are 100 ms or less. Longer times can be set (up to seconds), but what instruments do for these longer integration times is take several spectra of shorter integration times (i.e., 100 ms) and then signal average all of the spectra collected. This has the advantage of improving the S/N ratio. It makes sense that longer integration times lead to improved S/N because more time is spent actually making the measurement.

Diode array instruments, like single-beam instruments, would be susceptible to errors caused by fluctuations in the lamp intensity due to the time difference between measuring P and P_0. But as with single-beam instruments, diode array instruments can largely correct for this by using a beam splitter and a second detector to monitor the intensity of the sources.

The main advantage of diode array instruments over scanning instruments is their speed in collecting an entire spectrum. This makes them popular for kinetic studies because spectra can be collected rapidly and thus follow the kinetics of chemical reactions. They are also used as detectors for chromatographic separations because they can rapidly record the entire spectrum of each different compound as it elutes from the end of a column.

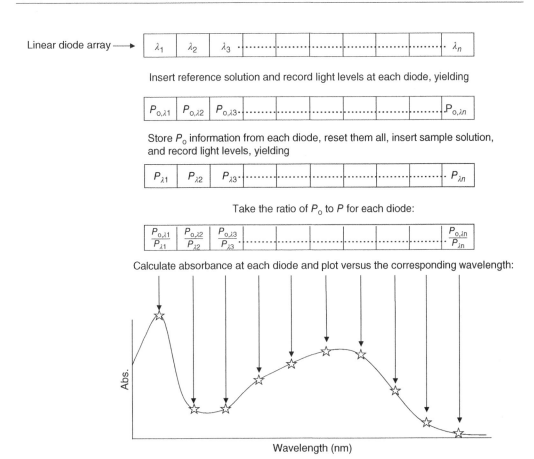

FIGURE 2.33 General measurement scheme of diode array detection. The linear array of boxes at the top of the figure represents the diode array, and each box within the array represents an individual sensing element (i.e., diode) within the array. Typical diode arrays have hundreds or thousands of diodes. Working your way through the figure from top to bottom explains how an entire spectrum is recorded with a diode array instrument.

The ability to collect multiple spectra quickly also presents the potential for signal averaging of multiple spectra, improving the S/N ratio of the spectra. Other advantages stem from the fact that the grating does not move. This provides good wavelength reproducibly and means that the instruments generally require little maintenance except for lamp replacement. Additionally, because all of the light from the sources is directed through the sample at once, the light intensity is high, which provides good sensitivity. The significant drawback to diode array instruments is the resolution that can be achieved. Resolution is limited by the physical proximity of the individual diodes in the array. Typically, this limits resolution to 1 nm, whereas scanning instruments can have resolutions as low as 0.05 nm.

Compare the technical information for the Agilent 8454 diode array system in Table 2.2 with the information about the PerkinElmer double-beam scanning instrument detailed above. Both instruments are designed for particular applications and have particular strengths (e.g., resolution vs. speed). Thus, while their figures of merit are not identical,

TABLE 2.2   Technical Specifications of the Agilent 8454 Diode Array Spectrometer

Wavelength range	190–1100 nm	
Slit width	1 nm	
Toluene/hexane limiting resolution (EP/BP and TGA test)	1.6	
Stray light	At 198 nm (12 g/L KCl, TGA, and BP/ EP method)	<1 %T
	At 220 nm (10 g/L NaI ASTM method)	<0.05 %T
	At 370 nm (50 mg/L NaNO$_2$)	<0.03 %T
Wavelength accuracy	<±0.5 nm (NIST 2034)	
	<±0.2 nm (at 486.0 and 656.1 nm)	
Photometric accuracy	<±0.005 A at 1 Abs (NIST 930e filters)	
	<±0.01 A potassium dichromate, EP method	
Wavelength reproducibility	<±0.02 nm	
Photometric noise (RMS)	<0.0002 A sixty 0.5 s scans at 0 Abs, 500 nm	
Photometric stability	<0.001 Abs/h, 340 nm after 1 h warm-up, 10 s SAT	

Source: Reproduced with permission of Agilent Technologies, Inc.

they may be equally good at performing the tasks for which they are designed. Here again, we see that the analyst must consider the information that is desired and select an appropriate instrument.

2.4.   THE PRACTICE OF SPECTROPHOTOMETRY

In the preceding sections, we established the following:

1. Molecules absorb electromagnetic radiation in the UV-visible part of the spectrum when the energy of the radiation matches an electronic transition within the molecule.
2. During the process of absorption, electrons within the molecule are promoted from lower energy to higher energy molecular orbitals.
3. Because molecules have different electronic configurations, different wavelengths are absorbed to different extents by different molecules.
4. Every absorbance measurement is actually the result of two measurements, P_0 and P, which are recorded using a reference and a sample.
5. Absorbance is proportional to concentration and path length (i.e., $A = \varepsilon bc$), and the proportionality constant is called the molar absorptivity, which reflects the likelihood that photons of a specific wavelength will be absorbed by the analyte.
6. Molar absorptivities are solvent dependent, a fact that has ramifications when preparing calibration curves.
7. A variety of spectrophotometer designs – single-beam, double-beam, and diode array – have been developed to take advantage of the light-absorbing properties of molecules.

8. Spectrophotometers have components that provide radiation (lamps), split the light into its constituent wavelengths (filters or monochromators), and detect the light as a function of wavelength either in space (diode array detectors) or over time (phototubes and photomultiplier tubes in scanning instruments).

9. UV-visible spectrophotometry can aid in structure elucidation by measuring an entire absorbance spectrum and in concentration determination by measuring absorbance at a fixed wavelength and comparing it to a calibration curve made with solutions of known concentrations.

Having established these fundamentals, we can discuss how spectrophotometry is practiced and some of its common applications. Topics that fall under this consideration include:

- The types of samples that can be analyzed.
- Preparation of calibration curves.
- Good practices in terms of experimental design and methodology.
- Cases in which deviations from Beer's law are likely to be encountered.
- Consideration of the accuracy and precision of spectrophotometric measurements.
- Common applications of spectroscopic techniques.

2.4.1. Types of Samples That Can Be Analyzed

The major limitation of UV-visible spectroscopy as an analytical technique is that, by definition, it requires that the analyte of interest absorbs light in the UV-visible range. Clearly, not all molecules do this. In particular, aliphatic compounds such as *n*-alkanes and alcohols (e.g., methanol, ethanol, etc.), and simple derivatives of these classes, either do not absorb at all or do not absorb strongly in the UV-visible range. Thus, UV-visible spectroscopy is "blind" to these molecules even when present at very high concentrations. Only as one approaches the extreme UV limit of 190–210 nm do these compounds start absorbing strongly. At this point, their absorbance often becomes more of a problem than a benefit because it limits their use as solvents. When present as a solvent (i.e., highly concentrated), even compounds with low molar absorptivities absorb significant amounts of light, making it impossible to measure the absorbance of analytes contained within them.

On the other hand, aromatic species and compounds that are highly conjugated are good candidates for analysis by UV-visible spectroscopy. Aromatic compounds tend to have a strong absorbance band between 250 and 300 nm, with additional conjugation shifting the position of maximum absorbance toward higher wavelengths (lower energy). Ketones and compounds containing nitro groups also absorb in the UV portion of the spectrum. Highly colored metal–ligand complexes are also well suited to UV-visible detection. In fact, a good rule of thumb is that if a solution of a compound appears colored to the naked eye, it absorbs light and can be analyzed by UV-visible spectroscopy. Even faintly yellow solutions like those resulting from nitroaromatic compounds are good candidates. The absence of color, however, does not necessarily mean that the species does not absorb because absorbance in the UV portion of the spectrum does not register with the human eye. A good example of this is benzene, which appears colorless but is amenable to analysis below 260 nm.

Compounds that do not absorb light on their own can sometimes be derivatized such that they do absorb light. The derivatization serves either to (1) alter the electronic structure such that the new molecule absorbs light or (2) simply attach a molecule that does absorb to one that does not. An example of the first case is the addition of ligands to metal ions, such as phenanthroline being bound to Fe^{2+} ions to develop a deep red color. In this case, the electronic distribution of both the iron atom and the phenanthroline are significantly affected by the development of charge-transfer bands. An example of the second case is the covalent binding of dansyl chloride to the amino groups of amino acids as shown in Figure 2.34. The majority of amino acids are not strongly absorbent, but the highly aromatic and conjugated dansyl chloride is. Thus, by attaching it to amino acids, the derivatized amino acids can be quantified.

2.4.2. Preparation of Calibration Curves

Typically, to use Beer's law to determine the concentration of an analyte in solution, five or six standard solutions of known analyte concentration are prepared in a solvent matrix that is as close as possible to that of the sample. This is often done by preparing a concentrated stock solution followed by serial dilution. In this case, it is critical that the stock solution is prepared and diluted carefully because the analysis results cannot be any more accurate than is the preparation of the standards. Ideally, the only thing that varies in these solutions is the analyte concentration – all other components (buffer components, salts, etc.) being kept constant. This is important because, as discussed earlier in the chapter, the molar absorptivity for compounds can be solvent dependent. If the compound is absorbing more or less light at a specific wavelength in the standard matrix compared to how well it absorbs the same wavelength in the sample matrix, the calibration curve will lead to artificially high or low (i.e., biased) concentrations for the unknown. A reference solution is also prepared. This should contain the same amount of all of the standard solution components except, of course, the analyte. This solution is ultimately used to determine P_0. Additional requirements for good analytical work are that (1) the concentrations

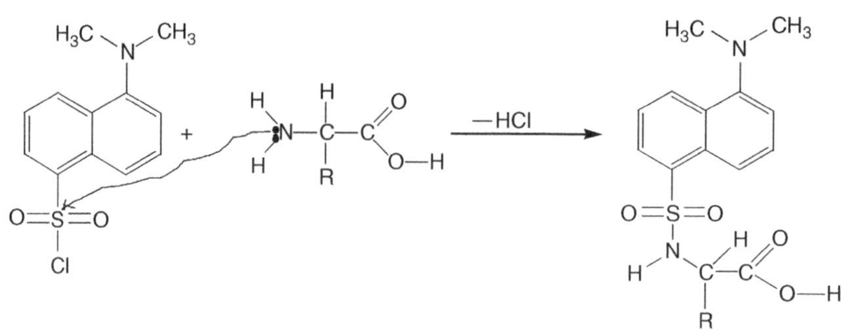

FIGURE 2.34   Reaction of dansyl chloride with an amino acid.

of unknown solutions fall near the middle of the range of standard concentrations and (2) the standard concentrations are equally spaced throughout the concentration range. *Furthermore, while a reference solution is an analytical necessity, a true "blank" of water should be the first solution inserted into the spectrophotometer, and the absorbance of the reference solution should be measured against it* [36]. This will call attention to any situation in which the constituents of the reference and standards themselves (solvent, buffer components, ligands, etc.) absorb light and thereby significantly decrease the intensity of the light reaching the detector.

After the solutions are prepared, the absorbance of each solution is measured, and a plot of absorbance versus known concentration is constructed. The best-fit slope and intercept are determined, and from these, the concentration of unknown solutions and their associated uncertainties can be determined.

When switching between the solutions in a cuvette, care must be taken to ensure that the old solution has been entirely removed or displaced by the new solution. If this is not done, the analyte concentration in the new solution will be altered by the previous solution. Consider an extreme situation. Imagine that 0.25 mL of a 5.00 mM standard solution of an absorbing species is left in the cuvette and that 0.75 mL of the next standard (7.5 mM) is added. The actual concentration of the more concentrated standard solution would be diluted by the presence of the less concentrated solution. In this case, it would drop to 6.875 mM and cause a significantly decreased absorbance compared to that which would result from the proper 7.5 Mm solution. The effect of previous solutions on the current one is known as "carry-over." Plotting this artificially decreased absorbance against 7.5 mM would introduce large errors in the calibration curve, leading to inaccurate or at least highly uncertain concentrations determined using it. Naturally, 0.25 mL is easily noticed, but nonetheless, the liquid that collects in the corners and remains on the side of cuvettes can change the concentration of subsequent solutions. For this reason, it is necessary to rinse the cuvette with the new solution a minimum of three times before taking its absorbance measurement.

It is commonly held that the standard solutions should be analyzed in ascending order of concentration (i.e., start with the most dilute and work up to the most concentrated). A better method is to build in some checks to ensure the reproducibility of the data. Table 2.3 shows a series of measurements that has been suggested in the literature [36]. Note that the standards are analyzed twice in this scheme. Measuring them twice provides a check on the reproducibility of the measurements, and measuring them in different orders incorporates a check for carry-over effects. If two measurements on the same solution disagree, then the source of the disagreement must be determined in order to have any confidence in the entire data set. If all of the absorbance readings are reproducible for all of the solutions, then the inclusion of such controls and reproducibility checks greatly increases the reliability of the data. Note that one solution (in this case, the control) is measured at the start, in the middle, and at the end of the data collection. Doing so helps ensure that the spectrophotometer is working in the same way throughout the time course of the experiment. This is important as spectrophotometers can drift with time, which leads to irreproducible measurements over time. Furthermore, some samples can change over time, either through degradation, reaction, or equilibrium processes such as metal–ligand binding. Designing checks for such variations into the experimental methodology

TABLE 2.3   Order for Analyzing Samples and Standards

Step	Solution
1	Control
2	Unknown
3	Standard 1 (least concentrated)
4	Standard 2
5	Standard 3
6	Control
7	Standard 4
8	Standard 5 (most concentrated)
9	Unknown
10	Standard 5
11	Standard 4
12	Standard 3
13	Control
14	Standard 2
15	Standard 1
16	Unknown
17	Control

Source: Used with permission from Vitha et al. [36], copyright © 2005, Division of Chemical Education, Inc.

is essential for quality assurance and provides much greater confidence in the data than when such checks are absent.

The "control" in the above scheme is a solution of known analyte concentration, ideally dissolved in the same matrix as the unknown. It is present to provide an independent measurement on a known system and thus to test the accuracy of the method. For purposes of data treatment, it is viewed as an unknown. In other words, its concentration is determined using the calibration curve. If the calibration curve does not produce the correct concentration for the control solution, then it is unlikely that an unknown concentration can be accurately determined. This serves as an accuracy check for the calibration curve.

After the data are collected, the absorbance values for each standard and the sample are averaged and the calibration curve is determined by plotting absorbance versus concentration. *It is important to make a plot of the data in order to visually check for any deviations or points that are significantly off the line.* Comparing Beer's law to the common generic equation for a line

$$A = \varepsilon bc + 0$$
$$y = mx + b$$

(2.24)

shows that a plot of absorbance versus concentration has a slope that is equal to the product of the molar absorptivity and path length (εb) and an intercept that should statistically be zero.

After plotting the data, the best-fit line is determined, yielding the slope and intercept and the uncertainty in each. Using these values, the analyte concentration in the sample and control, and the uncertainties in each, can be determined. This is done by substituting the absorbance into the equation for the best-fit line and by using the proper propagation of error equation [37].

2.4.3. Deviations from Beer's Law

The Beer–Lambert law (which we simply refer to as "Beer's law" for brevity from here on) is simple and incredibly powerful because it shows that when everything is in order, the absorbance of a solution is proportional to the concentration of the absorbing species in the solution. The direct proportionality between absorbance and concentration allows spectrophotometry to be used for a variety of important applications such as clinical tests, biochemical assays, real-time process monitoring, and kinetic measurements. As discussed above, in the most routine use of Beer's law, a set of five or six standard solutions containing varying concentrations of the analyte of interest is prepared. Their absorbance at a specific wavelength is measured and plotted against the known concentrations, yielding a straight line that passes through the origin. The absorbance of solutions with unknown concentrations of the analyte is also measured, and the equation for the best-fit line is used to determine the concentrations. In general, spectrophotometers and experimental methods are designed in such a way as to ensure that the direct proportionality holds. Because of this, it is easy to get lulled into complacency and assume that Beer's law always holds. This, however, is not true. Therefore, as practitioners we must be diligent to avoid situations that are known to cause deviations from Beer's law. Several situations where Beer's law is likely to fail, and thus produce curved plots of absorbance versus concentration, are discussed in the following sections.

2.4.3.1. *High Concentrations of Analyte or Other Species.* Calibration curves based on Beer's law rely on the molar absorptivity of the analyte being independent of its concentration. This generally holds if the analyte is very dilute [38] (mM or lower), meaning that every analyte molecule is entirely surrounded by solvent molecules. At higher analyte concentrations, however, the probability that analytes "see" and interact with one another increases. These interactions alter the electron distribution within the analyte molecules and subsequently alter their molar absorptivities. As the analyte concentration is increased, these interactions also increase, causing further alterations to the molar absorptivities. Thus, both ε and c in Beer's law ($A = \varepsilon bc$) are changing as a function of concentration, leading to nonlinear behavior. Similar effects can occur if the concentrations of other species in solution (e.g., electrolytes) vary. This effect is often not noticeable at analyte concentrations below 0.01 M, but certain situations make it more likely. For example, large organic species (e.g., polyaromatic compounds) in aqueous solutions tend to aggregate even at low concentrations – a manifestation of the hydrophobic effect. Participation in the aggregate alters the electronic structure of the individual molecules within the aggregate, giving them different absorption properties than the same molecules free in solution. In such cases, concentrations on the order of 10^{-6} M may be required before Beer's law is followed. Deviations from Beer's law based on analyte

concentrations may also arise if changes in the analyte concentration significantly alter the refractive index of the solution [38].

Such effects are easy to check for by simply taking a concentrated solution, measuring its absorbance, followed by significantly diluting the solution, say, by a factor of two or four, and measuring the absorbance of the diluted sample. The absorbance should decrease by the same factor (i.e., a twofold dilution should cut the absorbance in half). If not, then Beer's law is not being followed. The cause may not necessarily be analyte–analyte interactions because other factors may also be at work (see below), but at least the problem has been identified and possible solutions can be sought.

2.4.3.2. *Chemical Deviations.* Shifts in equilibria involving the absorbing species can cause deviations from Beer's law. Consider, for example, the following equilibria:

$$2CrO_4^{2-} + 2H^+ \rightleftharpoons 2HCrO_4^- \rightleftharpoons Cr_2O_7^{2-} + H_2O$$

The dichromate ion ($Cr_2O_7^{2-}$) absorbs in the visible region at 450 nm. Upon diluting a dichromate solution, the equilibrium shifts toward reactants (toward the side with more species to counteract the decrease in concentration brought about by dilution). Because the molar absorptivity of the chromate ion (CrO_4^{2-}) is much lower than that of the dichromate ion at 450 nm, the absorbance will decrease faster than it should if Beer's law were being followed. This is due to conversion of the absorbing dichromate ion into the nonabsorbing chromate ion. Such behavior leads to a plot like that shown in Figure 2.35. In this case, the nonlinearity as a function of concentration could be eliminated by maintaining a low pH in all solutions, thereby effectively driving the equilibrium entirely to the products and keeping all of the chromium in the dichromate form.

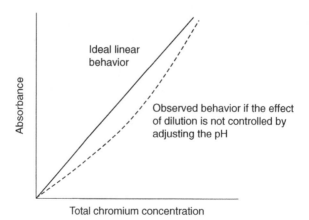

FIGURE 2.35 Deviation from Beer's law: ideal (linear) versus nonideal (curved) behavior of chromate/dichromate system if pH and ionic strength are not controlled.

Even in less complex equilibrium systems, such as that encountered with simple acid–base indicators, deviations will occur unless steps are taken to eliminate them. For example, consider the indicator phenol red. Its absorbance spectra as a function of pH are shown in Figure 2.36. The acid–base equilibrium for this indicator can be written as

$$HIn + H_2O \rightleftharpoons In^- + H_3O^+$$

where HIn and In⁻ are the protonated and deprotonated forms of phenol red, respectively. As the solution is diluted, the percentage of the total dye concentration that is in the deprotonated form increases. Figure 2.36 makes it apparent that the deprotonated form (see pH = 11.0) has a distinct absorbance spectrum compared to that of the protonated form (e.g., pH = 4.0). This difference in absorption properties leads to nonlinear behavior as the ratio of the protonated to deprotonated form changes with concentration. The absorbance properties (i.e., the wavelengths the indicator absorbs)

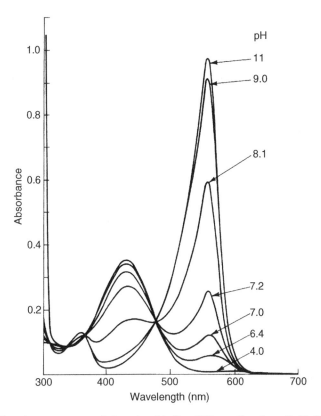

FIGURE 2.36 Absorbance spectra of phenol red (pK_a = 7.9) as a function of pH. Absorbance maxima are at 433 and 558 nm for the acidic (yellow) and basic (red) forms, respectively. Isosbestic points occur at 338, 367, and 480 nm. Source: From Willard et al. [12]. Reproduced with permission from Frank A. Settle, Jr.

change because protonating or deprotonating the indicator dramatically changes how the electrons are distributed within the molecular orbitals in the molecule, thus leading to changes in the energy difference between the highest occupied and lowest unoccupied orbitals, which ultimately means that the two different forms absorb different wavelengths of light.

Here again, the nonlinearity can be eliminated by carefully buffering each solution to the same pH (or by making all solutions extremely basic or acidic), thereby maintaining the same ratio of protonated to unprotonated species in all solutions. In addition, because acid–base equilibria are sensitive to ionic strength effects, the ionic strength should also be kept constant in all solutions via the addition of appropriate amounts of inert salts. Similar equilibria effects arise when studying complex ions (e.g., metal–ligand complexes).

In Figure 2.36, all of the spectra recorded at different pHs have the same absorbance at three distinct wavelengths (338, 367, and 480 nm). These points are known as isosbestic points and arise because the molar absorptivities of the protonated and deprotonated forms are equal at those wavelengths. At these wavelengths, the absorbance depends only on the total indicator concentration and not on the pH.

Because equilibria (acid–base, complex formation, etc.) are often involved in chemical systems, particularly in aqueous systems, it is important to keep them in mind as potential causes of nonlinear Beer's law behavior. If nonlinearity is observed, carefully consider the variables that control equilibria, including pH, temperature, concentrations of species involved in the equilibria, and ionic strength, and take appropriate steps to control the equilibria involved.

2.4.3.3. Instrumental Deviations.

Beer's law rigorously applies only in the case of monochromatic radiation. However, filters and monochromators never produce purely monochromatic radiation. Instead, a narrow band of wavelengths is passed through the sample. In certain situations, this can lead to nonlinear deviations from Beer's law. To understand how, consider the spectrum shown in Figure 2.37 and imagine simultaneously passing two wavelengths, 1 and 2, on opposite sides of point "a" through the solution.

Beer's law holds independently for each of the two wavelengths, such that we can write

$$A_1 = -\log\left(\frac{P_1}{P_{o,1}}\right) = \varepsilon_1 bc \quad \text{and} \quad A_2 = -\log\left(\frac{P_2}{P_{o,2}}\right) = \varepsilon_2 bc \tag{2.25}$$

These can be rearranged to

$$P_1 = P_{o,1} 10^{-\varepsilon_1 bc} \quad \text{and} \quad P_2 = P_{o,2} 10^{-\varepsilon_2 bc} \tag{2.26}$$

The detector does not "know" that two wavelengths are being passed simultaneously. It simply "sees" the total number of photons that reach it after being passed through the

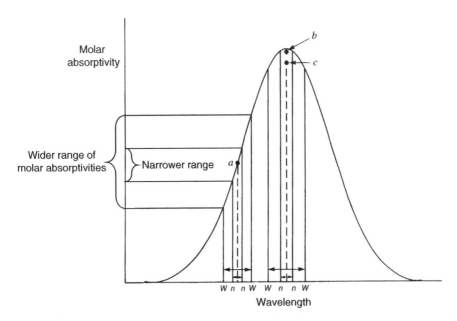

FIGURE 2.37 Effects of narrow (n–n) and wide (w–w) bandpasses. At point (a) with a wide bandpass, the range of molar absorptivities is significant as are the associated deviations from Beer's law. A narrow bandpass has a smaller range of molar absorptivities, but could still produce nonlinearities. Point (b) with a narrow bandpass is less prone to deviations because the molar absorptivity is nearly constant. With a wider bandpass, point (c), the average molar absorptivity decreases since more curvature, and thus a wider range of molar absorptivities, is incorporated in the range. Source: From Willard et al. [12]. Reproduced with permission from Frank A. Settle, Jr.

reference and sample solutions. The subsequent electronics still compute the absorbance based on these two measurements. In the case of two wavelengths, then the absorbance can be written as

$$A = \log \frac{\left(P_{o,1} + P_{o,2}\right)}{\left(P_1 + P_2\right)} = \log \frac{\left(P_{o,1} + P_{o,2}\right)}{\left(P_{o,1} 10^{-\varepsilon_1 bc} + P_{o,2} 10^{-\varepsilon_2 bc}\right)} \tag{2.27}$$

This can be rearranged to

$$A = \log\left(P_{o,1} + P_{o,2}\right) - \log\left(P_{o,1} 10^{-\varepsilon_1 bc} + P_{o,2} 10^{-\varepsilon_2 bc}\right) \tag{2.28}$$

If $\varepsilon_1 = \varepsilon_2$, the above equation collapses back to

$$A = \log \frac{\left(P_{o,1} + P_{o,2}\right)}{\left(P_{o,1} 10^{-\varepsilon bc} + P_{o,2} 10^{-\varepsilon bc}\right)} = \log \frac{\left(P_{o,1} + P_{o,2}\right)}{\left(P_{o,1} + P_{o,2}\right) 10^{-\varepsilon bc}}$$

$$= \log \frac{1}{10^{-\varepsilon bc}} = \log 10^{\varepsilon bc} = \varepsilon bc \tag{2.29}$$

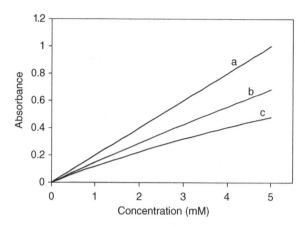

FIGURE 2.38 Plot of nonlinear compared to linear Beer's law behavior. The linear plot, a, is calculated in accord with the equations in this section of the chapter. It is based on two molar absorptivities, both equal to 200 cm^{-1}M^{-1}, and a path length of 1 cm. Plot b was obtained using the same path length but with two different molar absorptivities (100 and 200 cm^{-1}M^{-1}). Plot c was obtained using molar absorptivities of 50 and 200 cm^{-1}M^{-1}. It is clear that wider discrepancies in the molar absorptivities over the wavelength region passed through the sample lead to greater deviations from linearity.

which is Beer's law (i.e., $A = \epsilon bc$). If, however, the molar absorptivities are considerably different at the two wavelengths, the exponential terms cannot be combined and the equation does not collapse back to Beer's law. This leads to nonlinear behavior. While we have only treated the example of two wavelengths here, the same analysis holds for an entire band of wavelengths such as is typically passed by filters and monochromators. Furthermore, the larger the difference between the molar absorptivities, the larger the deviations. This can be observed in the plot in Figure 2.38, which is based on the equations derived above for two different molar absorptivities.

There are two practical consequences to the potential deviations caused by polychromatic radiation. First, we can better avoid nonlinearities by selecting a central wavelength that corresponds to the peak maximum of a broad UV-visible absorption band (e.g., point "*b*" in Figure 2.37). In this case, even if the bandpass is fairly wide, the variation in the molar absorptivity for surrounding wavelengths is quite small, and the approximation that they are all equal to one another is an acceptable one. The added advantage to choosing a wavelength corresponding to a peak maximum is that the molar absorptivity is highest at the peak, maximizing the sensitivity of the measurements. Note, however, that even at the peak maximum, a bandpass that is too wide can decrease the average molar absorptivity observed in the range and artificially depress the measured absorbances.

The second consequence of the non-linearities caused by polychromatic radiation is that users should avoid working at a central wavelength on the side of an absorption feature (e.g., point "*a*" in Figure 2.37) because here it is less likely that the molar absorptivities over a range of wavelengths are nearly equal, increasing the likelihood of nonlinear behavior. Furthermore, the sensitivity of the measurements is lower than at the peak maximum. Sometimes, however, it is not possible to make measurements at the wavelength corresponding to the peak maximum due to overlapping peaks caused by interferences. In these cases, the treatment above (and in particular Figure 2.38) shows that the smallest possible range in molar absorptivities will lead to the smallest deviations from linearity. The way to achieve the narrowest range in molar absorptivities is to use the smallest range of wavelengths (i.e., the narrowest bandpass) possible. For most scanning instruments, this means using exceptionally narrow slits. For filter-based instruments, it requires obtaining high-quality filters with narrow bandpasses. In both cases, less total light passes through the solution and reaches the detector, causing greater uncertainties in the measured absorbance values. However, this tradeoff between linearity and increased uncertainties is often made because fitting data to a straight line is much less complicated than nonlinear fitting.

These considerations of bandpass and wavelength reproducibility requirements ultimately lead to the common practice of *working near the peak maximum with as wide of a bandpass as can be used without inducing significant deviations from linearity*. The effect of slit width is easy to check for in systems with variable slits. Starting with deliberately very wide slits, they can be systematically narrowed until no significant changes in the spectra are recorded. This leads to the widest slits (i.e., best S/N) that do not distort the spectrum (i.e., the best accuracy). Broad, flat peaks that span tens or hundreds of nanometers allow for much wider slits, while narrow spectral features (10–50 nm) require much narrower slit widths. A rule of thumb is that deviations from Beer's law are generally minimal if the effective bandpass of the monochromator or filter is less than 1/10th the width of the absorption band at half-height. An excellent web-based applet that graphically shows the effect of slit width and the width of spectral features on Beer's law is available [39].

If a nonlinear calibration plot cannot be avoided after all of the above considerations are taken into account, then it is important to use a sufficiently large number of standards to characterize the calibration curve well enough to obtain accurate unknown concentration values.

2.4.3.4. Effect of Stray Radiation.

Both sets of instrumental performance characteristics provided above (Tables 2.1 and 2.2) list levels of stray radiation. Stray radiation leads to deviations from Beer's law and thus affects the accuracy of absorbance measurements [40–43]. Stray radiation is radiation that reaches the detector that is not in the desired wavelength range or that has not traveled through the solution. Principal sources of stray radiation are (1) scattering by dust or smudges on the surfaces of optical components and surface imperfections in mirrors, lenses, and cuvette walls, (2) diffraction of light at the slit edges, (3) reflections from interior surfaces of the monochromator, and (4) scattering by a grating that acts as a simple mirror instead of a perfect diffracting surface. The absolute intensity of stray radiation is constant at a given wavelength setting at any given time, but can increase as the instrument ages and the optical materials in the instrument deteriorate.

Stray radiation leads to positive deviations from Beer's law if the stray radiation is absorbed, and negative deviations if not. The latter is usually the case. Observed absorbances are reduced markedly as the amount of stray radiation increases and as the true absorbance of the solution increases. The apparent absorbance of a sample, assuming that none of the stray radiation, P_s, is absorbed, is given by

$$A_{apparent} = \log\left(\frac{P_o + P_s}{P + P_s}\right) \tag{2.30}$$

Consider, for example, an instrument in which the intensity of the stray radiation is 1% of that of the desired wavelength (i.e., $P_s = 0.01\ P_o$). In this case, the highest absorbance reading possible on the instrument (i.e., when $P = 0$) is

$$A_{apparent} = \log\left(\frac{P_o + P_s}{P + P_s}\right) = \log\left(\frac{P_o + 0.01P_o}{0 + 0.01P_o}\right) = \log\left(\frac{1.01}{0.01}\right) \approx 2 \tag{2.31}$$

Clearly the stray radiation is limiting the ability of the instrument to accurately measure high absorbances (this is one reason to be suspicious of absorbance readings above 2.0).

While the amount of stray radiation varies widely, with double monochromators, it is possible to achieve stray radiation levels of less than 10^{-6} relative to the true signal. In the past, stray radiation sometimes limited low-cost spectrophotometers to a dynamic range of 0.001–2.0 absorbance units (as illustrated by the calculation above). However, recent improvements in stray radiation filters and instrument designs have extended this range to 3.0 absorbance units or higher in some instruments. The ability to measure higher absorbance values with good accuracy helps to increase sample throughput because one calibration curve can cover a wider range of concentrations. This can eliminate the need to dilute more concentrated samples and thus save time. However, as a user, you should be leery of high absorbance values (2.0 and above), as the accuracy and reproducibility of absorbance measurements is generally better at lower absorbance values.

EXAMPLE 2.7

Suppose that for a particular analysis, signals of 6000 μA for the reference and 17 μA would be obtained if the instrument is working ideally (i.e., no stray radiation). What is the absorbance of this solution? Further suppose that due to stray radiation, both signals are increased by 15 μA, to 6015 and 32 μA, respectively. What is the percent error in the transmittance reading obtained in the presence of stray radiation?

Answer:

$$T_{ideal} = \left(\frac{P}{P_o}\right) = \left(\frac{17\,\mu A}{6000\,\mu A}\right) = 2.8 \times 10^{-3}$$

$$Abs_{ideal} = -\log T = -\log(2.8 \times 10^{-3}) = 2.55$$

$$T_{actual} = \left(\frac{P + P_s}{P_o + P_s}\right) = \left(\frac{32\,\mu A}{6015\,\mu A}\right) = 5.3 \times 10^{-3} \quad (abs_{actual} = 2.27)$$

$$\%\ error\ in\ T = 100 \times \left(\frac{5.3 \times 10^{-3} - 2.8 \times 10^{-3}}{2.8 \times 10^{-3}}\right) = 89\%\ error$$

Another question:
Repeat the calculations with the same reference signal and stray light level, but imagine that the sample signal would have been 4000 μA, ideally. Compare the results to those above. What can you conclude about the error caused by stray radiation and its dependence on sample absorbance?

Answer:

$$T_{ideal} = 0.6667$$
$$Abs_{ideal} = 0.1761$$
$$T_{actual} = 0.6675 \quad (abs_{actual} = 0.1755)$$

% error in $T = 0.12\,\%$. The percent error increases with decreasing transmittance and therefore with increasing absorbance. Thus, deviations from Beer's law due to stray radiation are most commonly observed at high absorbance values.

2.4.4. Precision: Relative Concentration Error

The above considerations dealt with the photometric accuracy of absorbance measurements – in other words, considerations of effects that could lead to biased absorbance values. Naturally, biased absorbance values lead to incorrect determinations of analyte concentrations. In this section, we focus on photometric precision [44], or the ability to reproduce absorbance values, which ultimately dictates the uncertainty in concentrations derived from absorbance measurements. Here again we must keep in mind that spectrophotometers do not actually measure absorbance, *per se*, but rather the intensity or power (P) of light striking a detector. Thus, the ability to reproducibly measure light intensity ultimately determines the reproducibility of concentration measurements.

Two main situations that can dictate the uncertainty of concentrations determined using Beer's Law are:

Case 1: The uncertainty in the intensity of light (here denoted as dP) is independent of the intensity of the light (i.e., dP is independent of P).

Case 2: dP varies with P because of shot noise – the statistical variation in the number of photons that reach the detector.

As discussed below, these different cases ultimately reveal optimum absorbance values that minimize uncertainty. In other words, they show the range of absorbance values in which uncertainties in concentrations are minimized and that, therefore, yield the most precise measurements.

2.4.4.1. *Case 1: Thermal Noise.*

Case 1 is primarily encountered with less expensive, simpler, and older instruments that use linear readout scales or digital readouts with limited resolution. It is a case in which thermal noise – random agitation of electrons within electrical components such as wires, resistors, and electrodes – is the main source of uncertainty. These random fluctuations create small, temporary voltages that manifest themselves as noise. Starting with a rearrangement of Beer's law

$$C = \frac{A}{\varepsilon b} = \frac{1}{\varepsilon b}\log\frac{P_o}{P} = \frac{-\log T}{\varepsilon b} \tag{2.32}$$

followed by differentiation, it can ultimately be shown that the relative uncertainty in concentration (dC/C) is given by

$$\frac{dC}{C} = \frac{-0.434}{A}\left(\frac{dP}{P}\right) \quad \text{or} \quad \frac{dC}{C} = \frac{0.434}{\log T}\left(\frac{dT}{T}\right) \tag{2.33}$$

Here, A, P, and T are the absorbance, power, and transmittance, respectively. Thus, the relative concentration error (dC/C) depends inversely on the product of the absorbance and transmitted power. This equation allows for the calculation of the transmittance at which the minimum relative concentration error will occur. This is done by taking the derivative of the above equation and setting it equal to zero. The result is that the minimum error is given by

$$\frac{dC}{C} = -2.72\,dT \tag{2.34}$$

Thus, it is reasonable to say that a 0.1% error in transmittance produces a 0.27% error in sample concentration. Figure 2.39 is a plot of the relative concentration error as a function of absorbance that results from the above calculation.

It is clear for the case considered here (i.e., thermal noise limited) that the graph is relatively flat between an absorbance of 0.3 and 0.6. Thus, it is desirable to adjust analyte concentrations, either by dilution or concentration, such that they result in absorbance values in this range. Doing so minimizes uncertainties in the calculated concentration. Note that the graph assumes a constant dT of 1%, but the shape will remain the same regardless of the actual percent transmittance error specified.

2.4.4.2. *Case 2: Shot Noise.*

Case 2, in which the relative uncertainty in concentration is dependent on the intensity of the power of the light, is encountered in modern UV-visible spectrophotometers where shot noise predominates. In this case, the shot noise arises from phenomena such as the random ejection of electrons from the

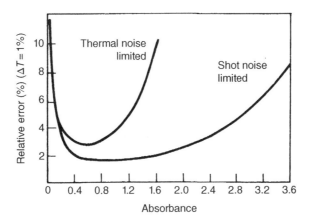

FIGURE 2.39   Relative concentration error, $\Delta C/C$, in percent, for a constant transmittance error of $\Delta T = 1\%$. Source: From Willard et al. [12]. Reproduced with permission from Frank A. Settle, Jr.

cathode of a photomultiplier tube. The noise (dP or dT) is proportional to the square root of the radiant power (i.e., $dT = k\sqrt{P}$, where k is a proportionality constant). Starting again with the rearranged form of Beer's law followed by differentiation and substitution of $dT = k\sqrt{P}$ yields

$$\frac{dC}{C} = -0.434k\left(\sqrt{P}\log\frac{P_o}{P}\right)^{-1} \tag{2.35}$$

The minimum of this function (i.e., where the derivative is equal to zero) occurs when $T = 0.135$, which corresponds to an absorbance reading of 0.868. The relative uncertainty in the concentration as a function of absorbance for the case in which shot noise is the dominant source of noise is plotted in Figure 2.39. For instruments in which shot noise predominates, the region of minimum error extends over a wide range of absorbance values from 0.3 to 2.0 (corresponding to $T = 0.5$–0.01). Note that the foregoing derivations apply to single-beam spectrophotometers. The calculations for the noise level of double-beam instruments are more complex, but the final results are identical. Thus, the take-away message is that absorbance values between 0.3 and 2.0 (some would put the upper limit at 1.5) are desirable because they minimize the instrumental contribution to uncertainties in concentrations.

2.4.4.3. *User-Induced Sources of Uncertainty.* The cases discussed above are related to the components of the instrument and are generally out of the user's control once a spectrometer has been purchased – although different instruments have different levels of photometric accuracy and precision, as can been seen in the performance characteristics provided earlier in this chapter.

While the instruments themselves influence the accuracy and precision of absorbance measurements, they are certainly not the only factor, and in many cases they make only minor contributions to inaccuracies compared to factors that are more directly in the user's control. Two such factors include the preparation of solutions and the cleanliness, condition, and positioning of the cuvettes. *If the sample and the solutions that are used to create the calibration curve are not prepared properly, then even the most expensive, highest-performing spectrophotometer will not provide accurate concentration determinations.* The careful preparation of standard solutions and the proper preparation of references and controls are entirely within the user's control and should thus be performed with great attention to the manipulations involved (weighing, transferring, pipetting, diluting, etc.).

Additionally, the condition and positioning of the cuvettes in the instrument can have a significant effect on the reproducibility of absorbance measurements. Because all cuvettes have slight imperfections, reflection and scattering losses change as different parts of the cuvette are exposed to the beam of radiation. Therefore, it is important to keep the cuvettes as clean as possible and to prevent scratching the surfaces. To clean cuvettes, dustless lens paper soaked in methanol should be used. When cuvette faces are cleaned in this manner, a film of methanol is left, which quickly evaporates and leaves the faces free of contaminants and stray dust particles that can scatter and reflect light. Cuvettes (and other optical components) can be easily scratched by rough paper. Using soft lens paper designed specifically for optical components minimizes this. Regarding the positioning of cuvettes within the instrument, leaving the cuvette in place and changing the solution by means of a syringe produces the maximum precision – of course, care must be taken to ensure that the old solution is completely replaced by the new so that none of the old solution mixes with the new and thus changes its concentration – an effect known as "carry-over," which was discussed above. Reproducibility is significantly improved when the cuvette is left in place compared to when it is removed prior to each measurement. With instrumentation capable of precisely reading a 0.0001 absorbance unit change, small differences are measurable.

2.4.5. The Desirable Absorbance Range

The calculations above for case 1 and case 2 errors in concentrations provide general guidance regarding the desirable range (i.e., lowest relative error) of absorbance values for standard and unknown solutions. Keep in mind that through sample manipulation (dilution, preconcentration, etc.), it is often possible to adjust the analyte concentration such that it falls within these desirable ranges. To provide further rationale for doing so, consider the definition of absorbance that we have repeated throughout the chapter:

$$Abs = \log\left(\frac{P_o}{P}\right) = -\log\left(\frac{P}{P_o}\right) \qquad (2.36)$$

The logarithmic nature of the definition of absorbance has important analytical consequences both at high and low absorbance values. Table 2.4 shows the ratio of P_o to P over a range of absorbance values. The same data is represented graphically in Figure 2.40. To highlight the analytical challenges of measuring high and low absorbance values, consider the P_o/P ratio at absorbances of 0.001 and 4.00. If P_o is arbitrarily taken to be 10 000 photons, at the low absorbance of 0.001, 9977 photons of the original 10 000 make it through the solution, meaning that only 23 out of the 10 000 photons have been absorbed.

TABLE 2.4   The Effect of the Logarithmic Definition of Absorbance on Ratio of P_o and P Assuming P_o = 10000 Photons

Absorbance	%T	P_o/P	P_o	P	Difference $(P_o - P)$
0	100.0	1	10000	10000	0
0.001	99.8	1.002	10000	9977	23
0.01	97.7	1.023	10000	9772	228
0.1	79.4	1.258	10000	7943	2057
0.5	31.6	3.162	10000	3162	6838
1.00	10.0	10	10000	1000	9000
2.00	1.0	100	10000	100	9900
3.00	0.1	1000	10000	10	9990
4.00	0.01	10000	10000	1	9999

As absorbance increases, P decreases significantly. Also, as absorbance increases from say, 3–4, compared to 1–2, the change in the difference in light intensity that the detector "sees" decreases substantially, making it difficult to accurately detect small changes in absorbance when absorbance is high.

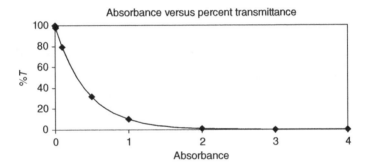

FIGURE 2.40   A plot of the absorbance versus percent transmittance data shown in Table 2.4. Note how low the percent transmittance is above an absorbance of 1.5 and how little it changes at higher absorbance values. This makes it difficult to detect small changes in absorbance of highly absorbing samples.

For solutions with low absorbance, then, it is hard to determine that any light has been absorbed at all. Conversely, at the high absorbance of 4.0, only 1 out of every 10000 photons is transmitted through the solution and reaches the detector, meaning that 9999 photons are absorbed. It is difficult to accurately measure such low light levels, much less the *difference* in such low light levels. Furthermore, at both extremes, the logarithmic nature of the definition of absorbance means that it is difficult to differentiate between solutions with similar concentrations because of small changes in the percentage of photons reaching the detector. For example, in going from 0.001 to 0.010 absorbance units, there is a difference of only 205 photons being transmitted through the solution, out of every 10000 that interact with the sample, and there is an even smaller difference in going from 3.0 to 4.0 absorbance units. However, in going from 1 to 2 absorbance units, there is a difference of 900 photons being transmitted to the detector out of 10000. Even more striking, the change from 0.5 to 1.0 results in a difference of 3162 out of every 10000 photons. These large differences in the amount of transmitted light are analytically easier to measure with accuracy and precision and thus dictate that absorbance measurements in the range of 0.5–2.0 are much more reliable than those at the extremes. In fact, for solutions with absorbance values over 2.0, the analyst should consider diluting the samples in half (at least) to get them into the more desirable range of absorbance values.

2.5. APPLICATIONS AND TECHNIQUES

It is rare in chemistry that a sample contains only a single analyte in a pure solvent. More commonly, samples contain multiple components, several of which might absorb in the UV-visible portion of the spectrum. This potentially complicates analyses because the absorbance spectra of the individual components might overlap and interfere with one another. Therefore, in addition to the relatively straightforward application of measuring the spectrum or the concentration of a pure analyte – or at least one free from interferences from other species – several useful variations of UV-visible spectroscopy exist. These techniques are discussed in the subsequent sections, followed by explanations of a few additional general uses of UV-visible spectroscopy. We end the section with a specific example of the use of UV-visible spectroscopy to measure enzyme kinetics.

2.5.1. Simultaneous Determinations of Multicomponent Systems

When no region can be found free from overlapping absorbance from two species, it is still possible to devise a method of concentration determination based on measurements at two or more wavelengths. Because their electronic structures are not the same, different molecules are likely to have different molar absorptivities at different wavelengths. If, therefore, measurements are made on an unknown solution at two wavelengths where the molar absorptivities of the two components are different, it is possible to set up two independent equations and solve them simultaneously for the two unknown concentrations:

$$A_1 = \varepsilon_{x,1} bc_x + \varepsilon_{y,1} bc_y \tag{2.37}$$

$$A_2 = \varepsilon_{x,2} bc_x + \varepsilon_{y,2} bc_y \tag{2.38}$$

Here, "x" and "y" are the two different species whose concentrations are sought. The subscripts "1" and "2" represent the two different wavelengths, and A_1 and A_2 are measured using the sample containing both compounds at the two different wavelengths. Separate solutions, each containing just one of the analytes at a known concentration, are used to determine the molar absorptivities of the two compounds at the two wavelengths of interest (i.e., to determine $\varepsilon_{x,1}, \varepsilon_{x,2}, \varepsilon_{y,1},$ and $\varepsilon_{y,2}$). The path length is set by the cuvette and is therefore known. Thus, there are two equations and two unknowns (c_x and c_y) allowing for the simultaneous determination of both concentrations.

It is preferable to choose wavelengths with the maximum difference in the molar absorptivities of the two species. This is depicted in Figure 2.41 at λ_1 and λ_2. Furthermore, it is possible, and in fact desirable if conditions permit, to make measurements at more

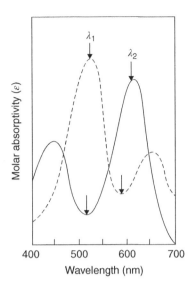

FIGURE 2.41   Simultaneous spectrophotometric analysis of a two-component system. Selection of analytical wavelengths is indicated by arrows. Source: From Willard et al. [12]. Reproduced with permission from Frank A. Settle, Jr.

wavelengths than there are unknowns. Such a set of equations is said to overdetermine the system. The best values for the unknown concentrations are then determined by the method of least squares [45]. In favorable situations, it is possible to determine a dozen or more unknown concentrations, provided that wavelengths can be found that are sufficiently distinct for each component and that Beer's law holds for each component at each wavelength.

Important Note: This method rests on a principle that has not yet been made explicit. The total absorbance at each wavelength (i.e., the absorbance measured by the instrument) is the sum of the absorbances of each component at that wavelength. Thus, for a four-component sample where the numbers 1, 2, 3, and 4 represent each component,

$$A_{\text{total}} = A_1 + A_2 + A_3 + A_4 = \varepsilon_1 b c_1 + \varepsilon_2 b c_2 + \varepsilon_3 b c_3 + \varepsilon_4 b c_4 \tag{2.39}$$

In other words, the spectrophotometer does not "know" how many absorbing species are present in solution. It only senses the total loss of light in the sample relative to the reference. It is up to the analyst to recognize the possibility that multiple absorbing species may be present in a sample and to devise clever means for making the chemical analyses that are desired.

2.5.2.   Difference Spectroscopy

In difference spectroscopy, two samples are placed in a double-beam spectrophotometer – one in the reference beam and the other in the sample beam. The recording, then, is the difference in transmittance (or absorbance) of the two samples. Common features of the two spectra cancel. Usually the concentration of absorbing material in the two samples is identical, but some solution parameter such as pH is different. Toxicology laboratories use this method for the analysis of drugs. Barbiturates, for example, have an absorption maximum at about 260 nm in 0.45 M NaOH and a maximum at 250 nm or below at pH 10.3 [46]. Using a pH 10 solution as reference, the spectrum from 230 to 280 nm is recorded. This spectrum is characteristic of the particular barbiturate and thus provides a means of identification. Biochemists frequently use difference spectroscopy to study the confirmation of globular proteins in solution by varying the solvent (adding alcohol) or changing the temperature of one sample. An enzyme-catalyzed reaction can be analyzed in the presence of nonenzyme-catalyzed reactions by placing the substrate in both beams. The nonenzyme reactions are canceled out, and the enzyme-catalyzed reaction produces a difference signal when enzyme is added to one sample.

2.5.3. Derivative Spectroscopy

As the name implies, in derivative spectroscopy, the first- or higher-order derivative of absorbance with respect to wavelength is recorded versus wavelength. This significantly increases the ability to detect and measure minor spectral features. This enhancement of characteristic spectral detail can distinguish between similar spectra and follow small changes in a spectrum [45, 47–50]. Thus, it is useful for finding minor impurities that contribute subtly different absorbance features to the major component. Purity analysis is critical in the pharmaceutical industry, where drugs that are being administered to animals and humans must be exceptionally pure and free from potentially harmful contaminants. Moreover, derivative spectroscopy can be of use in quantitative analysis to measure the concentration of an analyte whose peak is obscured by a larger overlapping peak and thus avoid having to use more time-consuming and expensive techniques like chromatography or mass spectrometry to make the measurements. An example of derivative spectroscopy is shown in Figure 2.42. The analyte band has been normalized so that it has a maximum absorbance of 1.00. In the second part of this figure, a broad band with twice the width and height of the analyte peak overlaps the left side of the analyte peak. The resulting curve is dominated by the overlapping peak. The analyte "peak" appears only as a small shoulder added on to the right side of the interfering peak.

To estimate the absorbance of the analyte peak in this case, one could draw the best guess tangent to the spectrum on the far right side in an effort to approximate the absorbance of the overlapping band at the wavelength of interest. Doing so allows one to measure the difference in absorbance values between the approximation (about 1.5) and the actual curve (1.9) in an attempt to determine the obscured band's absorbance. In this case, the reading turns out to be 0.4. This absorbance is far too low compared to the "true" absorbance of 1.0. Using the uncorrected absorbance of 1.9 is far too high. In the derivative spectrum in the lower right of the figure, the vertical distance between the adjacent maximum and minimum of the first derivative serves as the measure of the analyte absorbance. Now the estimate of the analyte absorbance is 0.88, only 12% low. Whenever the interfering band is broader than the analyte band by at least a factor of two, derivative spectroscopy usually proves advantageous.

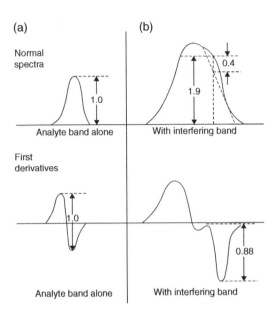

FIGURE 2.42 First derivative spectrometry for the quantitative measurement of the intensity of a small band (a) alone and (b) obscured by a broader overlapping band. See text for details. Source: Reprinted with permission from O'Haver [47]. Reproduced with permission of American Chemical Society.

2.5.4. Titration Curves

The change in absorbance of a solution may be used to follow the change in concentration of an absorbing species during a titration. As we have seen, for solutions that have only a single absorbing species, absorbance is directly proportional to the concentration of the absorbing species. A plot of absorbance versus the volume of titrant added consists, if the reaction is complete, of two straight lines that intersect at the endpoint of the titration. For reactions that are incomplete (as normally occurs near endpoints), extrapolation of the two linear segments of the titration curve establishes the intersection and thus endpoint volume. Possible shapes of photometric titrations are shown in Figure 2.43. Curve (a), for example, is typical of the titration where the titrant alone absorbs, as in the titration of arsenic(III) with bromate ions:

$$3As_2O_3 + 2BrO_3^- + 9H_2O \rightarrow 6H_3AsO_4 + 2Br^-$$

$$BrO_3^- + 5Br^- + 6H^+ \rightarrow 3Br_2 + 3H_2O$$

In this situation, the absorbance readings are taken at the wavelength where the bromine (Br_2) absorbs. As bromate is titrated into a solution containing arsenic(III), the bromate is consumed. This means that the second reaction cannot occur and no Br_2 is formed. Once all of the arsenic(III) is reacted, however, addition of more bromate generates Br_2 via the reaction with the bromide ion, which was formed via the first reaction. Thus, once the

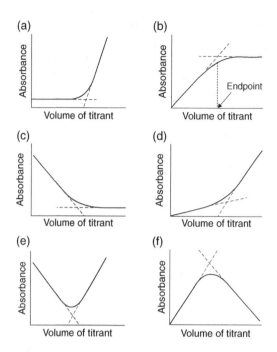

FIGURE 2.43 Possible shapes of photometric titration curves. See text for details. Source: From Willard et al. [12]. Reproduced with permission from Frank A. Settle, Jr.

endpoint has been reached, the absorbance – which was initially zero because no Br_2 was present – begins to increase and does so linearly with the increasing concentration of Br_2.

Curve "b" is characteristic of systems where the product of the reaction absorbs, as in the titration of copper(II) with EDTA. When the analyte is converted to a nonabsorbing product – for example, titration of p-toluidine in butanol with perchloric acid at 290 nm – curve "c" results. When a colored analyte is converted to a colorless product by a colored titrant – for example, bromination of a red dye – curves similar to "e" are obtained. Curves "d" and "f" might represent the successive addition of ligands to form two successive complexes of different absorptivity.

Photometric titrations have several advantages over a direct photometric determination. The presence of other absorbing species at the analytical wavelengths does not necessarily cause interference in a titration, because only *the change in absorbance* is significant. The additional absorbing species simply add a constant value to each of the absorbance readings in, say, curve "a," but do not affect the determination of the endpoint nor the concentration derived from it. However, the absorbance of these background species must not be intense, because if they are, the absorbance readings are limited to the undesirable upper end of the absorbance scale.

When using photometric titrations, keep in mind that by their very nature, titrations alter the volume of the sample being titrated. Absorbance readings, by virtue of their dependence on concentration (mole/volume), are inherently affected by the dilution caused by the addition of titrant. Thus, absorbance values must be corrected for dilution effects. This is done simply by multiplying the measured absorbance by the factor

$(V + \Delta V)/V$ where V is the initial volume of the sample and ΔV is the volume of titrant added up to each point in the titration. If the correction is not made, the lines are curved downward toward the volume axis and erroneous intersections are obtained. To minimize dilution effects, use a microsyringe to deliver small amounts of a highly concentrated titrant solution.

2.5.5. Turbidimetry and Nephelometry

Up to this point, we have focused on applications in which some species in the solution absorbs UV-visible radiation, which causes a decrease in the intensity of the light that would otherwise reach the detector. But the instruments do not "know" what the loss of light is due to – all they measure is how much light they detect. As a result, any phenomenon that causes a reduction in light intensity can theoretically be studied using spectrophotometry. One example is the measurement of the size or concentration of large particles in solution that cause scattering, and hence loss of light, rather than just absorbance of it. Most scattering is elastic, meaning that the incident and scattered light have the same wavelength. Turbidity is an expression of the optical properties of a sample that cause light to be scattered rather than transmitted in a straight line through the sample. It is caused by the presence of suspended matter in a liquid. A scattering particle is actually an optical inhomogeneity in an otherwise homogeneous medium. A simple macroscopic example is a piece of dust or lint, but under the right circumstances, an atom, molecule, cell, thermal density fluctuation, colloidal particle, or suspended particle can produce an optical inhomogeneity that results in scattering of radiation.

According to scattering theory, the intensity of the perpendicularly polarized component of scattered radiation and the parallel component is a function of the relative refractive index, the size of the scatterer (quantified via the size parameter), the angle of observation relative to the incident radiation, and the concentration of scatterers. The size parameter ($\alpha = 2\pi r/\lambda$) involves the radius of the scattering particle, r, and the incident wavelength, λ. When the size parameter is smaller than one-tenth the wavelength of the incident radiation and the refractive index of the particle is not greatly different from that of the surrounding medium, scattering occurs symmetrically in all directions and is called Rayleigh scattering. Ultimately, for particles larger than the wavelength of the incident radiation, the radiation intensity scattered by the particle depends in a complicated manner on the angle of scattering, but a large amount is scattered in the forward direction. This is known as Mie scattering. For very large particles, there is no wavelength dependence. Light-scattering theory is complicated by other sample parameters such as particle shape, molecular absorption, sample concentration, and size distribution of scatterers.

There are two methods for measuring the turbidity of a sample [51, 52] – turbidimetry and nephelometry. A turbidimeter measures the amount of radiation that passes through a sample in the forward direction, analogous to absorption spectrophotometry. A nephelometer measures the amount of radiation scattered by a turbid sample. These measurements are made at an angle relative to the incident beam, usually 45° or 90°, analogous to the fluorometers discussed in Chapter 3.

Clearly, these techniques are useful when the concentration of small particles is important. Applications include the analysis of large globular proteins, assessing the clarity of lakes and rivers, and measuring the purity of air samples.

2.6.   A SPECIFIC APPLICATION OF UV-VISIBLE SPECTROSCOPY: ENZYME KINETICS

As we have noted above, UV-visible spectroscopy can aid in structure determination (qualitative analysis) because the absorbance spectrum of a molecule depends on its structure. It is also suited for quantitative determinations via the application of Beer's law. Here we discuss an application in which UV-visible spectroscopy is used to monitor enzyme kinetics and to measure enzymatic rate constants. In general, UV-visible spectroscopy can be applied to relatively slow reactions that occur in the millisecond timescale or longer. Diode array spectrometers are particularly well suited for monitoring changes in spectra that occur as reactants are consumed and products are formed because the entire spectrum is recorded at one time and little time is required between spectral acquisitions. Fast-scanning spectrophotometers can also be used, but scan rates must be rapid (hundreds to thousands of nanometers per minute) in order to accurately capture spectral changes on the timescale of most reactions.

2.6.1.   Myeloperoxidase, Immune Responses, Heart Attacks, and Enzyme Kinetics

Myeloperoxidase (MPO) (Figure 2.44) is an enzyme found in neutrophils – a subset of white blood cells. The primary function of MPO is to catalyze the conversion of chloride ions to hypochlorous acid using hydrogen peroxide [54, 55]:

$$MPO + H_2O_2 \rightarrow H_2O + MPO\text{-}I\,(\text{a compound structurally related to MPO})$$
$$\underline{MPO\text{-}I + Cl^- + H^+ \rightarrow MPO + HOCl + H_2O}$$

Catalyzed rxn : $Cl^- + H_2O_2 + H^+ \rightarrow HOCl + 2H_2O$

The hypochlorous acid (the active ingredient in household bleach) is used by the white blood cells to kill foreign organisms such as bacteria, fungi, viruses, and parasites in the body. People with myeloperoxidase deficiency can, in extreme circumstances, have severely compromised immune responses, leading to fungal and bacterial infections. So having *too little* MPO can be bad. However, MPO is also linked to undesirable inflammatory responses and cardiovascular disease. Evidence suggests that MPO activity indirectly results in the oxidation of low density lipoproteins (LDL) and that oxidized LDL, rather than native LDL, causes the pathological events that lead to atherosclerosis (hardened arteries). In fact, an elevated blood plasma level of MPO is an early predictor for myocardial infarctions (heart attacks) [56]. So having *too much* MPO in your blood can also be a bad sign.

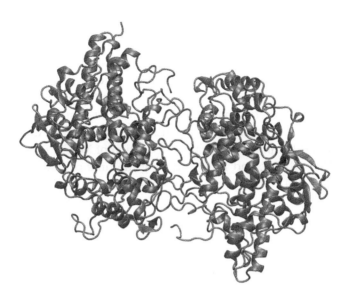

FIGURE 2.44 Representation of the structure of human myeloperoxidase from PDB entry 3F9P. Source: From Carpena et al. [53]. Reproduced with permission of The American Society for Biochemistry and Molecular Biology.

MPO is a heme protein. Heme is a large, highly conjugated ring system, meaning that it absorbs light in the UV-visible region. Because of this, MPO appears green and is responsible for the color of pus and phlegm.

2.6.2. Possible Mechanism for Myeloperoxidase Oxidation of LDL via Tyrosyl Radical Intermediates

Free tyrosine exists in the blood at levels ranging from 20 to 80 μM and is known to be a substrate for MPO. MPO can lead to oxidized LDL by catalyzing the reaction of hydrogen peroxide and tyrosine to create tyrosyl radicals. The tyrosyl radicals can then abstract hydrogen atoms from polyunsaturated fatty acids, which subsequently react with oxygen to form chain-propagating peroxyl radicals [57].

A general scheme [54, 55] for the production of free radicals by MPO catalysis is

$$MPO + H_2O_2 \rightarrow MPO\text{-}I + H_2O$$
$$MPO\text{-}I + AH_2 \rightarrow MPO\text{-}II + AH\cdot$$
$$\underline{MPO\text{-}II + AH_2 \rightarrow MPO + AH\cdot + H_2O}$$
$$Overall: H_2O_2 + 2AH_2 \rightarrow 2AH\cdot + 2H_2O$$

Here, MPO-I and MPO-II are intermediate forms of MPO, and AH_2 is a reducing agent, such as tyrosine. The first reaction is a two-electron oxidation of MPO. The second and third reactions are one-electron reductions of the MPO-I and MPO-II, respectively. The second and third reactions return the enzyme to its original state and, in the process, oxidize the substrate. Note that MPO is not consumed in this reaction, which is consistent with the catalytic nature of enzymes.

UV-visible spectroscopy has been used to measure the rate constants associated with tyrosyl radical production by MPO [54]. While these investigations are rather extensive because the mechanism has multiple steps and several mitigating factors, we focus here on one aspect of the reaction between MPO and tyrosine that demonstrates how UV-visible spectroscopy was used to study enzyme kinetics. Specifically, we examine a study published by Marquez and Dunford of the rate of reduction of MPO-I to MPO-II by tyrosine [54]. Spectra of MPO, MPO-I, and MPO-II during the time course of the reduction are shown in Figure 2.45 [54]. The spectra were collected using a commercially available rapid-scan instrument. Note in the caption that all of the spectra were recorded in 100 ms (about 10 ms per scan) and that tyrosine is in large excess of MPO, such that MPO is the limiting reagent. Note also that the first, second, and last scans were recorded at 2.5, 10, and 100 ms after mixing MPO with tyrosine and hydrogen peroxide. The first scan (the spectrum that seems like it does not "fit" with the rest), then, was taken when most of the MPO was still present in its original form, with a maximum absorbance around 430 nm. Quickly – in about 10 milliseconds – all of the MPO has been converted to MPO-I, which gives rise to the second spectrum from the top on the left side of the figure, with a maximum absorbance around 425 nm. As MPO-I is converted into MPO-II, new spectral features arise from 440 to 480 nm, with a maximum at 456 nm. They do so rapidly at first as the concentration of MPO-II quickly increases. This is evidenced by the large changes in absorbance at 456 nm. However, as the elapsed time nears 100 ms, the concentration of MPO-II levels off because almost all of the MPO-I has been consumed, so no more MPO-II

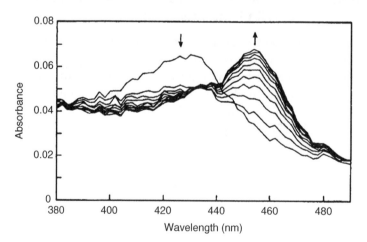

FIGURE 2.45 Rapid-scan spectra of MPO-I reduction to MPO-II by tyrosine. The arrows show the direction of absorbance changes with time. The first, second, and last scans were taken at 2.5, 10, and 100 ms after mixing a solution of 1.0 μm MPO and 50 μM tyrosine with another solution containing 20 μM H_2O_2. Source: Reprinted from Marquez and Dunford [54]. Reproduced with permission of The American Society for Biochemistry and Molecular Biology.

can be formed. It should be clear from this graph that UV-visible spectroscopy is useful for tracking the progress of the reaction as MPO-I is converted to MPO-II. Furthermore, because absorbance values are proportional to concentration, the changes in the absorbance at distinct wavelengths provide direct information about the change in concentration of MPO-I and MPO-II over time. Specifically, changes at 425 nm reflect the decrease in MPO-I and changes at 456 nm reflect increases in MPO-II. Clearly this relates to kinetics because kinetics is the study of the rate of change of concentration of products and reactants over time.

Figure 2.46 [54], and its important inset, shows how kinetic information is obtained from spectra such as those shown in Figure 2.45. The inset shows the absorbance at 456 nm (MPO-II maximum) versus time for a single concentration of tyrosine and MPO. The single exponential character of the plot in combination with the fact that tyrosine is in large excess in all of the reactions suggests that the reaction follows pseudo-first-order kinetics. The word "pseudo" is inserted to remind us that the overall order is not first order because it depends on both tyrosine and MPO-I, but when tyrosine is in large excess, its concentration does not change appreciably and therefore does not affect the rate of the reaction.

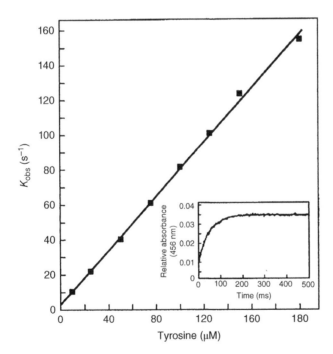

FIGURE 2.46 The inset shows a typical trace and curve fit (solid line) of the reaction followed at 456 nm used to determine the pseudo-first-order rate constants. Final concentrations were 0.25 μM MPO and 5 μM H_2O_2. The second-order rate constant was calculated from the slope of the plot of the observed pseudo-first-order rate constants versus the concentration of tyrosine. Source: Reprinted from Marquez and Dunford [54]. Reproduced with permission of The American Society for Biochemistry and Molecular Biology.

Recall from general kinetics that for a reaction, such as the second step in the three-step mechanism above, the rate is given by

$$\text{Rate} = k[\text{tyrosine}]^{a}[\text{MPO-I}]^{b} \tag{2.40}$$

where k is the overall reaction rate constant and the exponents "a" and "b" are the order of the reaction with respect to each reactant and must be determined experimentally. In the cases where the tyrosine concentration does not change appreciably during the reaction and thus appears constant, the rate equation can be rewritten as

$$\text{Rate} = k'[\text{MPO-I}]^{b} \tag{2.41}$$

Here, k' is a pseudo-first-order reaction rate constant. Integrating Eq. (2.41) shows that a first-order reaction (i.e., $b = 1$) results in a logarithmic increase in the concentration of the product like that shown in the inset. Plotting the logarithm of the concentration – in this case absorbance data – versus time yields a straight line with a slope equal to the pseudo-first order rate constant.

Furthermore, comparing the two equations above shows that under pseudo-first-order kinetics, $k' = k[\text{tyrosine}]^{a}$. This relationship, in the form of $y = mx + 0$, shows that if the experiment is repeated multiple times at different concentrations of excess tyrosine (i.e., multiple plots like that in Figure 2.45 and the inset in Figure 2.46 are conducted), then a plot of the measured k' values versus the respective tyrosine concentrations is linear if the reaction is first-order with respect to the tyrosine concentration (i.e., $a = 1$), and has a slope equal to the overall rate constant for the reaction. This series of experiments was conducted and yielded the plot shown in Figure 2.46 [54]. Ultimately, it was determined that the second-order rate constant, k, was $7.7 \pm 0.1 \times 10^{5}\,\text{M}^{-1}\text{s}^{-1}$. Thus, UV-visible spectroscopy was used to obtain a quantitative understanding of the kinetics of the reaction that contributes to the formation of plaque that clogs and hardens arteries.

The authors extended the work by examining other kinetic aspects of the oxidation of tyrosine by MPO and its physiological implications. A few of the experimental procedures noted in the article are worth discussing. First, in their paper, the authors explicitly state that they made sure that tyrosine does not absorb at any of the wavelengths they used to obtain concentration, and hence kinetic, data. They also conducted kinetic studies in the presence of chloride. Because chloride is a major substrate of MPO, it is possible that tyrosine could be effectively blocked from the active site when chloride is present and thus not be catalytically oxidized by MPO. This was not the case – tyrosine can effectively compete with chloride. The authors also investigated the formation of o,o'-dityrosine from tyrosine radicals and the kinetics of its oxidation by MPO. This was done using fluorescence rather than absorbance spectroscopy because the fluorescence signals were stronger and therefore more sensitive. In fact, throughout the article, the authors state that where options for making measurements existed (e.g., different wavelengths, different techniques, etc.), they routinely chose the method or technique that gave the largest changes in signal and thus the most reliable data. For example, after laying out two options related to the data in Figure 2.45, they state: "We chose to perform kinetic measurements for MPO-I reduction by tyrosine at 456 nm because the absorbance change accompanying MPO-II formation at this wavelength is larger [than the absorbance decrease at 442 nm associated with the disappearance of MPO-I] and leads to smaller experimental errors."

Overall, this was a thorough and carefully conducted set of experiments that probed many important aspects of the role of MPO-catalyzed oxidation of tyrosine. The researchers measured important rate constants, enzyme-binding constants, maximum velocities, and their dependence on variables such as pH from the spectrophotometric data they collected. As a result, by using UV-visible spectroscopy, they gained a better, more quantitative understanding of the mechanism of MPO catalysis.

2.7. SUMMARY

UV-visible spectrophotometry is based on the relatively simple premise that molecules absorb light when the energy of the light matches an electronic transition energy within the molecule. Both structural and quantitative information can be obtained via spectroscopy. Additionally, the absorbance of a solution is proportional to the concentration of the absorbing species. Absorbance is logarithmically related to the ratio of the power of light that reaches a detector in the presence and absence of analyte (i.e., P_o/P). Therefore, two measurements are made for every absorbance value that is produced by a spectrophotometer. The two measurements can be made sequentially as is done in single-beam and diode array spectrophotometers or simultaneously as is done in double-beam instruments. Furthermore, spectrophotometers must have some means of providing a continuous range of wavelengths, a means of isolating particular wavelengths from the continuous source(s), and a way to convert the light into an electrical signal. While UV-visible spectroscopy is relatively easy to conduct, potential sources of error and the limitations of low and high absorbance values must constantly be kept in mind. The major limitations of UV-visible spectroscopy are that it is generally not as sensitive as many other instrumental techniques and analytes must absorb in the UV-visible region of the spectrum in order to be detected. Nevertheless, UV-visible spectroscopy is widely used throughout all areas of science for both routine and highly specialized studies. The determination of kinetic constants, as discussed in detail in this chapter, is just one example of the application of UV-visible spectroscopy. Hundreds of other applications exist, and we trust that all scientists will encounter several of them throughout their careers.

PROBLEMS

2.1 (a) Calculate the energy of a mole of photons having a wavelength of 200 nm.

     (b) Compare this to the average C—C bond energy.

2.2 Convert the following percent transmittances into absorbance:

     (a) 0.5%

     (b) 1.0%

     (c) 15%

     (d) 30%

(e) 50%
(f) 75%
(g) 90%
(h) 99.9%
(i) Make a plot of percent transmittance versus absorbance.

2.3 Convert the following absorbances into percent transmittance:
(a) 0.003
(b) 0.05
(c) 0.18
(d) 0.30
(e) 0.80
(f) 1.2
(g) 2.0
(h) 2.5
(i) 3.0
(j) 3.5

2.4 (a) If the reference and sample signals in a double-beam-in-time instrument produce currents of 500 and 50 μA, respectively, what is the measured absorbance? What if the currents are:
(b) 7000 and 44 μA?
(c) 1.00 mA and 1.00 μA?
(d) 7.00 and 6.95 mA?

2.5 Suppose that if a spectrophotometer were acting ideally (i.e., no stray radiation), a particular analysis would yield signals of 8000 μA for the reference and 44 μA for the sample, but that due to stray radiation, both signals are increased by 20 μA to 8020 and 64 μA. What is the percent error in the absorbance reading obtained in the presence of stray radiation?

2.6 Using Eq. (2.30), calculate the absorbance values given the following powers (in watts) of light:

	P_o	P_{sample}	P_{stray}
(a)	100	50	0
(b)	100	50	2
(c)	100	2	0
(d)	100	2	2

(e) What is the percent error in the absorbance reading caused by the stray radiation in part (b) compared to (a)? For part (d) compared to (c)?
(f) Based on your calculations above, does stray light cause larger errors in higher or lower absorbance values?

2.7 At 510 nm with a 2.00 cm cuvette, the signal from a PMT for a reference solution was found to be 85.4 arbitrary units (i.e., $P_o = 85.4$ arbitrary units). With a 1.00×10^{-4} M solution of an absorbing species in the cuvette, the signal was 20.3 in the same arbitrary units as for the reference. Calculate the molar absorptivity of the species at 510 nm.

2.8   Suppose that a sample produces an absorbance of 0.185 relative to a reference solution in a double-beam spectrophotometer. If the reference and sample solutions were swapped (i.e., the sample used as the reference and the reference used as the sample), what would the absorbance reading be? After the switch, which would be greater, P or P_o?

2.9   The problem above suggests that negative absorbance values result when the reference solution absorbs more light than the sample. Considering that absorbance is directly proportional to concentration, does it make sense to have negative absorbance readings, or would a negative absorbance reading indicate that something is not correct?

2.10   When applying Beer's law, it is best to make absorbance measurements at a wavelength that corresponds to a peak in the spectrum of the analyte of interest. Give two reasons why this is so.

2.11   There are many instrument components and figures of merit mentioned in this chapter. Specify the following components and figures of merit for the commercial spectrophotometer described in Table 2.1:
(a) Grating mounting type
(b) Type of grating and number of lines
(c) Beam splitting device
(d) UV-visible detector
(e) IR detector
(f) UV-visible resolution
(g) Stray light at 370 nm
(h) Bandpass in UV-visible
(i) Photometric stability
(j) What does the acronym NIST stand for and what is the photometric accuracy using NIST 930D 1A filters?

2.12   One of the stated requirements for UV-visible radiation sources is stability over the time course of the measurement. Suppose a single-beam instrument is used to make an absorbance measurement by first inserting the reference solution into the cell holder, followed by the sample. If the lamp intensity is not stable over time, but rather decreases rapidly, will the measured absorbance value be artificially high or low? Explain your answer using the definition of absorbance.

2.13   (a) Calculate the first-order wavelength of light that an interference filter will pass if the thickness of the dielectric is 185 nm with a refractive index of 1.35.
(b) What other wavelengths will also be passed by this filter?

2.14   What is the width, in nm, of each groove of an echelette grating with 1180 grooves/mm?

2.15   For a spectrophotometer with a focal length of 0.50 m and a reciprocal linear dispersion of 1.6 nm/mm, if the instrument is set to a slit width of 5.0 nm (bandpass), (a) what is the physical (i.e., actual) width of the exit slit of the monochromator, and (b) what blaze width is required to achieve this reciprocal linear dispersion with this focal length using first order diffraction?

2.16 Determine whether wide or narrow slits would be desirable for the following analyses and briefly explain why.

(a) The determination of the concentration of a substance that has a broad absorption band using a standard calibration curve based on Beer's law.

(b) The accurate determination of the wavelength of maximum absorbance of a very sharp spectral feature of the substance of interest.

2.17 Sodium produces two sharp bands centered at 589.3 nm that are separated by 0.59 nm.

(a) What resolution is required to observe both peaks as distinct features?

(b) What is the minimum number of lines that a grating must have to resolve this doublet in the first order and the fourth order?

(c) What must the spectral bandpass be to achieve baseline resolution?

(d) What physical slit width is necessary if the monochromator has a reciprocal linear dispersion of 1.6 nm/mm?

2.18 The current generated in a diode is dependent on the wavelength of light striking it. This being the case, how can a diode array, which is just a linear grouping of individual diodes, be used to measure absorbance at different wavelengths as occurs in a diode array spectrophotometer? In other words, if the diode response is dependent on wavelength, then why isn't a UV-visible spectrum distorted by this wavelength dependence?

2.19 If a solution has an absorbance of 0.257 in a 2.00 cm cuvette at a concentration of 5.643×10^{-4} M at a particular wavelength, what will the absorbance of a 2.746×10^{-3} M solution be using a 1.00 cm cell at the same wavelength?

2.20 Iron(II) forms a 1 : 3 (Fe : ligand) complex with 1,10-phenanthroline to produce a colored species that absorbs light in the 500 nm region of the spectrum. If a 1.16×10^{-4} M solution of the complex has an absorbance of 1.289 with a 1.00 cm cuvette at 512 nm, how many milligrams of iron should be added to 1.000 L of a solution containing excess phenanthroline to yield a solution with an absorbance of 0.500 using the same path length and wavelength?

2.21 Consider the following protocol:

• 32.30 mg of p-nitroaniline is dissolved in 100.0 mL of methanol.

• 10.00 mL of this solution is transferred into a 100.0 mL volumetric flask and diluted to the mark with methanol.

• In subsequent dilutions, 5, 4, 3, 2, and 1 mL of this diluted solution are transferred into separate 10.00 mL volumetric flasks and diluted to the mark to create standard solutions for a calibration curve. The measured absorbance readings of these solutions are 1.832, 1.408, 1.102, 0.723, and 0.335, respectively, using a cell with a 1.00 cm path length and a wavelength of 375 nm.

(a) Determine the molar absorptivity of p-nitroaniline using the slope of the calibration curve.

(b) A solution with an unknown concentration of p-nitroaniline dissolved in methanol is diluted by taking 5.00 mL of the solution and diluting to 100.00 mL total with methanol. The resulting solution is then further diluted by taking 5.00 mL of it and diluting to a total volume of 25.00 mL. Use the best-fit line from the calibration curve determined in part (a) to determine the concentration of this diluted sample if its absorbance is found to be 0.681 with a 1.00 cm cell at 375 nm.

(c) Using your answer from part b, determine the concentration of the initial (i.e., undiluted) unknown solution of *p*-nitroaniline.

2.22 Suppose that the unknown *p*-nitroaniline solution in parts b and c in Problem 2.21 above were not dissolved in methanol but rather were dissolved in and diluted with cyclohexane, but that the calibration curve standards were prepared using methanol as the solvent. Based on the spectra of *p*-nitroaniline in both solvents presented in this chapter (and assuming that they were obtained in solutions of roughly equal molarity), if the calibration curve in methanol were used to determine the concentration of *p*-nitroaniline in a solution of cyclohexane and 375 nm were still used as the analytical wavelength, would the concentration determined using the calibration curve be higher or lower than the actual concentration? Explain your answer.

2.23 (a) Draw the structures of *p*-nitrophenol and m-nitrophenol.

(b) The molar absorptivities of *p*-nitrophenol at 294 and 400 nm are 1028 and 18 782 L/mol·cm, respectively, and those for m-nitrophenol at 294 and 400 nm are 4211 and 1485 L/mol·cm, respectively, all in a basic aqueous solution [37]. If a basic solution containing a mixture of the two compounds is found to have absorbances of 0.597 and 1.472 at 294 and 400 nm, respectively, what are the concentrations of the two species in the solution? The path length in all cases is 1.00 cm.

2.24 The absorbance maxima for the FD&C (food, drug, and cosmetic) dye Blue 1, Red 3, and Yellow 5 have absorbance maxima of approximately 625, 525, and 420 nm in water. Correlate these wavelengths with the visible colors of the spectrum, and explain why the dyes lead to the colors that they produce when incorporated into foods, drinks, and cosmetics.

2.25 (a) Make graphs of absorbance vs. pH at the specified wavelengths using the following data and determine the acid dissociation constant for *p*-nitrophenol.

pH	317 nm	407 nm
3.0	9720	330
4.0	9720	330
5.0	9720	330
6.0	9030	1660
6.2	8161	2280
6.4	8190	3990
6.6	7360	5140
6.8	6390	7220
7.0	5550	9160
7.2	4450	11650
7.4	3610	13400
7.6	2920	15000
7.8	2080	16900
8.0	1810	17500
9.0	1390	18330
10.0	1390	18330

(b) Explain why the pK_a of *p*-nitrophenol is significantly lower than that of *p*-methylphenol (pK_a = 10.26)

2.26 Many common organic solvents absorb electromagnetic radiation in the UV portion of the spectrum. For example, most alcohols begin to absorb strongly around 215 nm, with generally increasing absorbance at lower wavelengths. The wavelength below which a solvent begins to absorb is known as its cutoff wavelength (specifically, the cutoff wavelength corresponds to the longest wavelength that produces an absorbance of 1.00 in a 1.0 cm cuvette). This high background absorbance from the solvent can hinder or entirely prevent quantitative analysis of solutes dissolved in the solvent. Furthermore, below the cutoff wavelength, the spectra of the solutes can get exceptionally noisy – sometimes so much so that no spectrum can be observed. Explain the origin of the noise (i.e., why can't the spectrum of the analyte be observed?). Note: The cutoff wavelength for 1,2-dichloroethane is 250 nm. Because of this, the spectrum of Reichardt's dye in Figure 2.4 could not be recorded below 250 nm – it was simply wildly oscillating noise with no distinct chemical information.

2.27 The density of pure benzene is 0.874 g/mL at 25 °C and its molar absorptivity at 250 nm is 240 L/mol·cm.
(a) Determine the molar concentration of pure benzene.
(b) What absorbance would a solution of pure benzene produce if such an absorbance were measurable?
(c) What does this absorbance tell you about the amount of 250 nm light that makes it through the solution?
(d) If a solute was dissolved in benzene, could a quantitative analysis of the solute be conducted at 250 nm? Why or why not?

REFERENCES

1. Bayliss, N.S. and McRae, E.G. (1954). *J. Phys. Chem. 58*: 1002–1006.

2. Kamlet, M.J., Abboud, J.L., and Taft, R.W. (1977). *J. Am. Chem. Soc. 99*: 6027–6038.

3. Pavia, D.L., Lampman, G.M., and Kriz, G.S. (1979). *Introduction to Spectroscopy: A Guide for Students of Organic Chemistry*, 183–224. Austin: Saunders College Publishing.

4. Reichardt, C. (1988). *Solvents and Solvent Effects in Organic Chemistry*, 2e, 285–313. Weinheim: Wiley-VCH Publishing.

5. Jaffe, H.H. and Orchin, M. (1962). *Theory and Applications of Ultraviolet Spectroscopy*, 6–7. New York: Wiley.

6. Rao, C.N.R. (1967). *Ultra-Violet and Visible Spectroscopy Chemical Applications*, 2–3. New York: Plenum Press.

7. Strong, F.C. (1952). *Anal. Chem. 24*: 338–342.

8. Poole, R.K. and Kalnenieks, U. (2000). *Spectrophotometry and Spectrofluorimetry* (ed. M.G. Gore), 2–3. Oxford: Oxford University Press.

9. Kamlet, M.J. and Taft, R.W. (1976). *J. Am. Chem. Soc. 98*: 377–383.

10. Kumler, W.D. (1946). *J. Am. Chem. Soc. 68*: 1184–1192.

11. Taft, R.W. and Kamlet, M.J. (1976). *J. Am. Chem. Soc. 98*: 2886–2894.

12. Willard, H.H., Merritt, L.L. Jr., Dean, J.A., and Settle, F.A. Jr. (1988). *Instrumental Methods of Analysis*, 7e, 139. Belmont: Wadsworth, Inc.

13. Skoog, D.A., Holler, F.J., and Nieman, T.A. (1998). *Principles of Instrumental Analysis*, 5e. Philadelphia: Harcourt Brace and Company.

14. Buc, G.L. and Stearns, E.L. (1945). *J. Opt. Soc. Am. 35*: 458–464.

15. Hogness, T.R., Zscheile, F.P. Jr., and Sidwell, E.A. Jr. (1937). *J. Phys. Chem. 41*: 379–415.

16. Piunno, P.A.E. (2017). *J. Chem. Educ. 94*: 615–620.

17. Palmer, C. and Loewen, E. (2005). *Diffraction Grating Handbook*. New York: Newport Corporation.

18. Keliher, P.N. and Wohlers, C.C. (1976). *Anal. Chem. 48*: 333A–340A.

19. Flamand, J., Grillo, A., and Hayat, G. (1975). *Am. Lab. 7*: 47–51.

20. Ebert, H. (1889). *Wiedemann's Ann. 38*: 489.

21. Fastie, W.G. (1952). *J. Opt. Soc. Am. 42*: 641–647.

22. Fastie, W.G. (1952). *J. Opt. Soc. Am. 42*: 647–651.

23. Fastie, W.G. (1991). *Phys. Today 44*: 37–43.

24. Barry, D.J., Bagnuolo, W.G., and Riddle, R.L. (2002). *J. Astron. Soc. Pac. 114*: 198–206.

25. Czerny, M. and Turner, A.F. (1930). *Z. Physik 61*: 792–797.

26. Turner, L.W. (ed.) (1976). *Electronics Engineer's Reference Book*, 4e, 9–27. Boston: Butterworths & Co., Ltd.

27. Anderson, L.K. and McMurtry, B.J. (1966). *Appl. Opt. 5*: 1573–1587.

28. Sacks, R.D. (1981). Emission spectroscopy, Chapter 6. In: *Treatise on Analytical Chemistry*, 2e, part I, vol. 7 (ed. P.J. Elving, E.J. Meehan, and I.M. Kolthoff). New York: Wiley.

29. Talmi, Y. (1975). *Anal. Chem. 47*: 658A–670A. and 697A–709A.

30. Jones, D.C. (1985). *Anal. Chem. 57*: 1057A–1073A. and 1207A–1214A.

31. Sweedler, J.V. (1993). *Crit. Rev. Anal. Chem. 24*: 59–98.

32. Lytle, F.E. (1974). *Anal. Chem. 46*: 545A–557A.

33. Lott, P.F. (1968). *J. Chem. Educ. 45*: A182–A194. and A273–A292.

34. Lott, P.F. (1968). *J. Chem. Educ. 45*: A89–A111. and A169–A180.

35. Willis, B.G., Fustier, D.A., and Bonelli, E.J. (1981). *Am. Lab. 13*: 62–71.

36. Vitha, M.F., Carr, P.W., and Mabbott, G.A. (2005). *J. Chem. Educ. 82*: 901–902.

37. Harvey, D.T. (1999). *Modern Analytical Chemistry*. New York: McGraw-Hill.

38. Kortum, G. and Seiler, M.T. (1939). *Angew. Chem. 52*: 687–693.

39. Efstathiou, C.E. Educational applets: deviations from Beer's law: effect of polychromatic radiation. http://www.chem.uoa.gr/applets/AppletBeerLaw/Appl_Beer2.html (accessed 22 July 2017).

40. Cook, R.B. and Jankow, R. (1972). *J. Chem. Educ. 49*: 405–408.

41. Slavin, W. (1963). *Anal. Chem. 35*: 561–566.

42. Kaye, W. (1981). *Anal. Chem. 53*: 2201–2206.

43. Sharpe, M.R. (1984). *Anal. Chem. 56*: 339A–356A.

44. Rothman, L.D., Crouch, S.R., and Ingle, J.D. Jr. (1975). *Anal. Chem. 47*: 1226–1233.

45. Perkampus, H.-H. (1999). *UV-Vis Spectroscopy and Its Applications* (trans. C. Grinter and T.L. Threlfall), 58–68. Berlin: Springer-Verlag Publishing.

46. Sunshine, I. (1981). *Handbook of Spectrophotometric Data of Drugs*. Cleveland: Chemical Rubber Co.

47. O'Haver, T.C. (1979). *Anal. Chem. 51*: 91A–100A.

48. Hammond, V.J. and Price, W.C. (1953). *J. Opt. Soc. Am. 43*: 924.

49. Morrison, J.D. (1953). *J. Chem. Phys. 21*: 1767–1772.

50. Giese, A.T. and French, C.S. (1955). *Appl. Spectrosc. 9*: 78–96.

51. Surles, T., Erickson, J.O., and Priesner, D. (1975). *Am. Lab. 7*: 55.

52. Wendlandt, W.W. (1968). *J. Chem. Educ. 45*: A861 and A947.

53. Carpena, X., Vidossich, P., Schroettner, K. et al. (2009). *J. Biol. Chem. 284*: 25929–25937.

54. Marquez, L.A. and Dunford, H.B. (1995). *J. Biol. Chem. 51*: 30434–30440.

55. Furtmüller, P.G., Burner, U., and Obinger, C. (1998). *Biochem. 37*: 17923–17930.

56. Brennan, M.L., Penn, M.S., Van Lente, F. et al. (2003). *N. Engl. J. Med. 17*: 1595–1604.

57. Savenkova, M.I., Mueller, D.M., and Heinecke, J.W. (1994). *J. Biol. Chem. 269*: 20394–20400.

FURTHER READING

1. Gore, M.G. (ed.) (2000). *Spectrophotometry and Spectrofluorimetry*, 2–3. Oxford: Oxford University Press.

2. Ingle, J.D. Jr. and Crouch, S.R. (1988). *Spectrochemical Analysis*. Englewood Cliffs: Prentice Hall.

3. Jaffe, H.H. and Orchin, M. (1962). *Theory and Applications of Ultraviolet Spectroscopy*. New York: Wiley.

4. Loewen, E.G. and Popov, E. (1997). *Diffraction Grating and Applications*. Boco Raton: CRC Press.

5. Meehan, E.J. Chapters 1–3. In: *Treatise on Analytical Chemistry*, 2e, part I, vol. 7 (ed. P.J. Elving, E.J. Meehan, and I.M. Kolthoff). New York: Wiley.

6. Pavia, D.L., Lampman, G.M., and Kriz, G.S. (1979). *Introduction to Spectroscopy: A Guide for Students of Organic Chemistry*. Austin: Saunders College Publishing.

7. Perkampus, H.-H. (1999). *UV-Vis Spectroscopy and Its Applications* (trans. C. Grinter and T.L. Threlfall), 58–68. Berlin: Springer-Verlag Publishing.

8. Rao, C.N.R. (1967). *Ultra-Violet and Visible Spectroscopy Chemical Applications*, 2–3. New York: Plenum Press.

9. Reichardt, C. (1988). *Solvents and Solvent Effects in Organic Chemistry*, 2e, 285–313. Weinheim: Wiley-VCH Publishing.

10. Yadav, L.D.S. (2005). *Organic Spectroscopy*. Boston: Kluwer Publishers.

MOLECULAR LUMINESCENCE: FLUORESCENCE, PHOSPHORESCENCE, AND CHEMILUMINESCENCE

3

Just as molecules can absorb electromagnetic radiation, they can emit it, too. The emission is broadly called luminescence. Emission that follows the absorbance of a photon is called either fluorescence or phosphorescence, depending on the excited state from which it occurs. Chemiluminescence is the emission of light that results from a chemical reaction. This chapter describes the theory behind molecular luminescence and instrumental methods for measuring it. Analytical methods based on these phenomena can have exceptionally low detection limits, be quite selective, and have wide dynamic ranges. Luminescence spectroscopy has been used in applications such as DNA sequencing, single molecule detection, and fluorescence microscopy.

3.1. THEORY

In Chapter 1, we saw that molecules can absorb energy from electromagnetic radiation and in the process promote electrons from the ground state distribution into higher excited state distributions (known as S_1, S_2, etc.). Some of the energy is removed through rapid vibrational relaxation such that the molecule quickly returns to the lowest vibrational level in the lowest excited state (i.e., v_0 in S_1). These processes are summarized in the Perrin–Jablonski diagram [1–6] shown in Figure 3.1 and described in greater detail in Chapter 1.

A molecule in S_1 can either relax via nonradiative processes, relax by emitting a photon, or undergo intersystem crossing via an electron spin flip to enter a triplet state, T_1. Once in T_1, the molecule can transition back to the ground state nonradiatively or emit a photon. Emission from the S_1 state is known as fluorescence and emission from the T_1 state is known as phosphorescence. For our purposes, the only significant difference between fluorescence and phosphorescence is that phosphorescence requires that the molecule undergo a spin-forbidden transition from a singlet state to enter T_1 and another spin-forbidden transition back to the ground state. The practical implication is that because these transitions are unlikely and slow, most molecules relax back to the ground state via other faster processes. Therefore, most molecules do not phosphoresce, which limits the number of analytical methods that are based on phosphorescence. For those molecules that do phosphorescence, however, methods based on phosphorescence can be highly selective and sensitive.

Spectroscopy: Principles and Instrumentation, First Edition. Mark F. Vitha.
© 2019 John Wiley & Sons, Inc. Published 2019 by John Wiley & Sons, Inc.

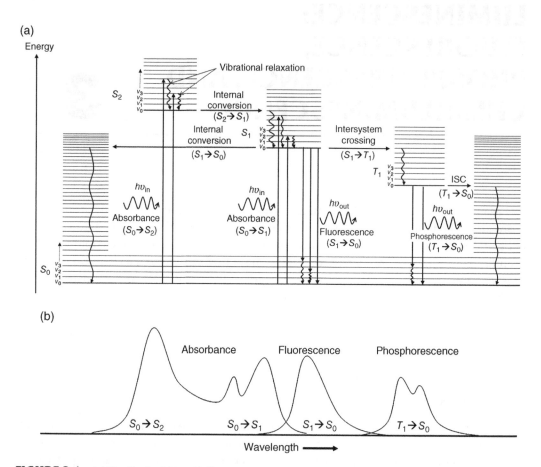

FIGURE 3.1 (a) The Perrin–Jablonski diagram summarizes a variety of processes related to the interaction of light with molecules. The arrow indicating that energy increases from top to bottom is an important feature of the diagram. In this chapter, we focus on the processes of fluorescence and phosphorescence. However, both of these processes require that the molecule first absorbs light. Other processes detailed in the diagram are explained in the text. Note that the energy involved in absorbance is greater than that in fluorescence, which is greater than that in phosphorescence. (b) Depiction of the shift in wavelengths of absorbance, fluorescence, and phosphorescence spectra. Source: Adapted in parts from Valeur and Berberan-Santos [6]. Reproduced with permission of Wiley-VCH Verlag BmbH & Co. KGaA.

Molecules that fluoresce are referred to as "fluorophores." This same word is also used to refer to a group of atoms within a molecule that is responsible for the fluorescence of a molecule. This chapter largely focuses on fluorescence because fluorescence is more common than phosphorescence and chemiluminescence. However, instruments that measure fluorescence can also be used to measure phosphorescence and chemiluminescence, so many of the principles apply to luminescence in general.

3.1.1. Absorbance Compared to Fluorescence

In absorbance measurements, the wavelength of electromagnetic radiation that goes into the sample is the same as that being detected after it passes through the sample. All that is measured is the loss of light of a particular wavelength due to it being *absorbed* by molecules. In contrast, in fluorescence, the intensity of *emitted* light is measured. Furthermore, the electromagnetic radiation emitted by fluorescent molecules has a longer wavelength than that which it absorbs. Fluorescence measurements also generally have significantly lower limits of detection and are more sensitive than absorption measurements. The cause of these differences is examined in this section.

Absorption occurs on a femtosecond (10^{-15} s) timescale. Because it is so rapid, only the electron distribution in a molecule changes during this time – the positions of the massive (on a relative scale) atoms in the molecule and in the surrounding solvent molecules do not. Therefore, immediately after absorption, the solvent "cage" that used to exist around the ground state still exists around the excited state molecule. This is shown in step 1 of Figure 3.2. Because the electron distribution of the fluorophore changes, the magnitude and direction of the molecule's dipole moment changes (see Figure 3.2), with the dipole moment of the excited state being greater than that of the ground state (i.e., $\mu_e > \mu_g$) for most molecules that are used to probe the polarity of their environment [6]. Because the solvent molecules have not moved, however, the solvent organization around the excited state is not optimal.

Over the next several picoseconds after absorption, the excited state molecule begins to relax. During this time, the solvent molecules begin to reorient to better interact with the excited state electron distribution. The increased interactions after solvent reorientation lower the energy of the excited state structure.

The molecule then fluoresces and returns to the ground state energy. But here again, as seen in Figure 3.2, the solvent cage does not have time to adjust to the new ground state dipole of the molecule during the emission process. Thus, the molecule returns to a ground state energy that is slightly higher in energy than it was before the absorbance and fluorescence processes took place. After emission has taken place, the solvent molecules then reorient back to the initial configuration best suited to solvate the ground state (i.e., the system returns to its lowest energy).

The practical implication of this solvent reorganization, combined with the rapid vibrational relaxation that occurs after excitation, is that the energy emitted via fluorescence is less than that which causes excitation. This means that fluorescence occurs at longer wavelengths than absorption (recall that $E \propto 1/\lambda$) and may be dependent on the polarity of the solvent, as seen in Figure 3.3. Furthermore, because fluorescence occurs predominantly from the S_1 state, fluorescence spectra often resemble the mirror image of absorption spectra, shifted to longer wavelengths as shown in Figure 3.4 [6, 9].

EXAMPLE 3.1

In which solvent, cyclohexane or acetonitrile, do you expect fluorescence to occur at lower energy for a molecule like that in Figure 3.3 for which $\mu_e > \mu_g$? [8]

Answer:
Acetonitrile as a solvent will cause a decreased energy of fluorescence. Because acetonitrile is polar, it can solvate the excited state better than cyclohexane. This lowers the excited state energy in acetonitrile. As a consequence, less energy is emitted as fluorescence in returning to the ground state.

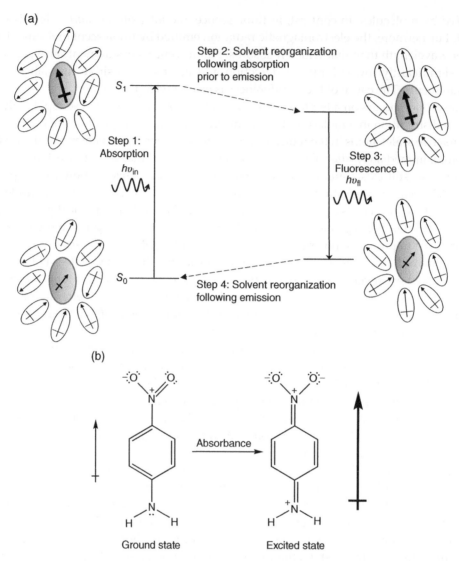

FIGURE 3.2 (a) Depiction of the change in dipole moment of a solute (shaded gray) upon absorption of a photon and the effect it has on the solvent molecules (open ovals) around the solute. In this case, the excited state dipole is greater than the ground state dipole (i.e., $\mu_e > \mu_g$), as indicated by the width and length of the dipole symbol. Step 1: No change in solvent organization during the absorption process. Note that the dipole moments of the solvent molecules are not oriented optimally around the excited state immediately upon absorption. Step 2: Solvent reorganization in response to the excited state electronic distribution that has a greater dipole moment than does the ground state. Step 3: Fluorescence – note that fluorescence, just like absorption, is so rapid that no solvent reorganization occurs during the event. Step 4: Reorganization of the solvent cage around the ground state dipole. (b) 4-Nitroaniline is an example of a molecule that has an excited state dipole greater than the ground state. Note that while the ground state is polar, the excited state is much more so. Electron density from the amine nitrogen is transferred toward the nitro group, causing the enhanced excited state dipole. Source: From Taft and Kamlet [7]. Reproduced with permission of American Chemical Society.

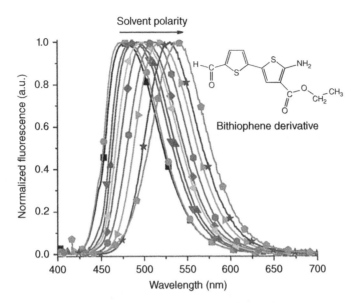

FIGURE 3.3 The emission spectra above were recorded for the derivative of the bithiophene shown in the figure in solvents with a wide range of polarity [8]. Solvents: toluene (■), ether (●), tetrahydrofuran (▲), ethyl acetate (▼), chloroform (◆), dichloromethane (◄), dimethylformamide (►), acetonitrile (◕), dimethyl sulfoxide (★), and ethanol (⬠). The polarity of the chemical environment of the fluorophore shifts the fluorescence spectrum toward longer wavelengths (lower energies). Source: Reproduced from Bolduc et al. [8] with permission of the PCCP Owner Societies. Reproduced with permission of Royal Society of Chemistry.

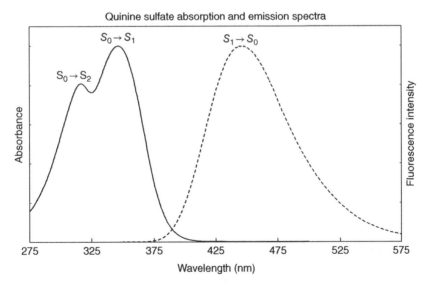

FIGURE 3.4 Normalized absorbance (solid line) and emission spectra (dashed line) of quinine sulfate in 0.25 M sulfuric acid. Note that fluorescence occurs at longer wavelength (lower energy), and while the absorbance spectrum reflects both the $S_0 \rightarrow S_2$ transition centered around 315 nm and the $S_0 \rightarrow S_1$ transition around 345 nm, the fluorescence only occurs from S_1 due to rapid internal conversion (445 nm peak).

EXAMPLE 3.2

(a) For the range of solvents shown in Figure 3.3, what is the effect on the wavelength of maximum fluorescence (i.e., how much is the wavelength shifted)?
(b) What is the difference in energy between the longest and shortest wavelength of maximum emission per photon? Per mole?

Answers:
(a) The wavelength of maximum fluorescence in toluene is approximately 480 nm and in ethanol is about 540 nm, for a total shift of 60 nm.
(b) To calculate the energy difference, we need to calculate the energy in each solvent:

$$\text{For toluene as solvent}: E = \frac{hc}{\lambda} = \frac{\left(6.626 \times 10^{-34}\ \text{Js}\right)\left(2.998 \times 10^{8}\ \text{m/s}\right)}{480 \times 10^{-9}\ \text{m}} = 4.14 \times 10^{-19}\ \text{J}$$

$$\text{For ethanol as solvent}: E = \frac{hc}{\lambda} = \frac{\left(6.626 \times 10^{-34}\ \text{Js}\right)\left(2.998 \times 10^{8}\ \text{m/s}\right)}{540 \times 10^{-9}\ \text{m}} = 3.68 \times 10^{-19}\ \text{J}$$

So the difference in energy is 4.14×10^{-19} J $- 3.68 \times 10^{-19}$ J $= 4.6 \times 10^{-20}$ J per photon, or about 28 kJ/mol.

The fact that fluorescence occurs at longer wavelengths than those absorbed by a molecule leads to a second significant difference between absorption and fluorescence spectroscopy, namely, that fluorescence is more sensitive than absorbance spectroscopy. The analytical challenge in absorption spectroscopy is to measure the *change* in intensity of a particular wavelength of light before and after it passes through a sample. In other words, one is trying to measure a decrease in the signal against a large initial intensity. In fluorescence spectroscopy, the task is to measure the *presence* of signal at a wavelength that is different than the excitation wavelength. The only way electromagnetic radiation at a longer wavelength is present is if a molecule emits it. So one is trying to measure the presence of signal against an essentially dark background, assuming no background fluorescence is present. It is much easier to measure the *presence* of light than it is to measure a *change* in the intensity of signal. This can be understood using an analogy. Imagine standing in the center of a stadium packed with tens of thousands of people, all of whom have flashlights. Absorbance measurement are like trying to figure out how many people turn off their flashlights after starting out with everybody having them turned on. In contrast, fluorescence is like standing in the middle of a totally dark stadium and measuring how many people turn on their flashlights starting with them all turned off. You can imagine it is much easier to see one or two flashlights turned on in a totally dark stadium (fluorescence) than it is to detect one or two flashlights being turned off while hundreds of thousands are still on (i.e., absorbance). This difference means that fluorescence measurements are much more sensitive.

3.1.2. Factors That Affect Fluorescence Intensity

If fluorescence is more sensitive than absorbance, it is reasonable to ask why absorbance spectroscopy is more common than fluorescence. The answer is that many molecules absorb electromagnetic radiation in the UV-visible region, but relatively few fluoresce and even fewer phosphoresce.

For those molecules that do fluoresce to some extent, the probability that they will is reflected in a quantity called the quantum yield, Φ_f. Specifically, quantum yield is the probability that a molecule in the excited state returns to the ground state via fluorescence:

$$\Phi_f = \frac{k_f}{k_f + k_{nf}} \tag{3.1}$$

where k_f is the rate constant for fluorescence and k_{nf} is the rate constant for all nonfluorescent relaxation processes, including internal conversion, intersystem crossing, and phosphorescence. The quantum yield can also be thought of as follows: Imagine a collection of *excited state* molecules of all the same type – for example, 100 pyrene molecules *that have already absorbed a photon*. The quantum yield is the ratio of the number of these excited state molecules that return to the ground state by emitting fluorescence relative to the total. If, say, 32 of the 100 lose their energy via fluorescence, then the quantum yield is 0.32. It is important to note that the quantum yield is *not* the ratio of molecules that fluoresce relative to the *total* number of fluorescent molecules present in a sample, but rather the ratio of those that fluoresce relative to those that absorbed a photon. For example, consider a particular molecule that does not absorb electromagnetic radiation at a particular wavelength very well. In other words, it has a low molar absorptivity. Even if you have a high concentration of that molecule, not a lot will be promoted to the excited state. However, if the vast majority of those molecules that do absorb ultimately fluoresce, then the quantum yield is high for that molecule. It is worth noting that while we have defined the quantum yield of fluorescence, a similar equation can be written for the quantum yield of phosphorescence.

For molecules with a high rate of fluorescence relative to nonfluorescent relaxation rates, Φ_f approaches 1.00, meaning that every molecule that absorbs a photon and is promoted to the excited state emits fluorescence. Molecules with relaxation rates for deactivation of the excited state that are competitive with k_f have much lower quantum yields. Clearly, high quantum yields are desirable when using fluorescence to detect and quantify molecules.

Quantum yields are affected by many variables, including:

1. Molecular structure
2. Solvent effects
3. Temperature
4. Solvent pH

which are explained in the following section [6, 10]. Quantum yields for several molecules are presented in Table 3.1.

TABLE 3.1 Quantum Yield of Common Fluorescent Molecules

Molecule	Quantum yield	Solvent	Source
Benzene	0.04	Ethanol	6
Naphthalene	0.21	Ethanol	6
Pyrene	0.65	Ethanol	6
Perylene	0.98	n-Hexane	6
Phenylalanine	0.024	Water	10
Tryptophan	0.13	Water	10

TABLE 3.1   (Continued)

Molecule	Quantum yield	Solvent	Source
Quinine sulfate	0.58	0.1 M H_2SO_4	10
Fluorescein	0.95	0.1 M NaOH	10
Rhodamine 6G	0.94	Ethanol	10

3.1.2.1.   Structural Effects.   As is evident from the structures in Table 3.1, fluorescence is expected in molecules that are aromatic or contain multiple conjugated double bonds. Saturated aliphatic compounds generally do not absorb in the UV-visible region, and because absorption is a prerequisite for fluorescence, they also do not fluoresce. However, just because molecules absorb light does not mean that they also fluoresce. Benzene, for instance, has a very low quantum yield. Increasing the number of fused rings increases the likelihood of fluorescence. Substituents that delocalize the π-electrons such as —NH_2, —OH, —F, —OCH_3, —$NHCH_3$, and —$N(CH_3)_3$ tend to increase fluorescence, whereas electron-withdrawing groups that contain —Cl, —Br, —I, —$NHCOCH_3$, —NO_2, or —COOH decrease or quench fluorescence completely [11]. Thus, aniline fluoresces but nitrobenzene does not.

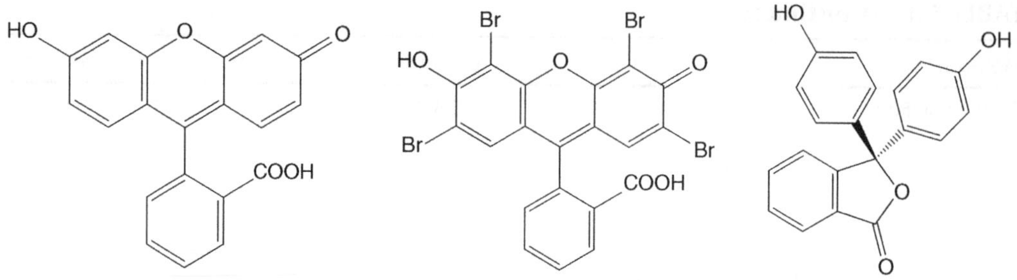

FIGURE 3.5   From left to right: fluorescein, eosin Y, and phenolphthalein.

Another factor that affects quantum yields is molecular rigidity. Rigidity reduces the interactions of the excited state with the surrounding solvent and thus reduces the rate of collisional deactivation (i.e., internal conversion). The reduced rate of deactivation by nonradiative processes leads to a greater probability of fluorescence (see Eq. 3.1). Given a series of aromatic compounds, those that are most planar, rigid, and sterically uncrowded are the most fluorescent. Specific examples of these effects include the fact that fluorescein and eosin (see Figure 3.5) are strongly fluorescent, but a similar compound, phenolphthalein, which is nonrigid and in which the conjugated system is disrupted, is not fluorescent despite its high molar absorptivity in the UV-visible region.

EXAMPLE 3.3

Draw the structures of aniline and nitrobenzene and show resonance structures for both. Which substituent ($-NH_2$ or $-NO_2$) is electron donating and which is electron withdrawing?

Answer:
Resonance structures of aniline

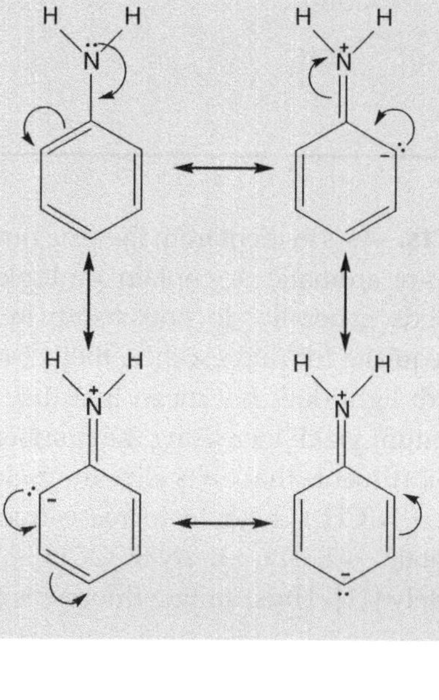

Resonance structures of nitrobenzene

The –NH$_2$ group is electron donating and the –NO$_2$ group is electron withdrawing

Another question:

Which molecule in each of the following pairs is likely to be more fluorescent?
  (a) Octanol or 1-aminonaphthalene
  (b) 2-Bromonaphthalene vs. 2-methoxynaphthalene
  (c) Pyrene or benzene

Answers:
  (a) 1-Aminonaphthalene
  (b) 2-Methoxynaphthalene
  (c) Pyrene

3.1.2.2. *Solvent Effects.*

Solvents capable of strong van der Waals interactions with the excited state species prolong the lifetime of collisional encounters, which leads to increased deactivation via nonfluorescent processes and decreased fluorescence (i.e., increased k_{nf} relative to k_f). Solvents that have molecular substituents such as —Br, —I, —NO$_2$, and —N=N— promote spin decoupling that leads to triplet state formation, thereby increasing intersystem crossing. Because fluorescence is defined as emission from a singlet state, intersystem crossing into the triplet state decreases fluorescence – although phosphorescence may be promoted as a consequence.

We also saw earlier that solvents can affect the wavelength of fluorescence depending on how well they stabilize the excited state electronic distribution. Greater stabilization, generally associated with greater solvent polarity and nonviscous solvents, shifts fluorescence to longer wavelengths compared to less polar, more viscous solvents. Viscosity matters in that viscous solvents restrict or slow the reorientation of the solvent molecules around the excited state dipole. If the reorientation takes a long time, the molecule is likely to fluoresce before the intermolecular interactions between the excited state fluorophore and the solvent molecules are optimized. Thus, fluorescence will occur from a higher energy state than it would if the solvent molecules move more quickly as they do in a less viscous medium.

The viscosity of the solvent also affects fluorescence quantum yields. This arises because viscous solvents restrict molecular motion, which in turn reduces the rate of collisions between the excited state fluorophore and solvent molecules, ultimately reducing the rate of nonradiative relaxation. Decreasing k_{nf} relative to k_f increases the fluorescence

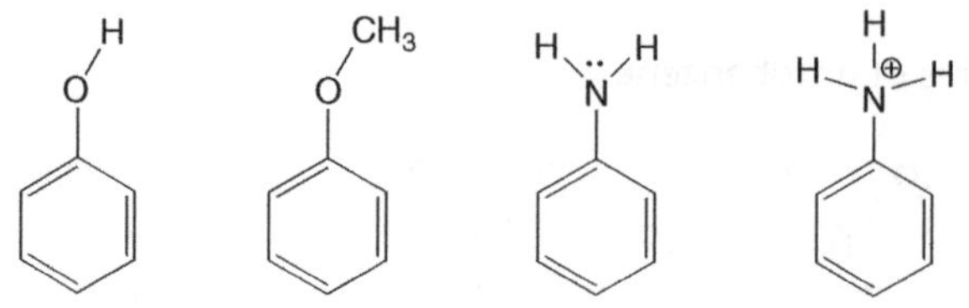

FIGURE 3.6   From left to right: phenol, anisole, aniline, and the anilinium ion.

quantum yield. Trapping fluorescent molecules in viscous gels, glasses, and solids leads to increased Φ_f values for the same reasons.

3.1.2.3. Temperature Effects. Lowering the temperature of a sample slows molecular motion and decreases the energy exchanged during molecular collisions. These effects generally lead to increased fluorescence.

3.1.2.4. pH Effects. Changes in the system pH that affect the charge on a fluorophore influence fluorescence. Both phenol and anisole (see Figure 3.6) fluoresce at pH = 7, but at pH = 12, phenol ($pK_a = 10.0$) is converted to a nonfluorescent anion, whereas anisole remains uncharged. Similarly, aniline ($pK_b = 9.4$) fluoresces in the visible range at pH = 7 through 12, but the protonated cation is nonfluorescent at pH = 2 [11]. These observations can be explained by comparing the resonance forms of anions and cations. For example, aniline in acidic solution has a positive charge fixed on the nitrogen atom, and the anilinium ion has only the same resonance forms as benzene, which fluoresces only in the UV region. In neutral or basic solution, however, aniline has three additional resonance structures, resulting in a more stable excited singlet state and a longer wavelength of fluorescent radiation [12].

3.1.3. Quenching

The phenomenon of quenching can be considered a solvent or medium effect, but we treat it here separately to call attention to its impact on Φ_f values. Quenching is the reduction or elimination of fluorescence because of interactions between the fluorophore and other species in solution. Oxygen (O_2) dissolved in solvents is a ubiquitous and well-known quencher. The most likely explanation of quenching by oxygen is that the paramagnetic nature of oxygen promotes intersystem crossing of a fluorophore's singlet state to a triplet state [10]. To reduce quenching by O_2, solutions must be degassed with N_2 or another inert gas before fluorescence is measured. Other efficient quenchers include solvents with amine groups and heavy atoms such as bromine and iodine. Quenching by such molecules relies on collisions between them and the fluorophores. It therefore depends on (1) the concentration of the quenchers, (2) the concentration of the fluorophore, (3) the viscosity of the solution, and (4) the temperature of the solution.

EXAMPLE 3.4

(a) As the viscosity of samples increases, do you expect the quenching caused by dissolved species to increase or decrease? Will fluorescence increase or decrease as a result?

(b) As the temperature of a sample is decreased, do you expect the quenching caused by dissolved species to increase or decrease? Will fluorescence increase or decrease as a result?

Answers:

(a) As viscosity increases, quenching by dissolved species decreases. This occurs because molecular diffusion is restricted and therefore slower in viscous solutions, so quenching species have a harder time reaching the excited state molecule before it fluoresces.

(b) Similarly, as temperature decreases, molecular motion slows, decreasing the rate of collisions between quenchers and fluorescing molecules. Because quenching decreases fluorescence, a decrease in the quenching rate results in an increase in fluorescence intensity as temperature decreases.

While quenching may seem like a negative effect because it reduces fluorescence, the specificity of some fluorophore–quencher complexes and the dependence on quencher concentration are the basis for some analytical measurements.

3.1.4. Quantum Yield and Fluorescence Intensity

The discussion above focused on the factors that affect the fluorescence quantum yield. These are important because higher quantum yields mean more molecules that absorb electromagnetic radiation emit it as fluorescence and hence contribute to the signal. But as we cautioned before, Φ_f is the ratio of molecules that emit a photon *relative to those that absorbed one.* Thus, the total amount of fluorescence observed also depends on the likelihood of a molecule absorbing a photon in the first place. So Φ_f is not the only variable controlling the total intensity of fluorescence. In fact, the intensity of fluorescence, P_f, is given by

$$P_f = \Phi_f \left(P_o - P \right) \tag{3.2}$$

where P_o is the power of the excitation beam incident upon the sample and P is the radiant energy emerging from the sample. The quantity $P_o - P$ is therefore the radiant power absorbed by the sample.

This equation makes it easy to see that fluorescence intensity is greatest for molecules that absorb the incoming radiation well (i.e., have high $P_o - P$) and that convert it efficiently into fluorescence (i.e., have high Φ_f).

Equation (3.2) can be manipulated to relate P_f to the concentration of the fluorescing molecule, which is clearly an analytically useful thing to do. Applying Beer's law ($A = \varepsilon bc = -\log(P/P_o)$) leads to

$$P_f = \Phi_f P_o \left(1 - 10^{-\varepsilon bc}\right) \tag{3.3}$$

where ε is the molar absorptivity of the fluorophore, c is its concentration, and b is the path length of the sample, which, as we discuss later, is ultimately related to the volume of solution illuminated from which fluorescence is emitted. Expanding Eq. (3.3) in a power series yields [11]

$$P_f = 2.303\Phi_f P_o \varepsilon bc \left[1 + \frac{2.303\varepsilon bc}{2!} + \frac{(2.303\varepsilon bc)^2}{3!} + \cdots + \frac{(2.303\varepsilon bc)^n}{(n+1)!}\right] \tag{3.4}$$

If absorbance is low (e.g., Abs < 0.05), only the first term in the series is significant (i.e., $1 \gg 2.303\varepsilon bc/2!\ldots$), and Eq. (3.4) can be written as

$$P_f \approx 2.303\Phi_f P_o \varepsilon bc \tag{3.5}$$

Equation (3.5) makes several important relationships clear:

1. The power (intensity) of fluorescence, P_f, is proportional to the quantum yield, Φ_f. This makes sense in that Φ_f is a measure of the likelihood that an excited state molecule fluoresces. The greater the probability, the more fluorescence will be observed.
2. The intensity of fluorescence, P_f, is proportional to absorbance (εbc) when absorbance is low. This means that the more excitation radiation that is absorbed by the sample, the more light that will be emitted. More specifically, the equation shows that *fluorescence increases linearly with concentration*, meaning that fluorescence can be used to make quantitative measurements, keeping in mind the restriction that the linear relationship holds only in the low absorbance region ($A < 0.05$ or more rigorously <0.02) [11]. Given the sensitivity of the technique, this requirement can often be met.
3. Lastly, Eq. (3.5) makes it clear that the amount of fluorescence emitted by a sample is proportional to the intensity of the *excitation* beam (P_o). This makes sense because the more photons that enter the solution, the greater the probability that the fluorophores in the sample are promoted to the excited state, from which they can then fluoresce. In other words, *more excitation photons in, more fluorescence out*. The practical consequence of this is that high intensity sources are used to increase fluorescence signals. Therefore, it is common to use lasers, which produce high photon fluxes, in fluorescence applications. High photon fluxes help to overcome relatively low fluorescence yields and low concentrations.

The proportionality between P_f and P_o is important to note, too, because source intensity is the main instrumental parameter in Eq. (3.5). The other parameters are largely set by the fluorophore and the sample characteristics. Technically, "b" – the path length – is also under the analyst's control, with longer path lengths leading to increased fluorescence intensity. However, the path length is often set or limited by the sample being analyzed (e.g., a cell wall in the case of fluorescence microscopy) or by the instrument configuration. It is worth noting, though, that in fluorescence, "b" is related to the volume of sample illuminated by the excitation beam because the greater the number of molecules that reach the excited state, the greater the number that fluoresce. Illuminating a greater volume excites a greater number of fluorophores, which increases the fluorescence signal. Wider slits also mean that a broader range of wavelengths (i.e., greater bandwidth) passes through the sample. For the same reasons as discussed in the UV-visible spectroscopy, this can lead to nonlinearities in calibration curves based on fluorescence. In practice, we will see that slit widths, which control the amount of solution that is illuminated, can be used to adjust the amount of signal obtained from a sample, with increasing slit widths generally associated with increasing signal.

3.1.5. Linearity and Nonlinearity of Fluorescence: Quenching and Self-Absorption

In accord with Eq. (3.5), plots of fluorescence (or phosphorescence for that matter), such as those shown in Figure 3.7 are often linear over three or more orders of magnitude of concentration. But there are some limiting factors as demonstrated by the curvature that exists in the plot at high concentration. Specifically, background fluorescence, quenching, and absorbance of exciting radiation by the solvent or by the analyte itself all affect the slope and linearity of such plots. The minimum detectable quantity of an analyte is generally limited by the signal arising from the blank, as solvent fluorescence and light scattering produce signals that may obscure the fluorescence of the analyte.

According to the general expression for fluorescence (Eq. 3.3), a sharp negative deviation from linearity is exhibited at high absorbance values. High absorbance is often correlated with high concentrations. At high concentrations, self-quenching and self-absorption may also contribute to negative deviations. Self-quenching arises when excited state molecules collide with other analyte molecules and lose their excitation energy by radiationless transfer. Collisional quenching, as discussed above, particularly arising from dissolved oxygen and other impurities in the sample, also causes decreased fluorescence due to radiationless deactivation.

Self-absorption arises when the fluorescence emitted by one analyte molecule is absorbed by another analyte molecule and thus goes undetected, causing a decrease in the anticipated signal. Figures 3.8a and b show why this is more of an issue at high rather than low concentrations. Self-absorption occurs because the wavelength of fluorescent light emitted by the analyte is based on a quantized excited-to-ground state transition. Other analyte molecules of the same type have exactly the same transitions, so the energy of the fluorescence emitted by one molecule perfectly matches the energy that can be absorbed by another. The presence of self-absorption can be detected by diluting the sample. For example, if a twofold dilution yields less than a twofold decrease in signal, then clearly the concentration is not in the linear range and self-absorption or other phenomenon is occurring.

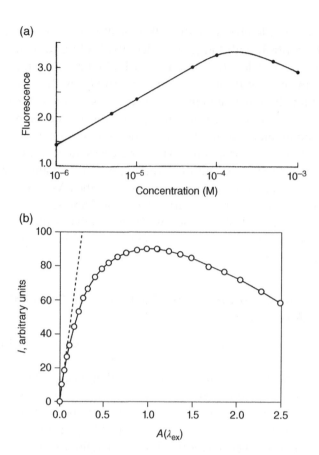

FIGURE 3.7 Examples of nonlinear fluorescence calibration curves. These plots depict calibration curves with linear and nonlinear portions (although note that the x-axis in (a) is a log scale). (a) A plot of fluorescence versus concentration for NADH in water. (b) A plot of fluorescence versus absorbance for quinine sulfate in 0.5 M H_2SO_4. The x-axis is the absorbance of the solution at the excitation wavelength. Because absorbance is typically linear with concentration, the graph is essentially a plot of fluorescence versus concentration. The solid line with open circle data points is the observed fluorescence. The dotted line is the tangent to the curve as the absorbance approaches zero, showing that the plot is linear until $A \gg 0.1$. Source: (a) From Willard et al. [11]. Reproduced with permission of Frank A. Settle, Jr.; (b) Reprinted from Credi and Prodi [13]. Reproduced with permission of Elsevier.

Another factor that can affect the linearity of fluorescence calibration curves is the presence of molecules within the sample that absorb either the excitation radiation or the emitted fluorescence. To minimize this effect, the sample can be diluted to reduce the interfering absorbance, provided that the fluorescence signal remains sufficiently detectable. The method of standard additions can also be used as the concentration of the interfering component remains constant.

If the analyte is too concentrated, or if another absorbing species is present at high concentration in solution, the first part of the solution absorbs a lot of the incoming excitation radiation. This means that as the excitation beam propagates through the solution, it gets less intense, as depicted in Figure 3.9a–d. The result is that the excitation beam is less intense when it reaches the solution at the back of the cuvette than it was at the front of the cuvette. Because fluorescence intensity is proportional to the excitation intensity, the

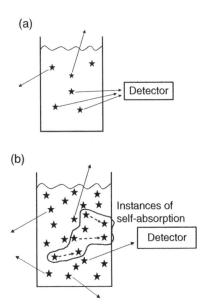

FIGURE 3.8   (a) A dilute solution of fluorophores (stars) emitting fluorescence (indicated by solid lines). (b) In a more concentrated solution of fluorophores, fluorescence emitted by one molecule can be absorbed by another (indicated by dashed arrows and circled). The second molecule can then nonradiatively relax, diminishing the total fluorescence observed. This phenomenon is more common at higher concentrations because the fluorescent photon is more likely to encounter another fluorophore before exiting the solution.

solution toward the back of the cuvette fluoresces less than it should given the concentration of analyte simply because the first part of the solution has absorbed so much of the excitation beam. If this occurs in the region of the solution from which the fluorescence is being collected and detected, the result is a nonlinear relationship between fluorescence and concentration. It is important in quantitative analyses to be aware of this problem because a given fluorescence level can correspond to two different concentrations. For example, a fluorescence level of 3.0 in Figure 3.7a corresponds to roughly 10^{-4} and 10^{-3} M, making it impossible to determine the true concentration of the analyte.

The effect of diminished beam intensity due to absorption and the effect of decreased fluorescence intensity reaching the detector due to self-absorption are collectively known as the inner filter effect. Both effects are accentuated at high concentrations of absorbing species in solution as shown in the graphs in Figure 3.7a and b. The inner filter effect can cause changes in the fluorescence spectrum as the analyte concentration is varied. It is also possible to check for nonlinear effects by diluting the solution by a known amount and seeing if the fluorescence decreases proportionally. If not, then the analyst knows she is not working in a linear region and inner filter effects may be the cause.

Another phenomenon that occurs with some solutes at high concentration is the formation of excimers, which are complexes between an excited state molecule and another molecule in the ground state. The two molecules in the excimer can be identical to one another, or they can be two different molecules (known as an exciplex). The term "excimer" is derived from the words "exited dimer" [6]. An example of a pyrene excimer is shown in Figure 3.10. After forming, the excimers may dissociate into two ground state molecules with emission of fluorescence but at wavelengths longer than that of single

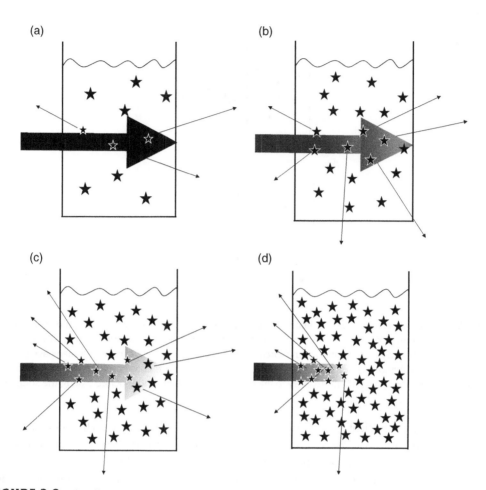

FIGURE 3.9   An illustration of the inner filter effect. In dilute solutions (a), the intensity of the excitation beam (solid black arrow) remains virtually unaltered as it propagates through the solution. This is important because the intensity of fluorescence is proportional to the intensity of the excitation radiation. As the solution concentration is increased, the fluorescence intensity should increase proportionally, as it does between the solutions depicted in (a) and (b). As the concentration continues to increase (pictures c and d), the inner filter effect attenuates the excitation beam. As a consequence, the fluorescence no longer increases linearly with concentration. In the final picture on the right, the attenuation is so severe that the excitation radiation does not even reach the back of the solution and thus no fluorescence is observed from fluorophores in this part of the solution. Also, keep in mind that as the solution concentration increases, the chances for self-absorption also increase, further reducing the amount of fluorescence observed.

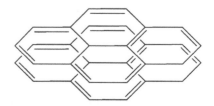

FIGURE 3.10   Pyrene excimer.

fluorescent molecules. Thus, fluorescence of the monomer would not be observed, and the signal would be lower than expected for the given concentration, causing a negative deviation to the calibration curve.

While here again quenching may seem problematic as it leads to nonlinear calibration curves, its dependence on the concentration of the fluorophore and other species in solution can also be used in a positive way to make quantitative measurements of a variety of species, even those that are not themselves fluorescent. Thus, quenching is actually the basis of many important analytical techniques such as FRET (fluorescence resonance energy transfer) and some immunoassays, the descriptions of which are beyond the scope of this discussion.

3.2. INSTRUMENTATION

This section focuses on instruments designed to measure fluorescence. To better understand the design requirements, we first briefly discuss two common applications of spectrofluorometers in order to provide a context for different designs and components used in instrumentation. Figure 3.11 shows an excitation spectrum and an emission spectrum. An *excitation spectrum* is a measure of the fluorescence intensity emitted at a particular wavelength as the wavelength of the excitation radiation is systematically varied. An *emission spectrum* is a measure of the fluorescence intensity emitted as a function of wavelength while exciting the sample with radiation of a fixed wavelength. These two types of spectra, summarized in Table 3.2, dictate the capabilities required in a fluorescence instrument. We discuss these types of scans and other applications in greater detail below, but it is helpful to have them in mind when reading about the design of spectrofluorometers.

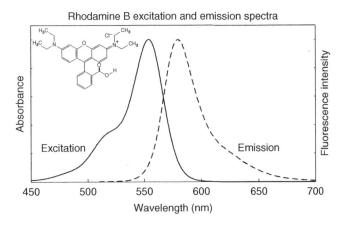

FIGURE 3.11 Excitation and emission spectra of rhodamine B.

TABLE 3.2 Summary of How Excitation and Emission Spectra Are Collected

Type of spectrum	Excitation wavelength	Emission wavelength
Excitation	Scanned	Fixed
Emission	Fixed	Scanned

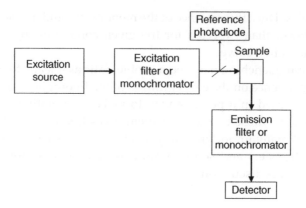

FIGURE 3.12   A general schematic of a fluorometer or spectrofluorometer. Electromagnetic radiation from the excitation source passes through a filter or monochromator. The nearly monochromatic light that emerges is then split by a beam splitter. Some of the light is sent to a reference detector in order to correct emission spectra for source intensity variations. The other part of the beam is sent to the sample. Molecules in the sample can absorb the excitation radiation and subsequently emit fluorescence over a range of wavelengths. The fluorescence is passed through the emission filter or monochromator. The nearly monochromatic light that emerges is then detected by the detector. Instruments that use filters are called fluorometers, and those that have monochromators are called spectrofluorometers. Both the excitation and emission monochromator can be scanned to collect excitation or emission spectra.

3.2.1.   Instrument Design

Figure 3.12 shows a general schematic of an instrument used to measure fluorescence and phosphorescence. It consists of:

1. An excitation source.
2. An excitation filter or monochromator with a reference detector.
3. A reference detector.
4. A sample cell.
5. An emission filter or monochromator.
6. A detector.
7. A readout device, typically a computer.

Simpler instruments that use filters instead of monochromators are referred to as fluorometers, whereas higher grade instruments with greater capabilities are referred to as spectrofluorometers.

3.2.2.   Sources

3.2.2.1.   *Xenon Lamps.*   Fluorescence requires absorbance, as was discussed in the underlying theory section. Therefore, instruments must deliver UV-visible electromagnetic radiation to the sample. The fact that fluorescence intensity is proportional to the excitation intensity suggests using intense sources. One of the most commonly used sources in spectrofluorometers is the high pressure xenon arc lamp. In these sources, xenon gas

is encased in a quartz tube that has electrodes at both ends of it (see Figure 3.13). A high voltage (20–40 kV) is applied across the electrodes, causing electrons to flow between them. The electrons ionize some of the xenon atoms. Recombination of electrons with xenon ions causes emission of a continuum of electromagnetic radiation from approximately 200 to 800 nm, with rapidly decreasing intensity in the 200–300 nm region as seen in Figure 3.14 for continuous arc lamps. In addition to the types of lamp shown in Figure 3.13, compact ceramic xenon arc lamps are also commercially available.

The output of continuous xenon lamps shown in Figure 3.14 features a broad continuum of radiation from 300 to 1000 nm. They also show very sharp lines in the high pressure xenon arc lamp emission spectrum. These arise from xenon atoms that are excited, but not ionized, and that emit specific wavelengths of light as they relax back to the ground state via quantum transitions [10]. Because the lamp intensity is high for these specific wavelengths, they can be used to maximize fluorescence signals. However, the fact that the intensity is not constant across all wavelengths complicates measuring true fluorescence spectra. We will come back to these discrete lines later in the chapter to illustrate some of the complexities of fluorescence spectroscopy relative to absorbance spectroscopy.

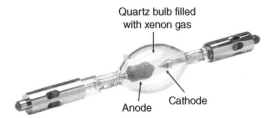

FIGURE 3.13 A high pressure xenon arc lamp. Source: Reproduced with permission from CHRISTIE®.

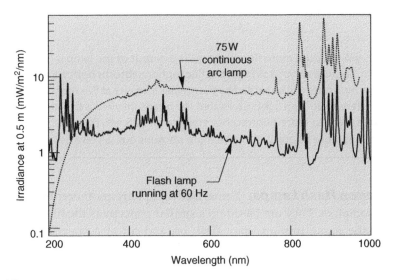

FIGURE 3.14 The spectral output of a xenon flash lamp. There is continuous output from 200 to 1000 nm. Spikes in the spectrum are due to specific electronic transitions of xenon atoms. Source: Reproduced with permission of Newport Corporation.

The lamps discussed above operate in a continuous manner, consume large amounts of power, and generate significant amounts of heat. Also, because they are continuously on, a sample sitting in the sample compartment is continuously illuminated unless the excitation beam is physically blocked when signal is not being recorded. This can cause photobleaching if the analyte is photolabile, which many fluorescent molecules are.

EXAMPLE 3.5

(a) Define photobleaching.
(b) If photobleaching occurs, would one expect fluorescence to increase or decrease over time as the sample is continuously illuminated?

Answers:
Photobleaching is the loss of fluorescence due to chemical changes in a fluorophore caused by illumination. As the molecules are continuously illuminated, bonds in the fluorophore can break, or the fluorophore can react with other species in solution, resulting in a structure that is no longer fluorescent. If photobleaching occurs, one would expect a decrease in fluorescence over time.

Another question:
Suppose a fluorophore absorbs both 600 nm light and 825 nm light equally well and that it has the same quantum yield for both wavelengths. If the xenon lamp that was used to record the spectrum in Figure 3.14 is used to induce fluorescence in the fluorophore at these two wavelengths, will the fluorescence intensity be the same with both excitation wavelengths, or will one wavelength generate more fluorescence than the other? If they produce different amounts of fluorescence, which excitation wavelength will generate more fluorescence? Why?

Answer:
Exciting at 825 will produce more fluorescence. The output of the lamp is greater at 825 nm than at 600 nm. Because we are told that the fluorophore absorbs both wavelengths equally well, then more molecules will be excited by the greater intensity at 825 nm than will be excited by the lower intensity 600 nm light. Given that the quantum yield is the same, then exciting more molecules leads to more fluorescence emission at 825 nm. Essentially, this boils down to the notion that more photons in create more photons out, all else being equal.

3.2.2.2. Xenon Flash Lamps. Xenon flash lamps are another common source used in spectrofluorometers. They are based on a similar concept as the high pressure xenon lamps, but as the name implies, they flash on and off at a specific frequency (e.g., 50 Hz). Each flash illuminates the sample and generates fluorescence, with multiple measurements being averaged together to generate a single data point. Flash lamps are more compact, have higher output in the UV region (see Figure 3.14), consume less power, and cause less photodegradation than continuous high pressure xenon arc lamps [10].

EXAMPLE 3.6

With a flash rate of 80 Hz, how many flashes occur in one minute?

Answer:
80 Hz is the same as 80 flashes per second. Given this:

$$\frac{80 \text{ flashes}}{\text{second}} \times \frac{60 \text{ seconds}}{\text{minute}} = \frac{4800 \text{ flashes}}{\text{minute}}$$

So a flash lamp operating at 80 Hz flashes 4800 times every minute.

3.2.2.3. Other Lamp Sources. While xenon arc and flash lamp sources are the most commonly used sources in spectrofluorometers, other lamps such as high pressure and low pressure mercury vapor discharge lamps have also been used. These lamps are useful in certain applications where specific, intense mercury lines overlap the absorption characteristics of the analyte and therefore generate greater signal than xenon lamps. Mixed mercury–xenon lamps have also been used, offering higher intensities in the UV region than xenon lamps coupled with the specific, intense mercury lines. While these lamps offer benefits for specific applications, they do not offer any for general-purpose instrumentation, and thus xenon arc and flash lamp sources are most commonly found in commercial instruments.

3.2.2.4. Lasers and Light-Emitting Diodes. As suggested earlier in the chapter, lasers are frequently used in fluorescence applications because of their high intensity. Also, because they can be focused on small areas, they find significant use in biological studies and in detection systems for instruments used in separations such as liquid chromatographs and capillary electrophoresis systems. Certain lasers can also be tuned to a variety of wavelengths and can be pulsed very rapidly – characteristics that are important in specific applications. However, because of their high cost relative to lamp sources, they are not commonly used in commercial instrumentation.

As an alternative to lasers, light-emitting diodes (LEDs) are used in some instruments [14]. LEDs produce radiation in a relatively narrow range of wavelengths, or they can be used in combination with filters to produce nearly monochromatic radiation. They also consume less power, have longer lifespans, and are more compact than other sources [10]. However, they are not typically as intense as lasers [15].

3.2.3. Filters and Monochromators

As we saw in the previous section, the sources used in spectrofluorometers emit a wide range of wavelengths. Similarly, the fluorescence emitted by a sample also often spans a wide range of wavelengths. Filters and monochromators are used to select monochromatic radiation that gets passed through the sample (excitation) or that gets detected (emission). The theory explaining how filters and monochromators work is discussed in Chapter 2.

The inclusion of filters versus monochromators differentiates fluorometers from spectrofluorometers. In general, fluorometers are simpler, less expensive instruments that use absorbance or interference filters to select excitation and emission wavelengths. Spectrofluorometers use monochromators to accomplish this task. In fact, some high-end research instruments use double monochromators for both excitation and emission in an effort to reduce stray light and thereby achieve remarkably low detection limits. The use of monochromators creates greater flexibility as it is possible to rapidly select any desired wavelength or to scan through a large wavelength range with subnanometer resolution, whereas filters limit the potential wavelength selection to the number of filters available in a laboratory and often require manually changing the filters in order to change the wavelength. The added flexibility and capability of spectrofluorometers, however, adds considerable expense relative to fluorometers.

3.2.4. Component Arrangement

Fluorometers and spectrofluorometers share an important characteristic, namely, that the emitted fluorescence is typically collected at 90° relative to the excitation beam as shown in Figure 3.12. This arrangement limits the amount of excitation radiation that enters the emission filter or monochromator, making it easier to detect just the emitted fluorescence free from the excitation radiation. However, because fluorescence is emitted at all angles, collecting it at 90° through narrow slits means that only a very small amount of the entire emitted fluorescence is detected. The technique is sensitive enough, however, that this is usually acceptable.

3.2.5. Fluorometers

In a fluorometer, electromagnetic radiation is split into a reference beam and a sample beam. Both beams pass through the excitation filter – called the "primary" filter – which selects the wavelength of radiation that is passed through the sample. The beams are focused using a series of lenses as shown in Figure 3.15. The reference beam is diverted to a reference detector, commonly a photomultiplier tube (PMT) or semiconductor detector (see Chapter 2 for a discussion of how these detectors work). A shutter allows light to be blocked from the sample until a measurement is made. This helps prevent photobleaching of the sample due to continuous exposure to electromagnetic radiation.

Fluorescence from the sample is collected by the emission optics, passed through the emission filter – also called the "secondary" filter – and focused on the detector, commonly a PMT. The reason for having two separate detectors is that the intensity of fluorescence scales with the lamp intensity (recall that $P_f \propto P_0$ at low absorbance). Therefore, the observation of a decrease in the fluorescence intensity for different samples may not be due to a lower concentration of analyte but may be due to a dying lamp intensity. The ratio of the fluorescence signal to the reference signal is taken to correct for variations in the fluorescence that arise due to lamp intensity variations.

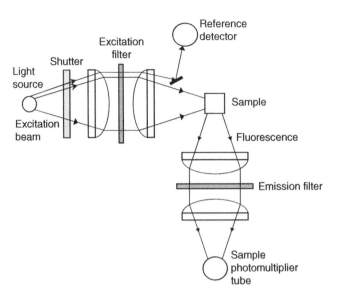

FIGURE 3.15 A bird's-eye view of fluorometer components. Electromagnetic radiation from the source is passed through the excitation filter to select the excitation wavelength. The beam is split by a beam splitter. A portion of the light is directed onto a reference detector to correct for fluctuations in the source intensity. The other portion of the beam is focused onto the sample. Molecules in the sample absorb the excitation radiation and can subsequently fluoresce. The fluorescence is collected at 90° by lenses, passed through the emission filter to select the emission wavelength that is monitored, and is then detected. While fluorescence is collected at 90°, it is worth noting that fluorescent light is emitted in all directions, so only a fraction of the actual light emitted is collected. Source: Reproduced with permission from Ruhle Companies, Inc., Farrand Optical Division and adapted from Willard et al. [11]. Used with permission from Wiley.

Because of the use of filters, fluorometers excite at one wavelength and observe at another – recalling that fluorescence occurs at lower energy/longer wavelength than excitation. Fluorometers cannot be used to scan through wavelength ranges to record entire spectra unless this is done manually. However, they are useful for routine, repetitive applications because they are relatively simple and inexpensive.

3.2.6.   Spectrofluorometers

Figure 3.16 shows a diagram for a spectrofluorometer. Monochromators replace the filters used in fluorometers for selecting both the excitation and emission wavelengths. Using excitation and emission monochromators allows for complete excitation and emission spectra of a sample to be recorded automatically, easily, and with subnanometer resolution. The ability to measure entire spectra aids in qualitative analyses because different analytes have different spectral characteristics.

3.2.6.1.   Excitation Spectra.   To record an excitation spectrum, the emission monochromator is fixed at one particular wavelength – a wavelength at which the analyte is known to fluoresce – and the excitation wavelength is systematically varied (see Figure 3.17). The equation $P_f \approx 2.303\Phi_f P_0 \epsilon bc$ shows that if the analyte does not absorb the incoming

(a)

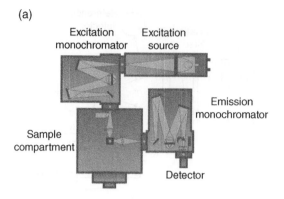

(b)

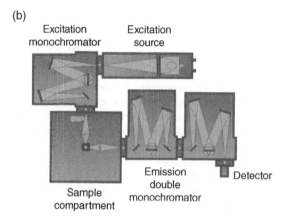

FIGURE 3.16 Bird's-eye view of the components of a spectrofluorometer. (a) Radiation from the excitation source enters the excitation grating monochromator, which selects the wavelength that is passed to the sample. Fluorescence from the sample is collected at 90° and passed through the emission grating monochromator to select the specific wavelength of fluorescence that is detected by the detector. Emission and excitation spectra are collected by scanning the respective monochromator while keeping the other fixed. (b) This diagram shows a spectrofluorometer with a double emission monochromator, which helps to decrease the stray light that reaches the detector. Source: Courtesy of HORIBA Scientific, Edison, NJ.

excitation wavelength (i.e., $\varepsilon = 0$), then no fluorescence is observed because molecules have to absorb a photon before they can emit fluorescence. Conversely, wavelengths that are associated with higher molar absorptivities for the analyte yield greater fluorescence. Thus, the excitation spectrum of an analyte closely resembles its absorbance spectrum. There are instrumental differences between UV-visible spectrophotometers and spectrofluorometers that lead to some variation, but it is generally correct to think of excitation spectra as reflecting the absorbance of the sample as a function of wavelength.

Figure 3.18 illustrates that one of the complications of fluorescence excitation spectroscopy relative to absorbance spectroscopy is that several sharp lines appear in the spectra (see "uncorrected excitation" spectrum) that would be absent in the absorption spectrum. These peaks, such as the ones around 470 and 490 nm, correspond to peaks in the spectrum of xenon, shown in Figure 3.14. Because fluorescence intensity depends on

Collecting an EXCITATION spectrum

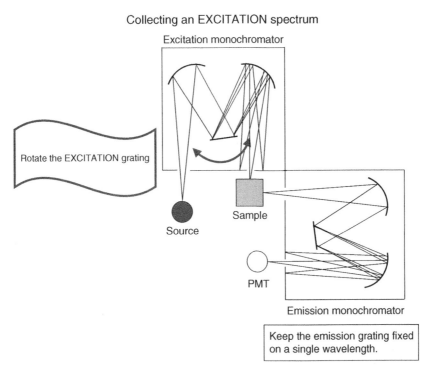

FIGURE 3.17 Collecting an excitation spectrum. To collect an excitation spectrum, the grating in the excitation monochromator is rotated in order to scan through the specified wavelength range. The emission monochromator is held constant at a wavelength where it is known that the analyte fluoresces.

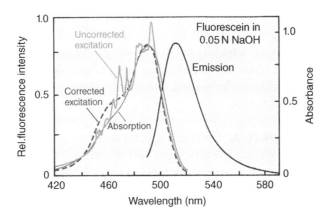

FIGURE 3.18 Uncorrected (solid spikey line) and corrected excitation spectra (dashed line) of fluorescein in 0.05 N NaOH compared to its absorbance spectrum (solid smooth line left) and its fluorescence spectrum (solid line right). The spikes in the uncorrected excitation spectrum are caused by variations in the intensity of the lamp at different wavelengths. There are other differences between the corrected and uncorrected spectra that arise from other instrumental components. The result is that even after correcting for lamp intensity variations as a function of wavelength, the corrected spectrum more closely resembles, but is still not identical to the absorbance spectrum. Source: From Lakowicz [10]. Reproduced with permission of Springer.

excitation intensity ($P_f \propto P_0$), and because the lamp output varies at different wavelengths, the spikes in the xenon lamp intensity lead to corresponding spikes in the fluorescence intensity emitted by the analyte.

To correct for this effect, the fluorescence intensity is normalized to a measure of the excitation intensity. Therefore, just as in fluorometers, a reference or compensating beam is split off from the excitation beam and its intensity measured. By taking the ratio of fluorescence to excitation intensity (P_f/P_0), spikes in fluorescence caused by spikes in the excitation intensity are eliminated. The spectrum labeled "corrected excitation" in Figure 3.18 reflects this. The spikes are no longer present after normalization. This shows why it is necessary to have a reference detector tracking the excitation intensity as an excitation spectrum is being recorded.

While the corrected excitation scan more closely resembles the absorption spectrum, there are still some clear differences, particularly around 460 nm. These arise from other effects such as variations as a function of wavelength in the response of the reference detector and the fluorescence detector (i.e., the detectors are more responsive to light at some wavelengths than others), the transmission efficiency of the excitation monochromator as a function of wavelength (i.e., monochromators do not pass light of all wavelengths equally well), and potential concentration effects if the sample is in a nonlinear range. These concerns are absent from absorbance spectroscopy because there, one is measuring the ratio of initial intensity relative to transmitted intensity of light *of the same wavelength*. So all of these effects get normalized. In fluorescence, however, because (1) there are two different monochromators that may have slightly different transmission characteristics, (2) the excitation and emission monochromators are not simultaneously passing the same wavelength of light, (3) there are two different detectors – each of which may have slightly different sensitivity that varies with wavelength – and (4) both detectors are not simultaneously sensing the same wavelength of light, these effects do not get normalized away. There are methods for measuring and correcting for such effects, but a discussion of them distracts from the thrust of our discussion here, which is the design of spectrofluorometers. Readers who are interested in learning more about obtaining fully corrected spectra are directed to commercial vendor manuals, information available online, and the treatment of the topic provided by Lakowicz [10].

3.2.6.2. Emission Spectra.

Emission spectra are recorded by fixing the excitation monochromator at one particular wavelength and measuring the fluorescence emitted from the sample as a function of wavelength (see Figure 3.19). Typically, the excitation wavelength that is chosen is one that is readily absorbed by the analyte of interest, meaning that the analyte has a high molar absorptivity at this wavelength, because greater absorbance leads to greater fluorescence at all wavelengths. There may be reasons for choosing other excitation wavelengths such as to avoid exciting other molecules besides the analyte in the solution or to minimize background fluorescence, but when possible, using the wavelength of maximum absorption results in the best sensitivity.

As an example, the emission spectrum of quinine sulfate is shown in Figure 3.20. As was discussed earlier in the theory section of this chapter, the fluorescence emission is

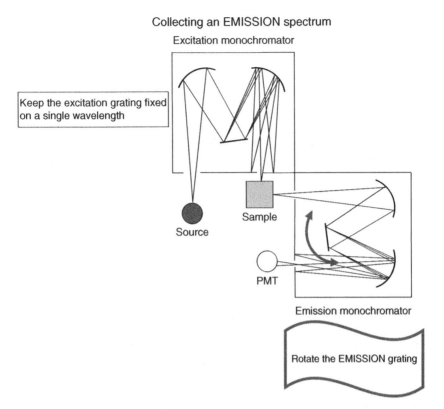

Collecting an EMISSION spectrum

Excitation monochromator

Keep the excitation grating fixed on a single wavelength

Source

Sample

PMT

Emission monochromator

Rotate the EMISSION grating

FIGURE 3.19 Collecting an emission spectrum. To collect an emission spectrum, the grating in the excitation monochromator is held constant at a wavelength where it is known that the analyte absorbs. The grating in the emission monochromator is rotated to scan through the wavelengths emitted by the analyte.

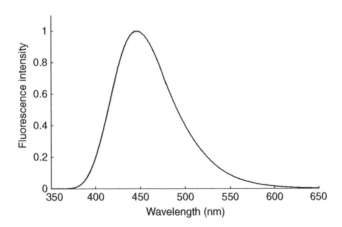

FIGURE 3.20 Normalized fluorescence spectrum of quinine sulfate in 0.25 N H_2SO_4, excited at 348 nm.

shifted to longer wavelength – lower energy – relative to the excitation wavelength, which in this case is 348 nm. It is important to note that the excitation wavelength generally only affects the *intensity* of the fluorescence, *not the shape* of the curve or the wavelength at which maximum emission occurs. Excitation wavelengths near an absorbance maximum lead to greater fluorescence intensity, assuming phenomena such as those that cause inner filtering are not occurring. The shape and the wavelength of the fluorescence maximum in the emission spectra, however, are dictated by the energy gap between the first excited state and the ground state, which are properties of the molecule (with some influence by the solvent as discussed above) – not properties that depend on the wavelength of excitation.

3.2.6.3. Fixed Excitation and Emission Wavelengths. Spectrofluorometers can also act like fluorometers simply by setting both the excitation and emission monochromators at fixed wavelengths. The fluorescence intensity of a series of standards can be measured to create a calibration curve, which is then used to measure the concentration of an analyte in unknown samples.

EXAMPLE 3.7

(a) To record an emission spectrum, which monochromator is scanned and which is fixed?
(b) Based on Figure 3.20, which emission wavelength would you select for quantitative measurements of quinine sulfate in order to obtain the greatest sensitivity?

Answers:
(a) The emission monochromator is scanned and the excitation monochromator is held constant to record an emission spectrum.
(b) Approximately 445 nm provides the maximum sensitivity.

3.2.7. Cells and Slit Widths

Just as in absorbance spectroscopy, 1 cm quartz cuvettes are commonly used in fluorescence and phosphorescence measurements. Other sizes and geometries, however, are also used in particular applications. Unlike cells used in absorbance spectroscopy, which typically have two opposite sides that are frosted for convenient handling, cells for luminescence measurements are not frosted on any side. This is largely because of the 90° angle between the excitation and emission collection. Frosted sides would interfere with either the excitation beam or the emitted luminescence.

The sample holders in commercially available instruments also offer the possibility of varying the temperature of the sample. It is useful to be able to cool the sample to increase the fluorescence because fluorescence intensity depends on temperature. Temperature control is also important when high reproducibility is required.

Monochromator slit widths influence the resolution of the spectra recorded, just as they do in absorbance spectroscopy. In luminescence measurements, the slits also control the amount of light delivered to the sample and the amount of emitted radiation collected by the emission monochromator. Increasing the widths of both the excitation and emission slits increases the number of excited state molecules created and the amount of fluorescence collected, assuming one is in a region where self-absorption and other quenching effects do not limit the signal.

While collection of fluorescence at a right angle relative to the excitation beam is common, some commercial instruments offer an option for "front-face" collection of fluorescence. As the name implies, fluorescence is collected from the front of the sample cell [16]. A diagram of such an arrangement is shown in Figure 3.21. This arrangement is particularly useful for detecting fluorescence from solids and opaque samples such as milk and hemoglobin in which most of the fluorescent light is reabsorbed before it escapes the sample. Collecting emission from the front face of the cell allows the fluorescence coming from the first "slice" of solution to be observed as depicted in Figure 3.21.

Many other cell geometries and sample holder options, including accessories to control temperature and to purge dissolved O_2, are available for specific applications.

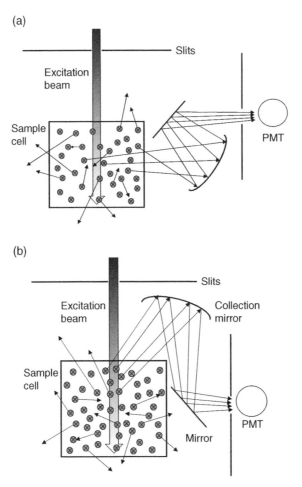

FIGURE 3.21 Bird's-eye view of front-face vs. 90° collection of fluorescence. (a) In 90° collection with opaque or concentrated samples, the potential for self-absorption or scattering exists such that light emitted by the excited molecules is lost before it can exit the cuvette and be detected. (b) Because there are fewer molecules between the fluorescent molecules and the front face, some fluorescence can "escape" and be detected before it is absorbed or scattered by other molecules.

3.2.8. Detectors

3.2.8.1. Photomultiplier Tubes: Analog Operation and Photon Counting.
Photomultiplier tubes (PMTs) are the most common detectors used in commercial spec-trofluorometers, the operating principles of which are described in Chapter 2. PMTs provide high sensitivity and rapid response, making them appropriate for fluorescence measurements.

Because PMTs can detect individual photons, they can be operated in either photon counting or analog mode. In photon counting, individual spikes in the anode current are recorded over time (see Figure 3.22) [17–19]. A threshold value is set so that only currents that rise above it are recorded as photons. This is done to discriminate signals arising from random ejection of electrons or other stray effects that cause a cascade of electrons starting somewhere in the chain of dynodes and therefore not providing the total current that arises when the initial photoemissive surface is struck by a photon. So in the simplest version of photon counting mode, the total magnitude of the current does not matter. Any current spike that is over the threshold value, regardless of how far over it is, is recorded as a single count. The number of signals above the threshold current is counted over a given period of time (i.e., counts/second is recorded rather than total integrated current) and is related to the fluorescence intensity emitted by the sample. Clearly, in dilute solutions, the greater the number of fluorescent molecules, the greater the number of counts/second. Hence, the count rate can be used to create a calibration curve for quantitative analysis.

There are limitations to photon counting, particularly at high levels of fluorescence when two or more photons may strike the PMT simultaneously or nearly so, yet would be counted as a single event because they appear as just a single spike (remember that the magnitude of the spike does not matter as long as it is over the threshold value). This situation is referred to as "pulse pile-up," meaning that the photons are arriving too quickly to be registered as individual counts, causing the signal or perceived number of events to flatten out even as the concentration of fluorescent species increases, ultimately

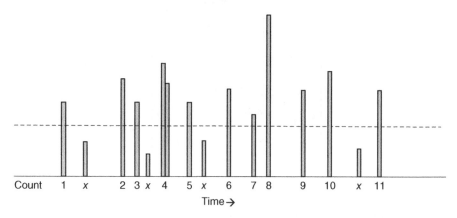

FIGURE 3.22   Single photon counting. Note that even though signals "4" and "8" are double in strength compared to most single counts, indicating that two photons arrived simultaneously or nearly so, they are still counted as just a single event.

leading to a nonlinear calibration curve. In other words, there is a limit to how fast the PMT can count, typically in the MHz range. Slit widths of the monochromators can be adjusted so that fewer molecules are illuminated by the excitation beam and hence fewer fluorescent photons strike the detector, but in some ways, this seems like "wasting" potential signal. There are limitations that occur at low fluorescence levels, too, largely arising from noise. These restrictions can limit the linear range of photon counting to as low as three orders of magnitude, although techniques exist for accounting for these effects and thus extending the linear range [20–23].

In analog mode, the individual current pulses are averaged together. So two photons arriving simultaneously contribute twice as much signal, and therefore both contribute to the overall average. This extends the linear range of calibration curves at high concentrations. At very low concentrations, however, the signal arising from fluorescence is lost within the average dark current of the PMT. Thus, single photon counting allows for lower detection limits than analog detection, but analog detection generally offers wider linear ranges and better precision because of higher average signal.

3.2.8.2. Charge-Coupled Devices (CCDs). As described in the UV-visible spectroscopy chapter, charge-coupled devices (CCDs) are two-dimensional arrays of sensors, or pixels, with excellent sensitivity and linear dynamic range. CCDs can have as many as a million or more pixels and thus offer high spatial resolution. Because of their 2D nature, they are used to take pictures of, or image, samples. They are commonly used in fluorescence microscopy, which is discussed in greater detail below, to image cells, organisms, and other biological systems. Images can also be recorded over time, creating videos of biological and chemical processes. They have grown in popularity as prices have fallen and quality increased. Fantastic colorized images are recorded with them, and the reader is encouraged to do an online search for such images. Simply searching terms like "fluorescence microscopy image" or "fluorescence microscopy organism" yields hundreds of striking images of interesting systems like endothelial cells, neurons, DNA-binding proteins, and even whole organisms like jellyfish and squid embryos.

3.3. PRACTICE OF LUMINESCENCE SPECTROSCOPY

3.3.1. Considerations and Options

In practice, there are many options and variables to consider when performing fluorescence and phosphorescence measurements. These include:

1. Do you want an excitation or emission scan?
2. What slit widths should be used?
3. What excitation wavelength to use for maximum fluorescence in an emission scan?
4. What emission wavelength to use for maximum fluorescence in an excitation scan?
5. Do you need to correct the spectra beyond taking the ratio of the fluorescence signal to the excitation beam?
6. Do you need to measure fluorescence at fixed excitation and emission wavelengths to ultimately create a calibration curve and do quantitative analyses?

7. What wavelengths should be used to achieve greatest sensitivity and selectivity?
8. What linear range is necessary to quantify the analyte?
9. Are quenchers present that are affecting the results?
10. Is cooling or temperature control required?
11. Should the sample be purged to eliminate O_2?
12. Is the concentration of the analyte too high such that self-quenching and self-absorption are causing nonlinear behavior?
13. Is there a better geometry or different cell to use for the sample?

The links between theory and the instrumental parameters that influence the answers to these questions are described in earlier sections of this chapter, but the questions themselves indicate the range of considerations required when conducting fluorescence and phosphorescence measurements. Given all of the options and variables, it is tempting to ask "why bother with fluorescence at all?" The answer to this is "sensitivity and selectivity." As noted earlier, because fluorescence occurs at wavelengths that are different than the excitation beam, the measurement amounts to trying to detect a little bit of light against a dark background – an easier task than detecting a slight decrease in intensity of light at the same wavelength. Furthermore, because most molecules do not fluoresce, fluorescent methods are selective for only those that do. On one hand, this is limiting in that the technique can only be applied for a small fraction of all known molecules. On the other hand, for those molecules that do fluoresce, it is a viable option that should be strongly considered, particularly for samples that are likely to be free of interfering signals from other fluorescent molecules.

In the sections below, we explain some variations on the types of measurements we have considered up to this point and that broaden the applications of fluorescence spectroscopy. These variations include fluorescence polarization, time-resolved fluorescence measurements, and fluorescence microscopy.

3.3.2. Fluorescence Polarization

In the discussions above, we have implicitly assumed that the light from the excitation source and that emitted by the sample are nonpolarized – or at the very least we have not explicitly considered the effects of polarization in fluorescence measurements. And for most fluorescence measurements, this is a perfectly acceptable situation.

However, explicitly using polarized radiation to excite the fluorophores in a sample and subsequently measuring the polarization of the emitted light leads to insights about the size, shape, fluidity, rigidity, and mobility of the chemical environment around the fluorophore [6, 10]. Such information is useful when studying things like enzyme/substrate binding, the effects of molecules like cholesterol on cell wall rigidity, and protein–protein interactions. These measurements are typically referred to as "fluorescence polarization" or "fluorescence anisotropy" and are the focus of this section.

In a typical fluorescence measurement, the excitation source is not polarized. But in fluorescence polarization measurements, a polarizing filter is placed in the path of the excitation beam prior to the sample as depicted in Figure 3.23 (see Chapter 1 for a general discussion of polarized radiation). Because the light that strikes the sample is polarized,

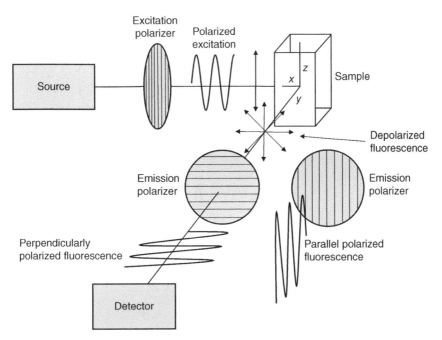

FIGURE 3.23 Arrangement of components for measuring fluorescence polarization. Polarizers are placed in the path of the excitation beam and the emitted fluorescence. After excitation, molecular rotation can cause the depolarization of fluorescence. Depending on the alignment of the emission polarization, components of the fluorescence that are parallel with or perpendicular to the excitation polarization can be selectively detected. In this figure as shown, the perpendicular component is being measured, and if the emission polarizer is rotated (image that is offset to the right), the parallel component is measured. Note excitation and emission filters or monochromators are still needed to select specific wavelengths but have been omitted from the figure for clarity.

only molecules that have a proper orientation and are thus able to absorb the polarized radiation are excited. Specifically, the transition moment of the fluorophore must be aligned parallel to the electric field of the excitation photons for maximum absorbance. Once molecules absorb the radiation, they undergo the typical relaxation processes, including fluorescence. Recall, however, that some time – about 1–10 nanoseconds – elapses between the initial excitation and the moment of fluorescence.

To understand fluorescence polarization and how it leads to information about the fluidity of the environment of the fluorophore, it is useful to consider two extreme conditions: one in which the excited molecules do not rotate at all prior to fluorescence and another situation in which they rotate very rapidly (see Figure 3.24).

In the first scenario, when the excited state molecules do not rotate at all in the time between excitation and fluorescence, the emitted fluorescence is polarized in the same plane as was the excitation radiation (i.e., all parallel). If, however, the excited state molecules rotate rapidly prior to emitting fluorescence, the polarization of the emitted light is randomized among all planes. This leads to a decrease in the polarization or a total depolarization of the emission [6].

Experimentally, measuring these effects simply requires placing one polarizing filter between the excitation source and the sample and another polarizing filter between the

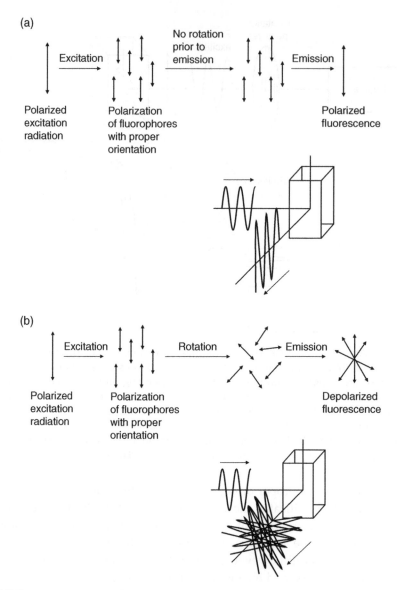

FIGURE 3.24 Emission following no rotation by the fluorescent molecules is polarized parallel to the excitation polarization (a). Emission after rotation is at least partially depolarized and thus has some component that is polarized parallel to the excitation polarization (b).

sample and the emission detector as shown in Figure 3.23. While keeping the excitation polarization constant, the intensity of the polarization of the emission that is parallel to ($I_\parallel$) and perpendicular to ($I_\perp$) the excitation polarization axis is measured. The polarization ratio, p, and the emission anisotropy, r, are calculated as shown in Eqs. (3.6) and (3.7):

$$p = \frac{I_\parallel - I_\perp}{I_\parallel + I_\perp} \tag{3.6}$$

$$r = \frac{I_{\parallel} - I_{\perp}}{I_{\parallel} + 2I_{\perp}} \qquad (3.7)$$

In both equations, the numerator evaluates the difference between intensities of the parallel and perpendicularly polarized fluorescence. If there is no difference between them (i.e., $I_{\parallel} - I_{\perp} = 0$), then both p and $r = 0$. In this case, the fluorescence is not polarized at all, meaning that the polarization of the emission has been totally randomized by molecular rotation. The denominator in the case of the polarization ratio is the total fluorescence intensity in the direction of the detector, whereas the denominator in the emission anisotropy equation accounts for the total intensity in all directions (x, y, and z planes). So both p and r effectively measure how different the perpendicular and parallel polarization intensities are relative to the total emission intensity.

We note here that in practice, vertically and horizontally polarized radiation are passed through the emission monochromator and detected with different efficiencies and that this difference is also wavelength dependent. For example, vertically polarized radiation may be passed with twofold greater efficiency than horizontally polarized radiation. If this is not accounted for, the polarization ratio based on uncorrected intensities is incorrect. Thus, correction factors are required in order to obtain true polarization ratio or emission anisotropy values. The procedure for measuring these correction factors and incorporating them into the calculation of true polarization ratios is discussed in detail in Ref. [10].

To understand the physical reality represented by these ratios, look again at the numerator. Recall that the excitation radiation is polarized and that if the excited molecules *do not rotate* prior to fluorescence, then the emission will also be polarized in the same plane as (i.e., parallel to) the excitation. So $I_{\parallel}$ is large and $I_{\perp}$ very small. In this case, the difference between $I_{\parallel}$ and $I_{\perp}$ is large, and consequently p and r will also be large.

In contrast, if the molecules rotate after excitation but prior to fluorescing, the intensity of the fluorescence that is perpendicular to the excitation beam increases and becomes comparable to that which is parallel. In this case, $I_{\parallel} - I_{\perp}$ is small. This means that the medium containing the fluorophores allows for rapid and free rotation.

The question then is, what conditions or what chemical systems restrict or enhance the rotation of molecules? Most often, fluorescence polarization is used to study biological systems where interactions between molecules are of great interest. For example, a molecule that is bound to a large enzyme will not be as free to rotate as it would if the molecule were free in solution. The polarization ratio thus provides a measure of its strength of binding and the degree to which its motion is restricted when bound. Fluorescence polarization is particularly useful in cases where the binding of a small molecule to a macromolecule does not lead to a change in the molecule's excitation or emission spectrum. In such cases, regular fluorescence measurements might make it look like no binding occurs. However, because of the binding, the molecule's motion is restricted, with greater binding leading to decreased mobility and a smaller change in the polarization ratio compared to unrestricted rotation. Thus, even in cases where regular fluorescence measurements are not sensitive to binding, polarization measurements are.

Fluorescence polarization is also used to study the fluidity of cell membranes [24–26]. The fluidity depends on a number of factors, including the ratio of saturated to unsaturated long-chain phospholipids. It also depends on the effects of additives like cholesterol. As the composition of the cell membrane changes, it becomes more or less rigid. Fluorescent molecules incorporated into the membrane can report on the fluidity of the membrane through changes in their polarization ratios. The membrane fluidity is important because it affects other cell wall structures such as membrane proteins that must undergo conformational changes to function. If the cell wall become rigid, protein conformational changes are restricted, which may inhibit protein functionality. Fluorescence polarization provides a useful and important method for monitoring such changes in the fluidity of membranes.

Other systems that are also studied using fluorescence polarization include protein–protein binding interactions, DNA–protein interactions, and antibody–antigen binding. In all cases, it is easy to imagine that bound molecules have a different mobility than those that are unbound and thus lead to changes in the polarization ratio.

3.3.3. Time-Resolved Fluorescence Spectroscopy

The chapter so far has focused on measurements that are classified as "steady state," meaning that they are conducted with constant illumination of the sample and constant collection of fluorescence. Most luminescence measurements are of this type. However, because fluorescence is a relatively long-lived process – fluorescence can last on the order of picoseconds to hundreds of nanoseconds, compared to femtosecond time-scales for absorption – it is also possible to perform "time-resolved" measurements in which the decay of fluorescence is measured over time. This makes it possible to apply a short pulse of excitation radiation to the sample, turn off the beam, and then detect the fluorescence and watch it fade away over a pico- or nanosecond timescale. Such measurements yield fluorescent lifetimes for excited state molecules. Variations in these lifetimes are related to the chemical environment of the fluorescent molecules. In fact, lifetime measurements are related to things such as (1) quencher concentration, (2) fluidity and rigidity of cell membranes and other media, (3) proximity of other molecules, and (4) the presence of multiple conformations of the molecules being studied [10, 27–29].

As a specific example of when time-resolved fluorescence is useful, consider the above section about fluorescence polarization. In that section, we implicitly focused on steady-state measurements. However, the measured anisotropy of a molecule depends not only on its mobility but also on its fluorescence lifetime (i.e., the time it takes for a molecule to fluoresce after it has been excited), with longer lifetimes leading to greater opportunity for rotation and hence greater depolarization of the fluorescence. So, if some change in the environment (for example, a polarity change) causes the fluorescence lifetime of a molecule to increase, the molecule has more time to rotate prior to fluorescence, and you would measure a smaller steady-state anisotropy, *even with no change in the fluidity of the environment*. Because of this, steady-state anisotropy changes alone do not prove that there has been a change in the fluidity of the environment.

To really prove it, one has to do time-resolved fluorescence anisotropy, which involves pulsing in the polarized excitation radiation and then following the decay of the anisotropy over time (R. Clarke, personal communication).

The details of making the necessary measurements on the timescales required for time-resolved fluorescence are beyond the scope of this chapter, but the fact that such measurements are not only possible but common demonstrates the power of fluorescence spectroscopy and the achievements in the area of fast pulsed sources and detection.

3.4. FLUORESCENCE MICROSCOPY

You are likely familiar with using microscopes to view small objects. Common laboratory microscopes, also known as bright-field microscopes, illuminate an object with light from a source. The light interacts with the sample (diffracts, refracts, scatters, gets absorbed, etc.) and is then collected through an objective lens and passed to a detector, which, commonly, is a human eye. To record 2D pictures, a digital camera can be added to the system.

Fluorescence microscopy is also used to visualize small objects, but rather than relying on the absorbance or scattering of the original light source to create the picture, fluorescence microscopes illuminate the sample with electromagnetic radiation that excites fluorescent molecules and then records the fluorescence emitted by those molecules. An objective lens captures the emitted light and focuses it on a charge couple device (CCD), creating a 2D image.

Diagrams of common fluorescence microscope configurations are shown in Figure 3.25. The output of a light source such as a light emitting diode (LED), xenon lamp, or mercury lamp passes through an excitation filter. The excitation filter determines which wavelength of light is used to excite the sample. The monochromatic radiation coming through the excitation filter then encounters a dichroic mirror (beam splitter), which directs the excitation light through the objective lens and onto the sample. Molecules in the sample are excited and emit fluorescence. The fluorescence is collected by the objective lens (the same lens used to focus the excitation light). Some of it is passed straight through the dichroic mirror and then through the emission filter (also called a barrier filter). The emission filter is selected such that it allows the desired fluorescence through it and blocks scattered background radiation – primarily from the excitation beam. The fluorescence then strikes the CCD and the image of the object is recorded. The excitation filter, dichroic mirror, and barrier filter are typically incorporated into a single component called a cube, and different cubes can be installed to vary the wavelengths of excitation and emission.

Illuminating the sample and capturing the fluorescence from the same face of the sample is known as epifluorescence and is commonly done. Two different microscope configurations are shown in Figure 3.25. The two configurations are called "upright" or "inverted." In the upright configuration, the objective lens is located above the sample, and in the inverted configuration, it is below the sample.

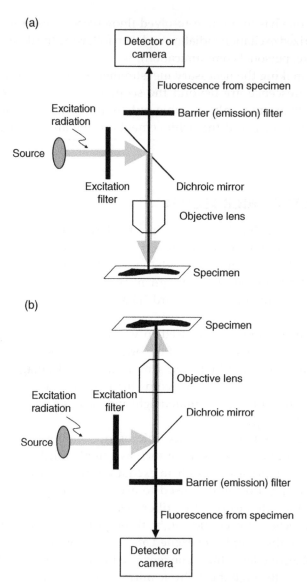

FIGURE 3.25 Diagrams of an (a) upright and (b) inverted fluorescence microscope. Excitation radiation from the source passes through the excitation filter, strikes a dichroic mirror (beam splitter), and is directed through the objective lens. The objective lens focuses the light onto the specimen. Fluorescence that is subsequently emitted from the specimen is collected by the objective lens, passes through the dichroic mirror and an emission filter, and ultimately strikes the detector or camera. In the upright configuration, the excitation radiation illuminates the specimen from above, and fluorescence is also captured above the specimen. In the inverted configuration, the excitation radiation strikes the specimen from below, and the fluorescence is captured from the underside of the specimen.

3.4.1. Fluorescence Microscopy Resolution

Naturally, when using a microscope, the hope is to magnify the specimen and be able to see structures and feature not visible with the naked eye. The extent to which fine details can be observed is dictated by the resolution of the microscope. Resolution is defined as the minimum distance by which two distinct objects or features must be separated in space (in an x–y plane) in order to appear as two separate entities in the enlarged image. The spatial resolution of conventional fluorescence microscopes is restricted by the diffraction limit. In practice, when dealing with wavelengths in the visible region of the spectrum as is typical in fluorescence microscopy, this amounts to a resolution limit of approximately 200–400 nm [6]. If the features are closer than this minimum distance, they blur together and appear as a single feature. Clever techniques such as scanning near-field optical microscopy (SNOM) and stimulated emission depletion (STED) can achieve resolution well below the diffraction limit, but detailed discussions of them are beyond the scope of this text.

3.4.2. Confocal Fluorescence Microscopy

Confocal fluorescence microscopy is similar to conventional fluorescence microscopy in that fluorescence that is emitted by the specimen is collected and focused onto a detector. However, in confocal microscopy, a laser is used as the excitation source. Additionally, a pinhole aperture is placed in front of the detector. The pinhole allows in only fluorescence that comes from the point in the sample where the excitation laser is focused. Emission from molecules in planes above or below the focal plane is physically blocked (see Figure 3.26).

The in-focus plane is scanned to develop a 2D image of that plane. Because the pinhole blocks fluorescence from other planes, only a very thin slice of the specimen is imaged. The fact that much of the fluorescence gets blocked and that which is detected comes from a very small volume of the sample (on the order of femtoliters) means that signals can be weak. For this reason, lasers are used as sources (recall that fluorescence intensity is proportional to excitation intensity and lasers are intense sources), and highly sensitive photoavalanche detectors are used to detect the fluorescence.

In addition to imaging a very thin (500 nm thick) plane, the specimen can be moved up or down and another plane imaged. By stacking multiple 2D images from different planes on top of one another, a 3D image of the sample is created. The ability to image planes from within the sample sets confocal microscopy apart from conventional microscopy. Of course, in either technique, the sample itself must have some inherent fluorescence, or specific molecules within the sample must be derivatized with fluorescent labels prior to collecting the images. Many interesting and visually appealing 3D images for a wide range of biological systems are available on the Internet, and we encourage readers to search for them and learn more about the capabilities of confocal fluorescence microscopy.

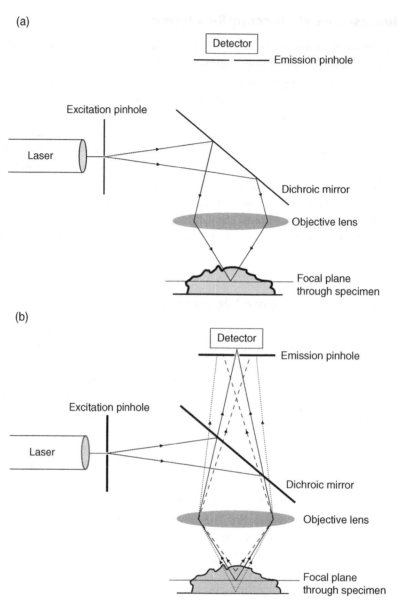

FIGURE 3.26 Figure of confocal fluorescence microscopy. (a) Depicts the illumination of the specimen. Electromagnetic radiation from the source (commonly a laser) strikes a dichroic mirror (beam splitter), passes through a lens, and illuminates the sample. (b) This image is similar to (a) except that it adds in the fluorescence caused by the excitation beam. Fluorescence is excited in many planes through the specimen. The fluorescence is collected by the lens and passes through the dichroic mirror toward the detector. Because of the pinholes located in front of the source and detector, only a small spot in a thin plane of the sample is focused on the detector. Light from other planes in the sample, represented by dashed and dotted lines in this figure, is not focused onto the detector and thus is not observed. A 2D image of the in-focus plane is obtained by scanning the plane, and 3D images are created by layering scans of multiple planes.

3.5. PHOSPHORESCENCE AND CHEMILUMINESCENCE

Before discussing the applications of luminescence techniques – with an emphasis on fluorescence – it is worth pausing here to comment explicitly on phosphorescence and chemiluminescence.

3.5.1. Phosphorescence

Recall from the examination of the Perrin–Jablonski diagram that phosphorescence is a longer-lived phenomenon than fluorescence, requiring two spin-forbidden transitions: one for a molecule to convert from an excited singlet state into a triplet state and one to return to the singlet ground state. Phosphorescence also generally occurs at longer wavelengths than fluorescence. However, both fluorescence and phosphorescence measurements require detecting the light emitted from a sample after excitation with electromagnetic radiation and thus measuring phosphorescence is essentially identical to measuring fluorescence. The only experimental difference besides the wavelengths at which they occur is that a synchronized delay between the excitation pulse and detection is usually required in order to measure the relatively long-lived phosphorescence free from any rapidly decaying fluorescence that is likely also emitted by the sample. The ability to cool the sample is also important in phosphorescence. Low temperatures reduce nonradiationless modes of deactivation, and thus increase the time molecules have to convert into the triplet state and phosphoresce. Given the similarities in the measurement requirements, the components of spectrofluorometers and their arrangement are suitable for phosphorescence measurements. Fluorescent measurements, however, are much more common as very few molecules of analytical interest phosphoresce.

3.5.2. Chemiluminescence

Chemiluminescence is different from fluorescence and phosphorescence in that it does not require the absorption of electromagnetic radiation prior to the emission of light. Rather, chemiluminescence derives from the energy created during a chemical reaction between the analyte and other reagents, which leads to the production of a molecule in the excited state that emits light as it relaxes to the ground state. Examples of such reactions are found in nature, most familiarly as the reaction that makes fireflies light up or that make beaches "glow" at night due to bioluminescent plankton.

Because no excitation source is needed, instrumentation used to measure chemiluminescence is relatively simple, requiring only a sample cell in which the reaction takes place, optics to focus the emitted electromagnetic radiation, and a detector. A wavelength selector such as a filter or monochromator can be used to select the wavelength of emitted light if desired, but is not necessary because the only source of light should be the reaction itself.

Typically, the signal initially rises and then falls as the chemiluminescent reaction takes place, as shown in Figure 3.27 for a reaction related to the bioluminescence of a crustacean found in the sea [30]. Time 0.00 on the graph is when the reactants – one of which is the analyte – are mixed together. The rise in signal is related to the time it takes for reactants to mix together and to begin creating the excited state product, and the decline reflects the consumption of reactants as the reaction proceeds. Typically, the signal is integrated for a

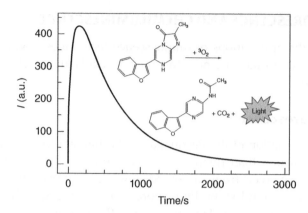

FIGURE 3.27 Chemiluminescence produced by the reaction shown in aerated diglyme–acetate buffer, pH = 5.6, at 25 °C. Time zero is when the reactants are mixed. The molecule in the reaction is an analog of a molecule responsible for blue-colored bioluminescence emitted by *Cypridina (Vargula)* – a species of ostracod crustacean. The native bioluminescence is produced by a luciferin–luciferase reaction. In this reaction, the analog reacts with triplet oxygen ($^3O_2$), undergoes a series of reactions (not shown), and ultimately creates the product shown while emitting a photon. Source: Adapted from Hirano et al. [30]. Reproduced with permission of Royal Society of Chemistry.

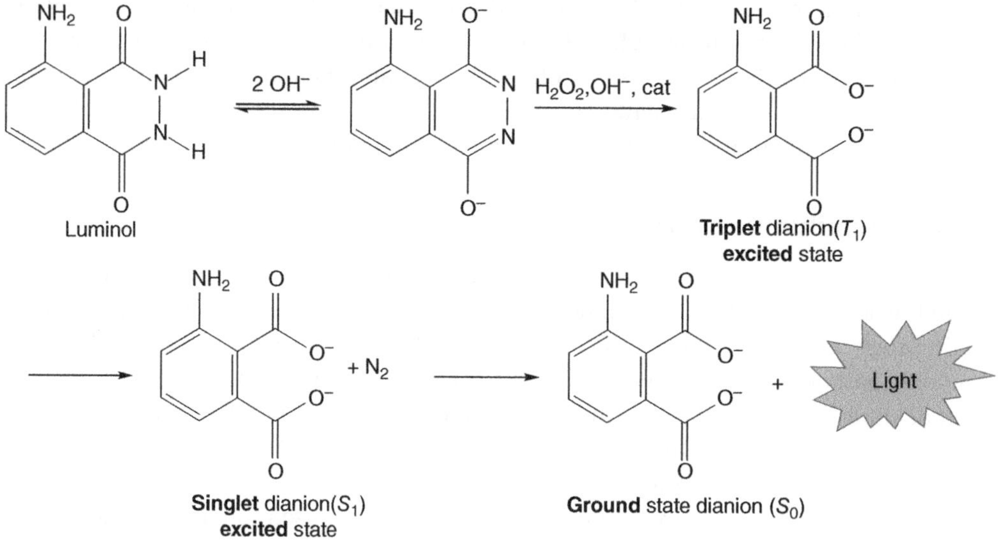

FIGURE 3.28 The luminol reaction. The reaction leads to the production of light due to the creation of high energy intermediates that relax back to the ground state via emission.

specified period of time and is proportional to the concentration of the analyte provided that other reagents are in excess. Because there is no background light to interfere with measurements, chemiluminescent assays can be quite sensitive at low concentrations.

An example of a chemiluminescent reaction is the well-known luminol reaction, which is used at crime scenes to detect blood because the iron in hemoglobin acts as a catalyst for the reaction shown in Figure 3.28 [31]. Chemiluminescence is also used in many clinical assays such as immunoassays, protein blotting, toxicological tests, and pharmaceutical tests [32]. It is also used to detect nitric oxide gas (NO) [33] for air quality purposes and was proposed as a technique for measuring the oxidation of elastomers in various types of rubber [34].

3.6. APPLICATIONS OF FLUORESCENCE: BIOLOGICAL SYSTEMS AND DNA SEQUENCING

Many of the numerous applications of fluorescence involve monitoring biochemical systems. Some biomolecules, such as tryptophan, tyrosine, phenylalanine, NADH (nicotinamide adenine dinucleotide), and riboflavin (see Figure 3.29), are fluorescent and can be used to directly monitor biochemical processes [10]. Biomolecules that are not natively fluorescent, or which have low quantum yields, can be derivatized with highly fluorescent molecules. For example, proteins are frequently labeled with fluorophores such as dansyl chloride, fluorescein isothiocyanate (FITC), and acrylodan (see Figure 3.30) [10]. Fluorescein, rhodamine, and BODIPY (boron dipyrromethene) and their derivatives are frequently used to label antibodies for use in fluorescence microscopy and immunoassays (see Figure 3.31). Lipids can also be derivatized with fluorescein and other fluorophores such as Texas Red and pyrene (see Figure 3.32). The derivatized lipids are then incorporated into biological membranes to study membrane potentials [10]. DNA can also be made fluorescent by incorporating purine bases that are analogs to natural bases such as adenine and guanine (see Figure 3.33) [10]. The ability to incorporate green fluorescent protein (GFP) and engineered mutants of it within cells has also been an important

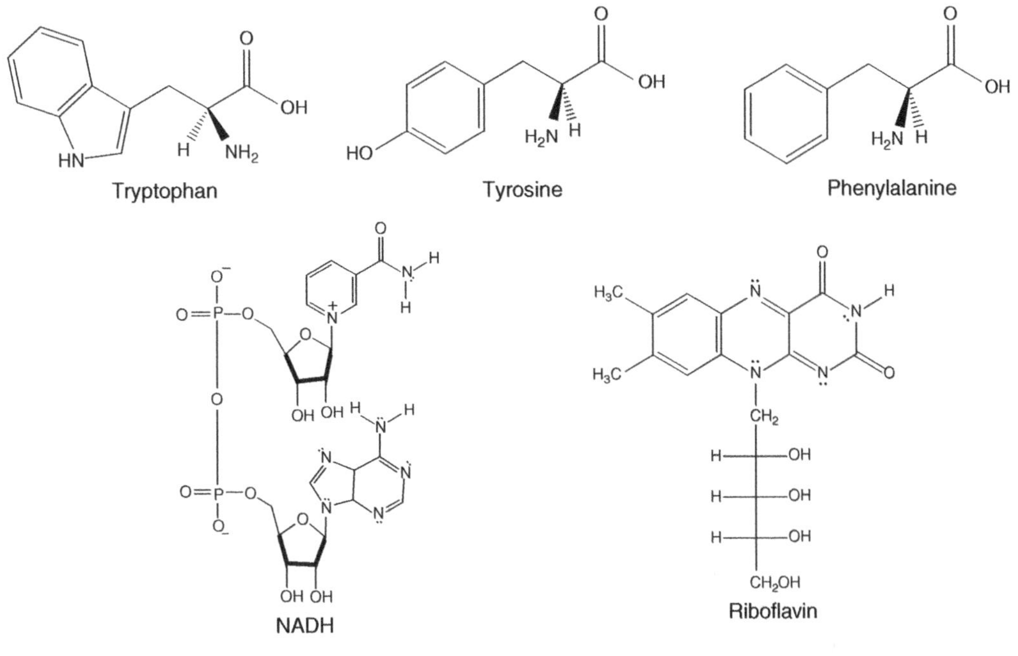

FIGURE 3.29 Examples of biomolecules with native fluorescence.

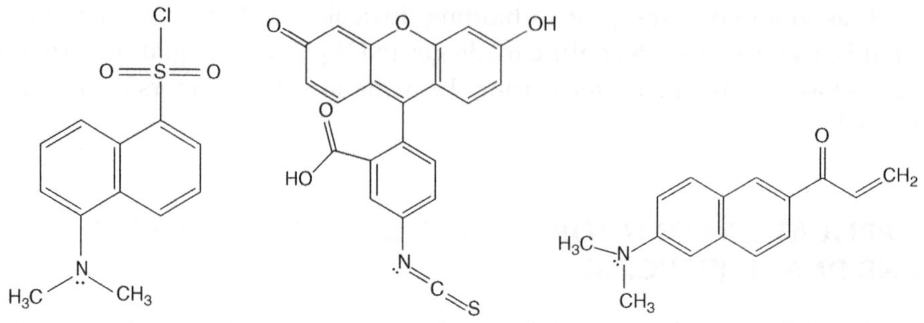

Dansyl chloride FITC Acrylodan

FIGURE 3.30 Examples of common fluorescent derivatizing agents.

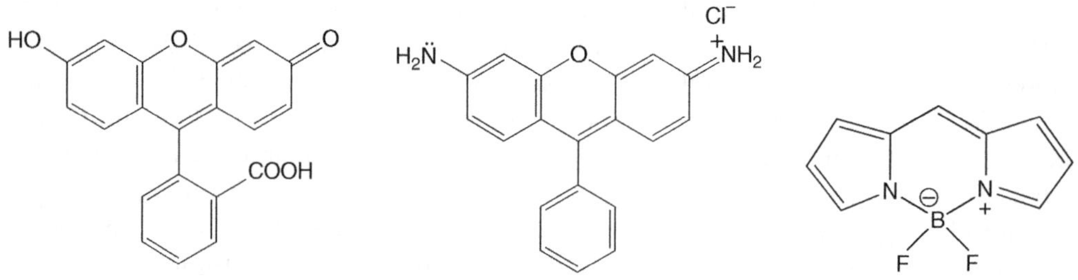

Fluorescein Rhodamine BODIPY

FIGURE 3.31 Examples of common derivatizing agents for fluorescence microscopy and immunoassays.

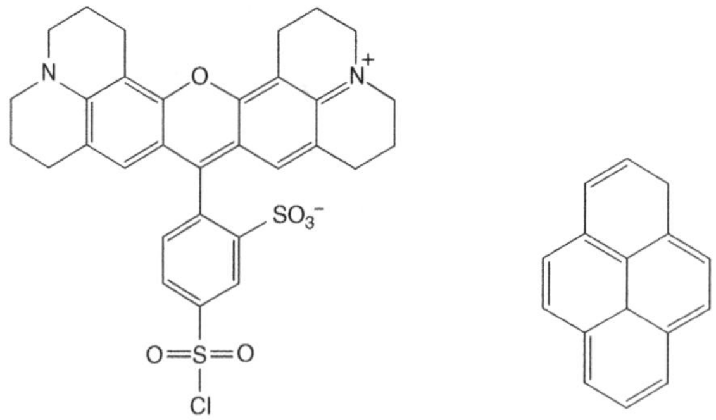

Texas Red Pyrene

FIGURE 3.32 Examples of dyes used to derivatize lipids.

advance in fluorescence-based imaging [10]. Applications of GFP include studying protein folding [35], protein transport [36, 37], and gene expression [38].

One of the most remarkable applications of fluorescence detection was its role in sequencing the human genome – and it continues to be used in modern sequencing systems. Many sequencing schemes are based on the Sanger method of sequencing [39]. In this method, a fragment of single-stranded DNA (ssDNA) is allowed to react with a

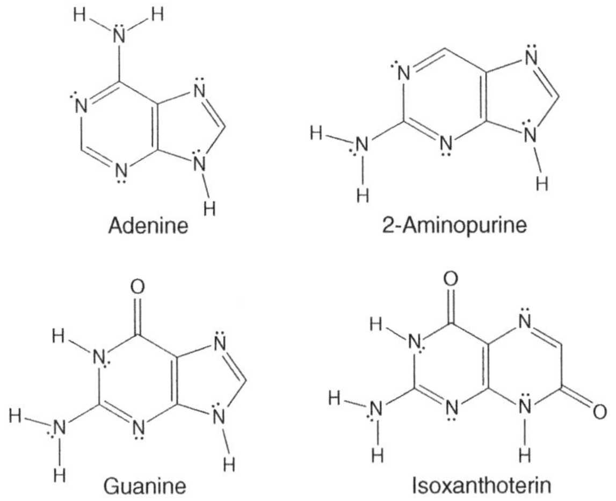

FIGURE 3.33 Examples of DNA bases and fluorescent analogs.

radiolabeled primer, which is a short chain of bases that is complementary to some portion of the fragment being sequenced. The primer anneals to the ssDNA. Base pairs that are complementary to those on the fragment are then added to the short primer using a polymerase enzyme. Some of the base pairs that are added to the reaction, however, are dideoxynucleotide analogs, meaning that they lack a critical –OH group at the 3' position. The absence of this group prohibits further extension of the growing DNA chain. This leads to the production of DNA fragments of different sizes, as illustrated in Figure 3.34a. Four separate preparations specific to each of the four bases ultimately create a whole series of fragments. The fragments in each separate mixture are then analyzed using slab gel electrophoresis as shown in Figure 3.34b, using one lane for each different base pair mixture and detected by autoradiography (i.e., by radioactivity) because the primer had been labeled with radioactive phosphorus. The DNA sequence is determined by "reading" the gel from the bottom up, assigning each subsequent base by seeing which lane had a band in it.

Because the original Sanger method uses slab gel electrophoresis, it is slow and labor intensive. And while methods using radiolabeled chemicals are quite sensitive, they require extra precaution and present potential safety concerns.

Smith et al. adapted the method by eliminating the radioactive phosphorus label and instead derivatized the primer sequence with one of four different fluorescent probes, shown in Figure 3.35 [40]. Extension of the primer with additional base pairs was done in four separate reaction vessels, one for each different labeled primer and each containing only one type of chain-terminating dideoxynucleotide. In this manner, each of the four fluorophores corresponded to or indicated that a specific base was at the end of the DNA fragment. For example, fluorescein isothiocyanate (FITC) was paired with chain-terminating adenine bases, and Texas Red was coupled with cytosine. This pairing of fluorophore and base was the key principle that ultimately allowed the DNA sequence to be read. Sequencing was accomplished by combining the four reaction mixtures into a single mixture and then using polyacrylamide gel tube

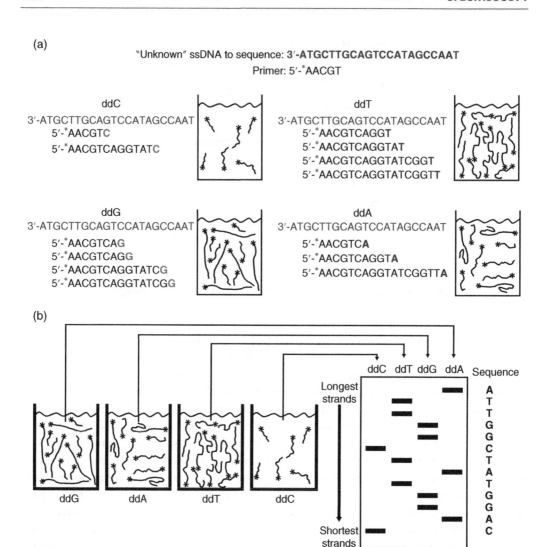

FIGURE 3.34 (a) The first step in the Sanger method of sequencing is to generate fragments of DNA that are radiolabeled (as indicated by the asterisks) and that terminate with dideoxynucleotides. Four separate reaction vessels are used, with one vessel for each of the four different dideoxynucleotides. (b) The ssDNA strands generated in each vessel are separated by slab gel electrophoresis, with a different lane for each reaction vessel. After the separation, the DNA sequence is read from the bottom of the gel to the top.

electrophoresis to separate the differently sized DNA strands, which emerged from the tube according to the number of bases they contained. As each strand reached the end of the tube, a laser excited the fluorophore at 488 or 514 nm. The emission was collected and passed through filters to distinguish the different fluorescent wavelengths arising from each dye.

Ultimately, the plots in Figure 3.36 were obtained [40]. The vertical lines in the plot show that it is possible to "read" the peaks from left to right and determine which base

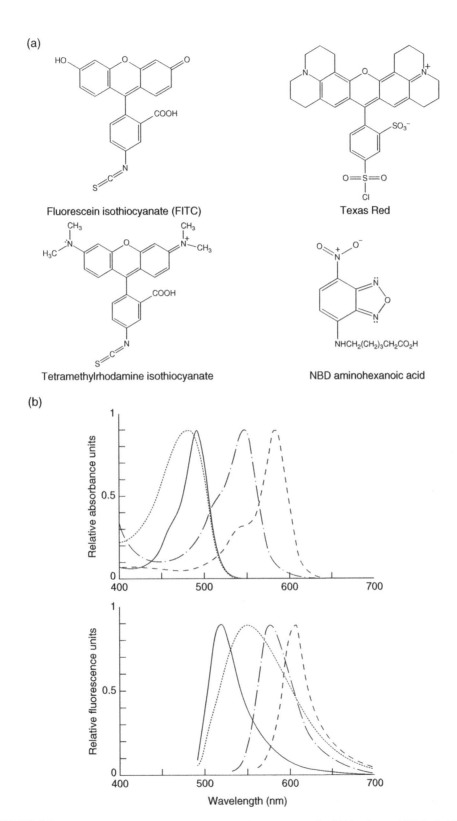

FIGURE 3.35 (a) Fluorescent dyes used by Smith et al. to derivatize the DNA primers. NBD is 4-chloro-7-nitrobenzo-2-oxa-1-diazole. (b) The absorbance and emission spectra of fluorescein (solid line), NBD (dotted line), Texas Red (dashed line), and tetramethylrhodamine (dashed/dot line). Source: Reprinted by permission from Smith et al. [40]. Reproduced with permission of Nature Publishing Group.

(a)

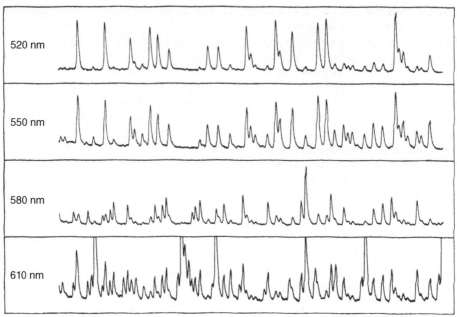

(b)

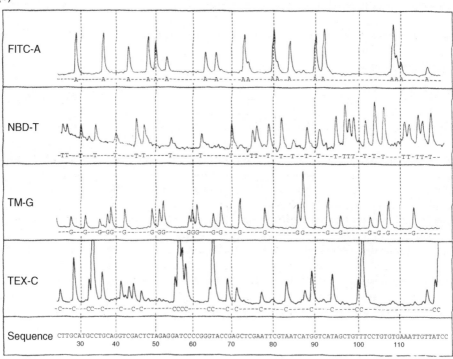

FIGURE 3.36 Partial electropherogram from which the DNA sequence of the fragment Smith et al. analyzed was read. (a) The electropherogram collected at the four different wavelengths used to detect the DNA. (b) Electropherogram after adjusting for migration shifts caused by the primers showing how the sequence can be read from the peaks (note the sequence at the bottom of the plot). (c) Enlargement of part of the electropherogram (1000 data points) to make it clearer how the sequence is read from the peaks. Source: Reprinted by permission from Smith et al. [40]. Reproduced with permission of Nature Publishing Group.

(c)

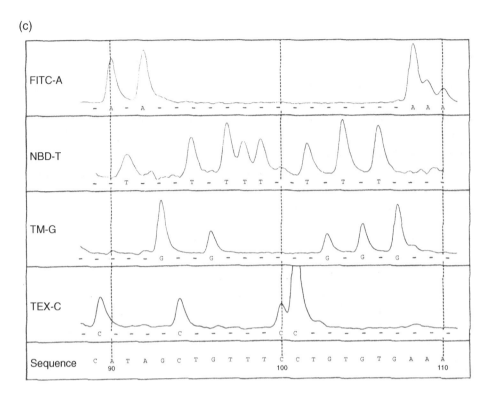

FIGURE 3.36 (Continued)

came next in the sequence. For example, the first vertical line shown corresponds clearly to T, the next also to T, and the third to A. The letters underneath the individual plots (--A------A--) were then assembled into the entire sequence of the 200-base pair fragment of DNA with which they were working.

The entire human genome, which has over three *billion* base pairs, was sequenced in essentially the same manner by fragmenting the DNA into many pieces, sequencing the fragments, and then combining all of the data from individual fragments to create the entire sequence. The speed of capillary electrophoresis, including advances in parallel analyses using multiple capillary arrays, coupled with the sensitivity associated with laser-induced fluorescence and the specificity obtained by using fluorophores with distinct wavelengths of maximum emission, made it possible for this rather remarkable feat to be accomplished.

Modern commercially available sequencers, while employing modified DNA synthesis strategies and improved fluorescent probes, are still based on the same principles as those just explained. Different fluorophores are used in different sequencers to avoid some of the complications associated with the ones used by Smith et al. Nevertheless, the "genome revolution" was fostered by and continues to rest heavily on the speed, sensitivity, and selectivity of fluorescence spectroscopy.

3.7. SUMMARY

Fluorescence is a sensitive and selective method that offers the potential for exception-ally low detection limits and wide dynamic ranges. It is based on measuring the emission of electromagnetic radiation by molecules returning to their ground state following excitation. Phosphorescence and chemiluminescence offer similar advantages but are less frequently used because fewer molecules phosphoresce or chemilumi-nesce relative to the number that fluoresce. Fluorometers and spectrofluorometers use bright light sources to increase the amount of fluorescence observed. Filters and mono-chromators are used to select excitation and emission wavelengths, and PMTs or CCDs are used to detect fluorescence. Many factors affect the probability that a molecule will fluoresce, including molecular structure, solvent viscosity, the presence or absence of quenchers, and sample temperature and pH. Fluorescence intensity can be increased by increasing the intensity of the excitation beam, increasing the slit widths, and decreasing the temperature of the sample. Because of its sensitivity and selectivity, fluorescence is used in many applications, particularly in biochemical studies, including fluorescence imaging, immunoassays, single molecule detection, membrane studies, and DNA sequencing.

PROBLEMS

3.1 The structure of laurdan, a fluorescent molecule, is shown below. Circle the part of the molecule that is responsible for the fluorescence (i.e., circle the fluorophore). What types of biological systems could laurdan be used to study?

3.2 The absorbance and emission spectra for tryptophan are shown below.
   (a) To obtain the maximum sensitivity for a calibration curve based on the fluores-cence of tryptophan, which wavelengths for excitation and emission should be used?
   (b) The emission spectrum shown was collected using an excitation wavelength of 260 nm. If 275 nm is used instead, will the fluorescence maximum shift to a different wavelength, and if so, will it shift to longer or shorter wavelengths? Assume all else stays the same. Explain your answer. Also, if 275 nm is used for excitation, will the emission intensity change? If so, will it increase or decrease? Assume all else stays the same. Explain your answer.

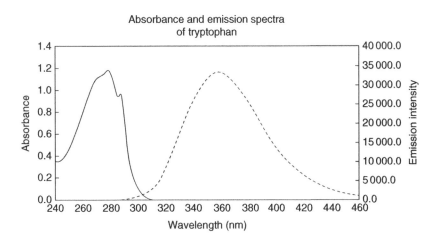

Absorbance and emission spectra of tryptophan

3.3 Does fluorescence intensity increase or decrease with the following changes, assuming all other conditions are constant? Briefly explain why.

(a) Decreasing molar absorptivity of the analyte.

(b) Wider slits.

(c) Weaker source.

(d) Higher sample temperature.

(e) Decreasing solvent viscosity.

(f) Using ethanol as a solvent instead of bromoethanol.

(g) Exciting at a wavelength that is not the wavelength of maximum absorption relative to exciting at the wavelength of maximum absorption.

3.4 Why is fluorescence intensity increased (and thus detection limits potentially decreased) when using a laser instead of a lamp as the excitation source?

3.5 Explain why increasing the temperature of a sample generally decreases fluorescence intensity.

3.6 The molecule di-8-ANEPPS is a voltage-sensitive dye used to study cell membranes.

(a) Look up and draw the structure of di-8-ANEPPS. Identify the lone and π-electrons that are involved in the absorption and emission of electromagnetic radiation.

(b) This molecule has a larger dipole moment in the excited state than in the ground state. In which direction will the emission spectrum shift if it is recorded in benzene instead of ethanol?

3.7 The wavelength of maximum absorbance of the compound in Figure 3.3 is around 420 nm in all of the solvents studied. Look at Figure 3.3. Which solvents lead to the smallest and largest differences between the wavelength of maximum absorbance and the wavelength of maximum fluorescence? Explain why this is the case.

3.8 How is intersystem crossing different from internal conversion?

3.9 (a) From the spectra for phenanthrene and naphthalene shown on the next page, devise a method of analysis for each component assuming a spectrofluorometer is available (hint: multiple measurements are possible with different combinations of excitation and emission wavelengths). (b) Would using phosphorescence help in this analysis?

Problem 3.9 Excitation and emission spectra for phenanthrene (top) and naphthalene (bottom). A, excitation; F, fluorescence; P, phosphorescence.

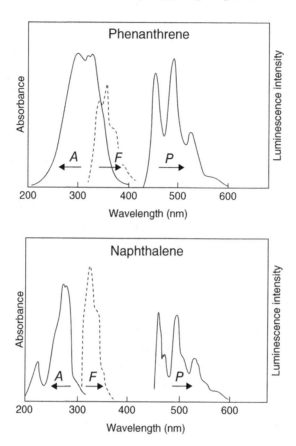

3.10 (a) Explain the difference between emission and excitation spectra. In both cases, be sure to identify which monochromator is varied and which remains fixed.

    (b) Which spectrum, emission or excitation, more closely resembles an absorbance spectrum?

3.11 What makes analytical methods based on fluorescence more sensitive than those based on absorption?

3.12 3,4-Benzopyrene is a potent carcinogen often found in polluted air. A sensitive method for its determination is to measure its fluorescence in sulfuric acid solution. The excitation wavelength is 520 nm and the emission wavelength is 548 nm. Twenty liters of air were drawn through 10.0 mL of dilute sulfuric acid solution to capture the 3,4-benzopyrene in the air. One milliliter of this sulfuric acid solution gave a reading of 33.3 in a fluorometer. Two different standard solutions of 3,4-benzopyrene containing 0.750 and 1.25 μg per 10.00 mL of sulfuric acid gave readings of 24.5 and 38.6 when 1.00 mL was placed in the same cell used for the unknown. A blank sample with no 3,4-benzopyrene gave a reading of 3.5. Calculate the weight of 3,4-benzopyrene in 1 liter of air.

3.13 In January of one particular year, a laboratory measured a calibration curve for quinine sulfate in aqueous solutions. Because this was all done on the same day, the source intensity did not vary appreciably, and the ratio of fluorescence to source intensity was not used – just raw fluorescence intensity was measured. The equation of the line was found to be: fluorescence intensity = 1.35×10^{11} [quinine sulfate (M)] $- 1.52 \times 10^3$.

The same laboratory workers then analyzed an unknown solution that produced a fluorescence intensity of 13,425 fluorescence units.

(a) What is the concentration of quinine sulfate in solution?

(b) Now, suppose that four months later, the lab is still performing quinine sulfate assays and still using the same calibration curve established in January. However, over that time, the lamp output has decreased to 83.7% of what it was when the calibration curve was established. Suppose that a solution of exactly the same concentration as calculated in part (a) is made and analyzed. What would the fluorescence intensity read (assuming that the ratio of the emission intensity to the excitation intensity is NOT taken into account)? Using the calibration curve and this uncorrected intensity, what quinine sulfate concentration would be calculated for the sample?

(c) What percent error resulted by not accounting for changes in lamp intensity?

3.14 Tyrosine absorbs at 280 nm with a molar absorptivity of 1280 $M^{-1}cm^{-1}$ and fluoresces at 303 nm. Suppose that on a given instrument, for a very dilute solution of tyrosine, the fluorescence intensity at 303 nm is found to be 3.69×10^5 arbitrary fluorescence units when excited using 280 nm.

If the sample is excited at a different wavelength where the molar absorptivity is 930 $M^{-1}cm^{-1}$ and the source intensity is 2.6 times greater at this new wavelength, what is the new fluorescence reading?

Assume the quantum yield and path length remain constant.

3.15 The following data were obtained for standard solutions of a fluorescent dye at a fixed wavelength of excitation and emission.

Concentration (µM)	Fluorescence signal
0.1000	90.0
0.5000	340.0
1.000	700.0
5.000	3810
10.00	7210
30.00	23500
50.00	38300
80.00	57200
100.0	70200
125.	69300
150.	62800
175.	51800
200.	38000

(a) Suppose that an unknown solution has a signal of 51 500 fluorescence units. What is the concentration of the dye in this solution? Plot the data and look carefully at the plot before answering.

(b) If a 10-fold dilution is performed and results in a signal of 13 520 fluorescence units, what is the concentration of the original solution?

3.16 It took Smith et al. about 13 h to sequence 200 base pairs of DNA. Using just a single capillary electrophoresis instrument and making no improvements in the method, how many years would it take to sequence the human genome (about three billion base pairs)? Compare this to the actual time it took (13 years; 1990–2003) – what do you conclude from this comparison?

3.17 What advantages are offered from front-face collection of fluorescence compared to 90° collection? In what situations might it be useful?

3.18 The excitation spectrum of Fluorescent red 610 (a fluorescent tag used to derivatize biological molecules) was recorded from 350 to 600 nm using a mercury lamp as the excitation source. The emission monochromator was set at 670 nm. The excitation spectrum was uncorrected and no reference detector was used. The spectrum showed spikes in the fluorescence intensity at 365, 405, 436, 546, and 578 nm on top of an otherwise smoothly changing spectrum. What is the cause of the spikes?

3.19 Under ideal conditions, Eq. (3.5) suggests fluorescence intensity is directly proportional to analyte concentration. This is generally true at low concentrations, but deviations from linearity are observed at higher concentrations. Identify two sources of deviations that arise at high analyte concentration, and explain how they lead to deviations.

3.20 In a sample containing 1 femtomole of a fluorescent compound, if one-tenth of the molecules are excited by the excitation beam, and 4.3×10^7 molecules fluoresce as a result, what is the fluorescence quantum yield?

3.21 Why are analyses based on phosphorescence much less common than analyses based on fluorescence?

3.22 What components that are in a spectrofluorometer used for fluorescence measurements are not needed in an instrument used to detect chemiluminescence? Why is this?

REFERENCES

1. http://photobiology.info/HistJablonski.html (accessed 2 September 2017).
2. Perrin, F. (1929). *Ann. Phys. (Paris) 12*: 169–275.
3. Perrin, J. (1922). *Trans. Faraday Soc. 17*: 546–572.
4. Jablonski, A. (1933). *Nature 131*: 839–840.
5. Jablonski, A. (1935). *Z. Physik 94*: 38–46.
6. Valeur, B. and Berberan-Santos, M.N. (2013). *Molecular Fluorescence: Principles and Applications*, 2e. Weinheim: Wiley-VCH Verlag BmbH & Co. KGaA.
7. Taft, R.W. and Kamlet, M.J. (1976). *J. Am. Chem. Soc. 98*: 2886–2894.

8. Bolduc, A., Dong, Y., Guérin, A., and Skene, W.G. (2012). *Phys. Chem. Chem. Phys. 14*: 6946–6956.

9. http://www.olympusmicro.com/primer/techniques/confocal/fluoroexciteemit.html (accessed 2 September 2017).

10. Lakowicz, J.R. (2006). *Principles of Fluorescence Spectroscopy*, 3e. New York: Springer.

11. Willard, H.H., Merritt, L.L. Jr., Dean, J.A., and Settle, F.A. Jr. (1988). *Instrumental Methods of Analysis*, 7e. Belmont: Wadsworth Publishing Co.

12. https://chem.libretexts.org/Core/Physical_and_Theoretical_Chemistry/Spectroscopy/Electronic_Spectroscopy/Fluorescence_and_Phosphorescence (accessed 9 February 2018).

13. Credi, A. and Prodi, L. (1998). *Spectrochim. Acta A 54*: 159–170.

14. Resch-Genger, U., Hoffmann, K., and Pauli, J. (2014). *Comprehensive Biomedical Physics, Vol. 4: Optical Molecular Imaging* (ed. F. Alves and D. Kiessling). Amsterdam: Elsevier.

15. Puiu, A., Fiorani, L., Menicucci, I. et al. (2015). *Sensors 15*: 14415–14434.

16. Hercules, D.M. (1966). *Fluorescence and Phosphorescence Analysis: Principles and Applications*. New York: Wiley.

17. https://www.hamamatsu.com/resources/pdf/etd/PMT_handbook_v3aE-Chapter6.pdf (accessed 4 September 2017).

18. Malmstadt, H.V., Franklin, M.L., and Horlick, G. (1972). *Anal. Chem. 44*: 63A–76A.

19. https://www.azom.com/article.aspx?ArticleID=10886 (accessed 4 September 2017).

20. Kissick, D.J., Muir, R.D., and Simpson, G.J. (2010). *Anal. Chem. 82*: 10129–10134.

21. Taguchi, K., Frey, E.C., Wang, X. et al. (2010). *Med. Phys. 37*: 3957–3969.

22. Becker, W. (2005). *Advanced Time-Correlated Single Photon Counting Techniques*. Berlin: Springer-Verlag.

23. O'Connor, D.V. and Phillips, D. (1984). *Time-correlated Single Photon Counting*. London: Academic Press, Inc.

24. Dawaliby, R., Trubbia, C., Delporte, C. et al. (2016). *J. Biol. Chem. 291*: 3658–3667.

25. Scott, B.M., Vu, M.P., McLemore, C.O. et al. (2008). *J. Lipid Res. 49*: 1202–1215.

26. Harris, F.M., Best, K.B., and Bell, J.D. (2002). *Biochim. Biophys. Acta 1565*: 123–128.

27. Jameson, D.M. (2014). *Introduction to Fluorescence*. Boca Raton: CRC Press.

28. Berzin, M.Y. and Achilefu, S. (2010). *Chem. Rev. 110*: 2641–2684.

29. Otosu, T., Nishimoto, E., and Yamashita, S. (2010). *J. Biochem. 147*: 191–200.

30. Hirano, T., Takahashi, Y., Kondo, H. et al. (2008). *Photochem. Photobiol. Sci. 7*: 197–207.

31. https://en.wikipedia.org/wiki/Luminol#/media/File:Luminol_chemiluminescence_molecular_representation.svg (accessed 4 September 2017).

32. Rodriguez-Orozco, A.R., Ruiz-Reyes, H., and Medina-Serriteno, N. (2010). *Mini Rev. Med. Chem. 10*: 1393–1400.

33. Bates, J.N. (1992). *Neuroprotocols 1*: 141–149.

34. Forsström, D., Kron, A., Mattson, B. et al. (1992). *Rubber Chem. Technol. 65*: 736–743.

35. Tsien, R.Y. (1998). *Annu. Rev. Biochem. 67*: 509–544.

36. Stadler, R., Lauterbach, C., and Sauer, N. (2005). *Plant Physiol. 139*: 701–712.

37. Kaether, C. and Gerdes, H.H. (1995). *FEBS Lett. 369*: 267–271.

38. Komorowski, M., Finkenstadt, B., and Rand, D. (2010). *Biophys. J. 98*: 2759–2769.

39. Sanger, F., Nicklen, S., and Coulson, A.R. (1977). *Proc. Natl. Acad. Sci. U.S.A. 74*: 5463–5467.

40. Smith, L.M., Sanders, J.Z., Kaiser, R.J. et al. (1986). *Nature 321*: 674–679.

FURTHER READING

1. Becker, R.S. (1969). *Theory and Interpretation of Fluorescence and Phosphorescence*. New York: Wiley.
2. Hercules, D.M. (1966). *Fluorescence and Phosphorescence Analysis: Principles and Applications*. New York: Wiley.
3. Jameson, D.M. (2014). *Introduction to Fluorescence*. Boca Raton: CRC Press.
4. Kubitscheck, U. (2017). *Fluorescence Microscopy: From Principles to Biological Applications*, 2e. Weinheim: Wiley-VCH Verlag BmbH & Co. KGaA.
5. Lakowicz, J.R. (2006). *Principles of Fluorescence Spectroscopy*, 3e. New York: Springer.
6. Perkampus, H.-H. *Encyclopedia of Spectroscopy*, Grinter, H.-C. and Grinter, R. (Trans.); VCH Verlagsgesellschaft mbH: Weinheim, 1995.
7. Sauer, M., Hofkens, J., and Enderlein, J. (2011). *Handbook of Fluorescence Spectroscopy and Imaging: From Ensemble to Single Molecules*. Weinheim: Wiley-VCH Verlag BmbH & Co. KGaA.
8. Sharma, A. and Schulman, S.G. (1999). *Introduction to Fluorescence Spectroscopy*. New York: Wiley.
9. Valeur, B. and Berberan-Santos, M.N. (2013). *Molecular Fluorescence: Principles and Applications*, 2e. Weinheim: Wiley-VCH Verlag BmbH & Co. KGaA.
10. White, C.E. and Argauer, R.J. (1970). *Fluorescence Analysis: A Practical Approach*. New York: Marcel Dekker, Inc.

INFRARED SPECTROSCOPY

4

This chapter continues our discussion of the analytical uses of spectroscopy that started with the chapter on the fundamental characteristics of electromagnetic radiation (EMR) and continued into the chapters about UV-visible and fluorescence spectroscopy. Here we focus on the infrared (IR) region of the spectrum. Two of the most common tasks for chemists are verifying the structure of reaction products and determining the identity of unknown compounds. IR spectroscopy provides information about the functional groups present in molecules by probing the nature of the vibrational energy of bonds. It is therefore frequently used to assist in structure elucidation. Because of its use in structural analysis, IR spectroscopy is generally associated with qualitative analyses, but it is also used for quantitative analyses, with recent advances in instrumentation expanding such uses.

In the sections below, we

1. Discuss the fundamental bond vibrations upon which IR spectroscopy is based and the theories used to explain why different bonds vibrate with different frequencies.
2. Describe the instrumentation designed to measure the absorption of IR radiation and how we gain chemical information from such measurements.
3. Explain a variety of sampling methods for analyzing solids, liquids, and gases.
4. Examine a number of important applications that illustrate how the entire IR region of the spectrum, including the near-, mid-, and far-IR regions, are used in quantitative and qualitative chemical analyses.

4.1. THEORY

In the UV-visible spectroscopy chapter, we saw that molecules absorb electromagnetic radiation when the energy of the radiation matches a ground to excited state electronic transition in the molecule. The amount of light absorbed, A, is given by

$$A = -\log\left(\frac{P}{P_o}\right) = -\log T \qquad (4.1)$$

Spectroscopy: Principles and Instrumentation, First Edition. Mark F. Vitha.
© 2019 John Wiley & Sons, Inc. Published 2019 by John Wiley & Sons, Inc.

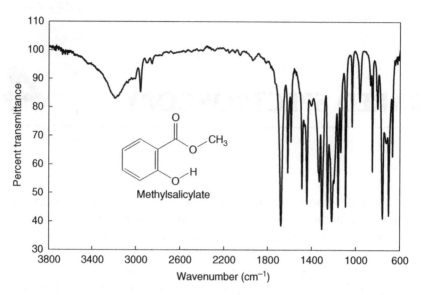

FIGURE 4.1 A typical IR spectrum taken of methyl salicylate as the analyte. Methyl salicylate is used to make candy "wintergreen" flavored and gives candy a terrific minty smell. Note that the axes are transmittance vs. wavenumber ($\bar{v}$ in cm^{-1}), compared to UV-visible spectra that are typically plotted as absorbance vs. wavelength (in nm). It is possible, however, to convert between transmittance and absorbance ($A = -\log T$) and between wavelength and wavenumber ($\lambda = 1/\bar{v}$).

where P_0 is the power of radiation transmitted to the detector when a reference is present and P is power transmitted to the detector when the sample is present. The transmittance, T, is equal to P/P_0, and is the fraction of light transmitted through the sample. This relationship is also true in IR spectroscopy. However, the fundamental molecular change that causes the absorption of radiation is different. UV-visible spectroscopy probes electronic transitions, while IR spectroscopy probes the vibrational motions of bonds in a molecule. A typical IR spectrum is shown in Figure 4.1. Each of the peaks arises from bond vibrations. Infrared spectra such as the one in this figure for a wide range of compounds can be found in Ref. [1].

Vibrational energies are roughly 10–1000 times smaller than electronic transition energies (e.g., the S_0–S_1 transition), as depicted in Figure 4.2. Thus, to probe vibrations, electromagnetic radiation with much lower energy than that used in UV-visible spectroscopy is needed. Infrared radiation spans the appropriate range of required energies.

The IR region of the spectrum is associated with electromagnetic radiation in an energy range from approximately 2.60×10^{-19} J (780 nm) to 2.0×10^{-22} J (1000 μm). This range is frequently subdivided into three regions:

1. Near-IR
2. Mid-IR
3. Far-IR

The mid-IR region is most frequently used by chemists for structural analysis purposes and is the focus of this chapter. However, due to advances in instrumentation and

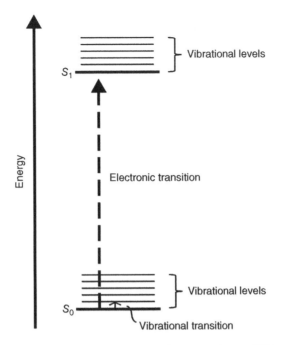

FIGURE 4.2 Depiction of electronic transitions from the ground state (S_0) to the first excited state (S_1) state such as those observed in UV-visible spectroscopy compared to a transition from the lowest vibrational level to the first vibrational level like those probed by IR spectroscopy. Note that energy increases from the bottom to the top of the image. It is clear that vibrational transitions are much lower in energy than electronic transitions.

data analysis techniques, applications in the near-IR and far-IR are growing rapidly and are also discussed in this chapter.

The difference in the energy of the radiation used in mid-IR spectroscopy compared to UV-visible spectroscopy requires different instrumental components for recording IR spectra. It also requires understanding the atomic-level changes taking place when molecules absorb infrared radiation, which is the focus of the next section.

EXAMPLE 4.1

(a) Methyl salicylate absorbs at 306 nm in the UV portion of the spectrum. Calculate the energy, in joules, associated with this $S_0 \rightarrow S_1$ electronic transition. Hint: Recall Planck's relationships $E = h\upsilon = hc/\lambda = hc\bar{\nu}$. Careful attention to units will help you get the correct answer.

(b) The peak in the IR spectrum of methyl salicylate that is located at 1695 cm^{-1} (i.e., $\bar{\nu} = 1695$ cm^{-1}) arises from the carbonyl bond. Calculate the energy, in joules, of the bond vibration associated with this wavenumber. Careful attention to units will help you get the correct answer. Hint: To make working with units easier, it might be helpful to rewrite 1696 cm^{-1} as $\dfrac{1695}{cm}$. Then work with units and $E = hc\bar{\nu}$ to get cm^{-1} converted to joules.

(c) Calculate the ratio of the energy of the UV transition relative to that of the vibration in the IR region (i.e., how many times larger is the UV transition energy compared to that of the IR?).

Answers:

(a) $E = \dfrac{hc}{\lambda} = \dfrac{\left(6.626 \times 10^{-34}\ \text{Js}\right)\left(2.998 \times 10^{8}\ \text{m/s}\right)}{306 \times 10^{-9}\ \text{m}} = 6.49 \times 10^{-19}\ \text{J}$

(b) $E = hc\bar{v} = \left(6.626 \times 10^{-34}\ \text{Js}\right)\left(2.998 \times 10^{8}\ \text{m/s}\right)\left(\dfrac{1695}{\text{cm}}\right)\left(\dfrac{100\ \text{cm}}{\text{m}}\right) = 3.367 \times 10^{-20}\ \text{J}$

(c) $\text{Ratio} = \dfrac{6.49 \times 10^{-19}\ \text{J}}{3.367 \times 10^{-20}\ \text{J}} = 19.3$

Thus, the energy of the UV transition is nearly 20 times greater than that of the IR vibration.

Another question:
Calculate the energy, in joules, associated with the broad feature in the IR spectrum of methyl salicylate centered around 3200 cm⁻¹.

Answer:
6.357×10^{-20} J (or 6.4×10^{-20} J to two significant figures because we are limited by the estimate of 3200 cm⁻¹, which has only two significant figures).

4.1.1. Bond Vibrations

Molecules are collections of atoms held in a geometric relationship to one another by Coulombic attractions (see Figure 4.3). Fundamentally, bonds form because sharing electrons between atoms lowers the energy of the system relative to the atoms existing apart from one another. This decrease in energy is seen in the classic plot shown in Figure 4.4, which is perhaps one of the most important plots in all of chemistry because it rationalizes why bonds and molecules, rather than just separate atoms, exist. In this plot, two atoms separated far apart from one another (far right side of the figure) are taken as having zero energy. As the distance between them decreases, the electronic orbitals overlap and electrons (negatively charged) are allowed to interact with *two* nuclei (positively charged). If this attraction between positive and negative charges overcomes the electron/electron repulsions and nucleus/nucleus repulsions that are also introduced, the energy of the system decreases and a bond forms. As the atoms continue to get closer and closer together, the positive/positive repulsive forces arising from the two nuclei continue to increase, resulting in an increase in the energy of the system (far left side of the figure).

As you likely already know, the lowest part of the curve is associated with the bond energy (y-axis value) and bond length (x-axis value), with different bonds (for example, C—C, N—H, C=O, etc.) having different energies and lengths.

This plot is important in IR spectroscopy because it shows that the atoms do not have to stay at a fixed distance apart in space, but rather can move back and forth relative to one

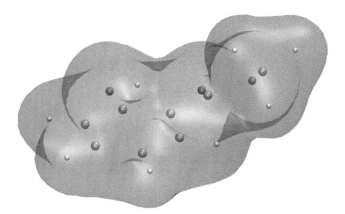

FIGURE 4.3 A depiction of methyl salicylate (for which the IR spectrum was shown in Figure 4.1). Balls represent atomic nuclei and the gray-shaded region represents the electron density. Lines representing bonds between atoms have been deliberately omitted to emphasize the fact that Coulombic forces, not sticks or springs, hold the atoms in space relative to one another. Source: Image courtesy of Dr. Matthew Zwier.

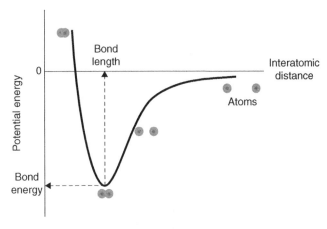

FIGURE 4.4 Starting at the right edge, this curve shows the potential energy as two atoms get increasingly closer to each other. The lowest energy point is referred to as the bond energy. The separation of the two nuclei at this point is known as the bond length. The energy initially decreases as favorable positive/negative interactions between protons located in the nuclei and electrons increase. At very short distances (far left), the energy increases and eventually becomes positive (i.e., unfavorable) as the repulsive forces between the two positive nuclei outweigh the favorable positive/negative interactions. The actual values of bond distance and bond energy change depending on the numbers of protons and electrons in the specific atoms involved.

another and still be bonded (i.e., have lower energy than when they are apart), as depicted in Figure 4.5 for a carbonyl bond. As the atoms move back and forth, attractive and repulsive forces act to keep the atoms separated by an average distance, which we call the bond length. This constant motion of moving apart and coming back together is called a bond vibration. It is the energy, or frequency, of this motion, along with other types of movements of atoms relative to one another in space, that is probed in infrared spectroscopy.

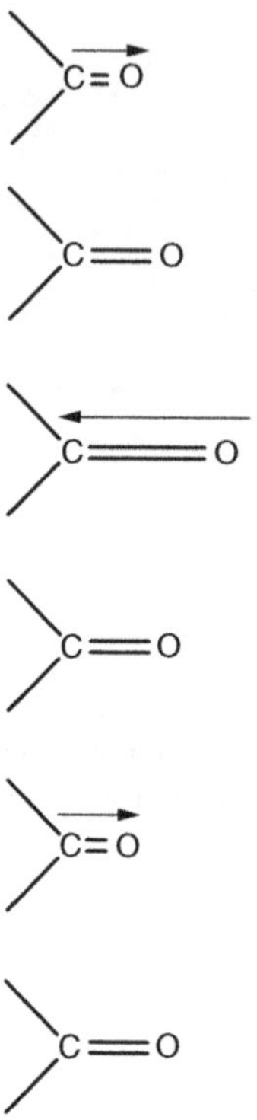

FIGURE 4.5 Vibration of a carbonyl bond. From top to bottom: atoms close to one another repel each other, and the atoms begin to move apart. As the bond lengthens, restoring forces bring the atoms back together again. The process continues with a frequency that is characteristic of the atoms involved and the strength of the bond.

4.1.2. Other Types of Vibrations

Before continuing, we pause here to acknowledge that, while we have been talking about vibrations involving just two atoms, in reality, vibrational modes in polyatomic molecules involve movement of atoms throughout the molecule [2]. Informative animations and videos of such motions are readily available on the Internet [3–10]. Accurate animations depict the entire molecule jiggling and dancing in a periodic manner. In many cases, the bond of interest experiences the vast majority of the movement, which is why we tend to focus on particular bond vibrations and bending motions, but at the atomic level, the motions are distributed over a wide number of atoms.

Vibrations that involve changes in the distance between atoms along the bond axis are called stretching vibrations, whereas those that involve changes in the angle between bonds are called bending vibrations (see Figure 4.6).

Some vibrations lead to a change in the dipole moment of the molecule and some do not. This is important because *only those vibrations that cause a change in the dipole moment produce peaks in IR spectra*. Such vibrations are referred to as "IR active." Those that do not produce peaks are called "IR inactive."

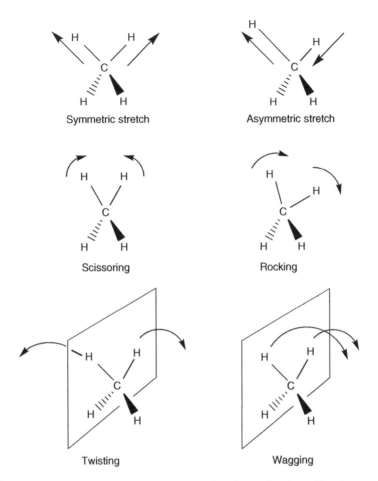

FIGURE 4.6 Examples of common stretching and bending vibrations. The top two images are stretches, and the middle and bottom images depict bending vibrations. Note that stretching vibrations result in a change in bond lengths, whereas bending vibrations result in changes in bond angles [11].

Recall that the formula for the dipole moment of two equal but opposite charges separated by some distance in space is given by

$$\mu = qr \tag{4.2}$$

where μ is the dipole moment, q is the magnitude of the charge, and r is the distance between the charges. As bonds vibrate, the distance between charges changes, and thus the dipole moment can change. However, the symmetry of the molecule and of the particular vibration may cause the changes to offset, producing no net change in the dipole during the vibration. Figure 4.7 shows several vibrations for CO_2 and H_2O to help visualize vibrations that do and do not produce changes in the dipole moment.

Within a molecule, stretching and vibrating motions have regular, periodic frequencies that depend on the particular atoms and bond energies involved. The change in dipole of

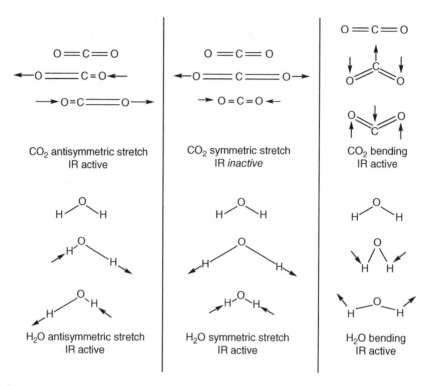

FIGURE 4.7 Vibrational modes of CO_2 and H_2O. In all instances except the symmetric stretch of CO_2 (upper middle), the stretching or bending motions result in a change in the dipole of the molecule and are therefore IR-active.

IR-active modes oscillates with the same frequency as the vibrational motion. The electric field of electromagnetic energy also oscillates with a given frequency. When electromagnetic radiation of the same frequency as the oscillating dipole interacts with the molecule, the molecule can absorb it. Recall that $E = h\upsilon$, such that the frequency of the radiation really corresponds to an energy and that absorbing radiation really means absorbing energy. The energy goes into increasing the *amplitude* of the vibration, leading to a larger displacement of the atoms relative to one another. Vibrations that do not have an associated oscillating dipole (i.e., oscillating electric field) cannot interact with the oscillating electric field of the incoming radiation and thus they do not absorb the radiation.

4.1.3.   Modeling Vibrations: Harmonic and Nonharmonic Oscillators

4.1.3.1.   The Simple Harmonic Oscillator and Hooke's Law. A ball and spring model is often used to understand bond vibrations. In this model, atoms are represented by solid spheres having mass, m, connected by a spring with a force constant, k. It is easiest to develop this model starting with a single mass suspended from a ceiling by a spring as shown in Figure 4.8.

The situation in which the ball is at rest is associated with zero displacement and zero potential energy. Now imagine pulling the ball straight down, displacing it by some

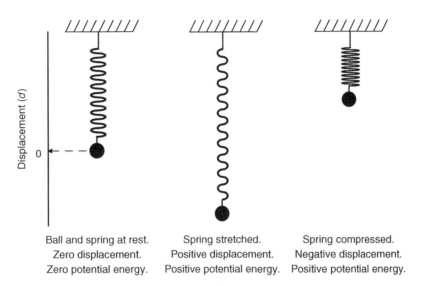

Ball and spring at rest.   Spring stretched.   Spring compressed.
Zero displacement.   Positive displacement.   Negative displacement.
Zero potential energy.   Positive potential energy.   Positive potential energy.

FIGURE 4.8 Ball and spring model for vibrations. When the system is at rest, it has no potential energy. Displacements of the ball from the rest position due to stretching or compressing the spring are measured relative to the rest position. Displacements in either direction result in the system having potential energy.

distance, d. By doing work on the system, you have given the system some potential energy. If you want to displace the ball by a greater distance, you have to put even more work into the system. Thus, larger displacements mean the system (i.e., the ball and spring) has greater potential energy. According to Hooke's law as applied to simple harmonic oscillators, the energy of the system is given by

$$E = \frac{1}{2}kd^2 \tag{4.3}$$

where k is the force constant, or stiffness, of the spring (Hooke's law is $F = -kd$, and the equation above can be derived from it). The dependence on the force constant makes sense. If the spring is difficult to displace – like the stiff springs in a car's suspension system (high k value) – then you have to put a lot more energy (i.e., higher E) into stretching the spring a given distance than you would compared to a flexible, loose spring with a low force constant.

Compressing the spring by pushing the mass in Figure 4.8 up (movement in the $-d$ direction) also adds energy to the system. The coiled spring is poised to push back against the compression and thus has potential energy, again given by Eq. (4.3). Squaring that d-term indicates that displacement in either direction from the rest position results in positive potential energy for the system. In fact, graphing E vs. displacement produces the plots in Figure 4.9. Notice that (1) higher k values produce higher energies for the same displacement, (2) greater displacements (meaning more stretching or compression) lead to greater potential energies, and (3) both positive and negative displacements lead to positive potential energy.

It is easy to imagine that letting go of the ball after displacing it results in the ball oscillating up and down with a regular, periodic motion characterized by a particular frequency, v. The frequency with which it oscillates is given by

$$v = \frac{1}{2\pi}\sqrt{\frac{k}{m}} \qquad (4.4)$$

where m is the mass of the ball. Here again, the equation matches intuition. Lighter objects (i.e., smaller m) produce higher frequencies than heavier objects, as do springs with higher force constants.

To make this model more applicable to atoms that are in a chemical bond, we now imagine two balls of mass m_1 and m_2, attached by a spring (see Figure 4.10) – keeping in mind again that atoms are not really attached to one another with springs but that it is a convenient model to use. When the spring is at rest, the distance between the balls is analogous to the bond distance, and the bond strength (i.e., bonded energy) is associated with the force constant of the spring. When the balls are moved further away from or closer to one another and then let go, they oscillate with the characteristic frequency given by

$$v = \frac{1}{2\pi}\sqrt{\frac{k}{\mu}} \qquad (4.5)$$

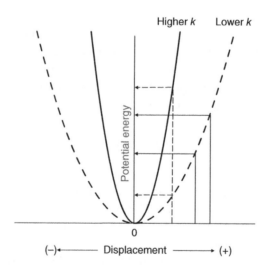

FIGURE 4.9 The potential energy versus displacement for two different springs: one with a higher force constant, k (solid curve), and one with a lower force constant (dashed line). Note that the spring with the higher force constant has higher potential energy for the same displacement of both springs (follow the dashed vertical line). Similarly, a greater displacement of either spring results in a higher potential energy as indicated by the vertical solid lines.

FIGURE 4.10 Two balls of different mass (m_1 and m_2) attached to a spring. Left: resting or equilibrium position. Middle: spring is stretched. Right: spring is compressed. When let go after being stretched or compressed, the system would begin to vibrate with a characteristic frequency depending on the force constant of the spring and the masses of the two balls.

where μ now takes the place of the single mass in Eq. (4.4) and is known as the reduced mass [12]. The reduced mass is equal to

$$\mu = \frac{m_1 m_2}{m_1 + m_2} \tag{4.6}$$

such that

$$\upsilon = \frac{1}{2\pi} \sqrt{\frac{k(m_1 + m_2)}{m_1 m_2}} \tag{4.7}$$

When applied to chemical bond vibrations, m_1 and m_2 are the masses of the two atoms in a bond, and k is the bond energy.

This equation can be used to estimate the frequencies of infrared radiation absorbed by specific bonds within a molecule, provided that adequate values of the bond energies are known. However, this model is only approximate for several reasons, three of which are:

1. Atoms are not connected by springs but rather are held together by Coulombic forces. Thus, the model breaks down under certain conditions.
2. There is an important selection rule that pertains to IR spectroscopy, namely, that in order for a particular vibration to absorb electromagnetic radiation (i.e., to be an "IR-active" vibration), the movement of the atoms involved in that vibration must result in a change in the dipole moment of the molecule, as explained earlier.
3. This model is based on classical physics and completely hides the fundamental quantized nature of vibrational transitions in molecules. We saw in UV-visible spectroscopy that the energy gaps between *electronic* states are quantized; so, too, are the *vibrational* levels.

4.1.3.2. Quantum Mechanical Harmonic Oscillator Model.

Taking a quantum mechanical approach to develop a model for bond frequencies starts with the energy of a harmonic oscillator given by

$$E = \left(\upsilon + \frac{1}{2}\right) \frac{h}{2\pi} \sqrt{\frac{k}{\mu}} \tag{4.8}$$

where υ is the vibrational quantum number that can only be positive integer values (including zero) and h is Planck's constant (6.626×10^{-34} Js) [13].

Quantum transitions generally entail changes from one quantum state (e.g., $v = 0$) to another (e.g., $v = 1$). Calculating the quantized energy difference, ΔE, between $v = 0$ and $v = 1$ yields

$$\Delta E = \frac{h}{2\pi} \sqrt{\frac{k}{\mu}} \tag{4.9}$$

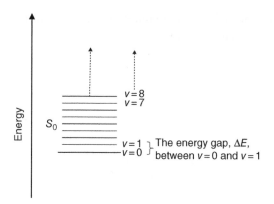

FIGURE 4.11 Vibrational energy levels within the ground electronic level (S_0). In this image, the difference between consecutive energy levels is approximately the same. This approximation does not hold true at high vibrational levels for actual molecules as discussed below, but we use the approximation here for illustrative purposes.

Applying Planck's equation, $\Delta E = h\upsilon$, results in

$$\upsilon = \frac{1}{2\pi}\sqrt{\frac{k}{\mu}} \tag{4.10}$$

which is exactly the same result given by the classical physics model of two balls on a spring. This result is general for any quantum change of one vibrational level; however, the $v = 0$ to 1 transition (see Figure 4.11) is most relevant because, as we saw in the fundamentals of spectroscopy chapter, the Boltzmann distribution shows that the large majority of molecules are in the lowest vibration state (i.e., $v = 0$) at typical experimental temperatures.

Note that the frequencies that are predicted by the equations above are easily converted into wavelengths, λ, or wavenumbers, $\bar{\nu}$, via the relationships we have seen in previous chapters, namely,

$$E = h\upsilon = \frac{hc}{\lambda} = hc\bar{\nu} \tag{4.11}$$

where h is Planck's constant and c is the speed of light.

EXAMPLE 4.2

(a) Calculate the expected vibrational frequency of the ^1H—^{35}Cl bond. The force constant for this bond is 481 N/m. Hint: 1 N = kg·m/s². Atomic mass of ^{35}Cl = 34.969 g/mol, ^1H = 1.008 g/mol.

(b) Would you expect the vibrational frequency of D—^{35}Cl (i.e., ^1H—^{35}Cl) to be higher or lower than for ^1H—^{35}Cl? Explain.

(c) Calculate the expected vibrational frequency of D—^{35}Cl using the same force constant as above. Atomic mass of deuterium = 2.014 g/mol. Does the calculation match your prediction?

Answer:

(a) First calculate the reduced mass, keeping in mind that it will be convenient to use units of kilograms because the force constant has kilograms embedded in it. Hint: We also must keep in mind we are dealing with *individual atom* masses, *not moles* of atoms. What constant (i.e., number) relates moles to individuals?

$$\text{Mass of } ^{1}\text{H} = \frac{1.008 \text{ g}}{\text{mol}} \times \frac{1 \text{ mol}}{6.022 \times 10^{23}} \times \frac{1 \text{ kg}}{1000 \text{ g}} = 1.674 \times 10^{-27} \text{ kg}$$

$$\text{Mass of } ^{35}\text{Cl} = \frac{34.969 \text{ g}}{\text{mol}} \times \frac{1 \text{ mol}}{6.022 \times 10^{23}} \times \frac{1 \text{ kg}}{1000 \text{ g}} = 5.807 \times 10^{-26} \text{ kg}$$

$$\mu = \frac{m_1 m_2}{m_1 + m_2} = \left(\frac{1.674 \times 10^{-27} \text{ kg} \times 5.807 \times 10^{-26} \text{ kg}}{1.674 \times 10^{-27} \text{ kg} + 5.807 \times 10^{-26} \text{ kg}} \right) = 1.627 \times 10^{-27} \text{ kg}$$

Then use the equation based on Hooke's law while continuing to pay attention to units and their cancelation:

$$\upsilon = \frac{1}{2\pi} \sqrt{\frac{k}{\mu}} = \frac{1}{2\pi} \sqrt{\frac{481 \text{ N/m}}{1.627 \times 10^{-27} \text{ kg}}}$$

$$= \frac{1}{2\pi} \sqrt{\frac{481 \text{ kg} \cdot \text{m/s}^2 \text{m}}{1.627 \times 10^{-27} \text{ kg}}} = 8.65 \times 10^{13} \text{ s}^{-1} \text{ (or Hz)}$$

(b) Because deuterium is twice as heavy as ^{1}H due to the presence of a neutron in the nucleus, we expect a larger reduced mass. A larger reduced mass is expected to decrease the frequency of the vibration:

(c) $$\mu = \frac{m_1 m_2}{m_1 + m_2} = \left(\frac{3.344 \times 10^{-27} \text{ kg} \times 5.807 \times 10^{-26} \text{ kg}}{3.344 \times 10^{-27} \text{ kg} + 5.807 \times 10^{-26} \text{ kg}} \right) = 3.162 \times 10^{-27} \text{ kg}$$

$$\upsilon = \frac{1}{2\pi} \sqrt{\frac{k}{\mu}} = \frac{1}{2\pi} \sqrt{\frac{481 \text{ N/m}}{3.162 \times 10^{-27} \text{ kg}}} = \frac{1}{2\pi} \sqrt{\frac{481 \text{ kg} \cdot \text{m/s}^2 \text{m}}{3.162 \times 10^{-27} \text{ kg}}} = 6.21 \times 10^{13} \text{ s}^{-1} \text{ (or Hz)}$$

Compared to $n = 8.65 \times 10^{13}$ s^{-1} for ^{1}H—^{35}Cl, we do see a decreased frequency of 6.21×10^{13} s^{-1} for D—^{35}Cl.

Another question:

Convert the frequencies calculated above into wavenumbers. The values you calculated are where you would predict a peak to appear in an IR spectrum.

Answers:

2885 cm^{-1} for ^{1}H—^{35}Cl

2071 cm^{-1} for D—^{35}Cl

While Eq. (4.10) is used to predict the frequency (or wavenumber) of the electromagnetic radiation absorbed by a given bond vibration, the chemical environment of a particular bond influences the actual frequency of radiation that it absorbs. This occurs because individual bond vibrations are not truly isolated from the rest of the molecule and the vibration depends on the atoms bonded to the group of interest. For example, the carbonyl stretch from $C=O$ bonds appears at 1640–1690 cm^{-1} in amides, 1720–1740 cm^{-1} in aldehydes, 1705–1725 cm^{-1} in acyclic ketones, and 1735–1750 cm^{-1} in esters [14]. They are all clustered in a general range from 1640 to 1750 cm^{-1}, but the exact location depends on what, specifically, is bonded to the carbonyl group. In this way, the exact location of a peak in the IR spectrum provides additional information about other groups bonded to the carbonyl moiety (e.g., $-H$ vs. $-O-$, vs. $-CH_2-$, etc.).

4.1.3.3. Anharmonic Oscillators. Like all models, the quantum model of a harmonic oscillator is an approximation and does not entirely account for the actual behavior of molecules. Specifically, the harmonic oscillator model breaks down for the following reasons:

1. Bonds in molecules can break, meaning that the atoms can move far enough apart that they overcome the restoring force that tries to keep them together.
2. As nuclei move close to one another, their charges cause repulsions that result in an increase in energy that is steeper than predicted by masses that do not repel one another.
3. According to the quantum mechanical model selection rules, only transitions of $\Delta v = \pm 1$ vibrational levels are allowed, but we frequently observe transitions between more than one vibrational level (i.e., $\Delta v = \pm 2$ or ± 3).
4. In addition to only allowing transitions of $\Delta v = \pm 1$, the harmonic oscillator model predicts that all of these transitions have the same energy difference regardless of the starting vibrational level (e.g., $E_{v=0\,to\,1} = E_{v=7\,to\,8}$; see Figure 4.11). This means that a particular vibrational motion would have only a single frequency associated with it and therefore would produce only one peak in the IR spectrum. However, we know this is not the case.

A more complete model, based on quantum mechanical wave equations, provides better approximations to the behavior of actual molecules by employing an anharmonic oscillator model. The development of this model is not detailed here, but the results are shown in Figure 4.12. We see in this figure that the more complete model shows (1) the possibility for bond dissociation at large interaction distances, (2) a steeper rise in energy at smaller distances related to electrostatic repulsion between nuclei, and (3) a decrease in the difference in energies as the vibrational levels increase.

The decreased energy gap between levels allows for transitions of $\Delta v = \pm 2$ or ± 3 to be observed in spectra. Such transitions are referred to as overtones. Because they are still roughly twice or three times greater in energy than the fundamental $\Delta v = \pm 1$ transition, they occur at frequencies (i.e., energies) that are approximately two or three times greater than the energy of the fundamental vibration. So, for instance, a $C-O$ stretch that appears around 1100 cm^{-1} could have overtones around 2200 and 3300 cm^{-1}. Transitions between

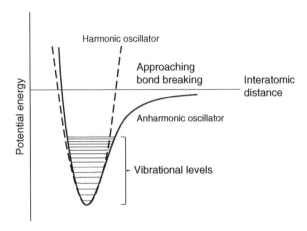

FIGURE 4.12 Depiction of the harmonic (dashed lines) and anharmonic (solid line) oscillator models. Near the potential energy minimum, the harmonic oscillator model approximates anharmonic behavior fairly well. However, when the interatomic distance is very short (far left of the diagram), the potential energy of a bond increases more rapidly than predicted by the harmonic oscillator model because of the repulsion between positively charged nuclei – an effect not present with neutral balls and springs. At long distances between atoms (far right of the diagram), bonds can break, which is also not accounted for in the harmonic oscillator model. Also note that the energy difference between vibrational levels decreases at the higher vibrational levels.

multiple vibrational levels (e.g., $\Delta v = \pm 2$ or ± 3) are "forbidden" according to quantum selection rules. While the word "forbidden" seems to suggest that such transitions cannot occur, they actually can for anharmonic oscillators, but their occurrence is of low probability. The result is that spectral features arising from overtones are weak and difficult to observe, if they are observed at all. Furthermore, some overtones are at frequencies (wavenumbers) not explored in typical IR spectra.

Related to this is one final phenomenon that leads to peaks at unexpected frequencies, and that is the effect known as "coupling." In coupling, the energy levels of multiple vibrations mix together to produce peaks not associated with any single specific bond. The probability of this occurring is greatest when adjacent bonds have similar frequencies. This means that coupling is likely for C–C, C–O, and C–N stretching and C–H wagging and rocking motions in bonds that comprise the skeletal backbone of a molecule [15]. While peaks arising from coupling generally cannot be assigned to any one particular bond, they do contribute to the "fingerprint" pattern for a particular sample and, in this way, provide useful information that can be compared to spectra of known molecules to assist in structure determinations.

4.1.4. The 3N–6 Rule

The number of peaks in an IR spectrum that can be observed for a molecule containing N atoms is given by $3N-6$. So for a molecule with eight atoms, 18 vibrational modes are possible. The purpose of this section is to explain the origin of this well-known rule of thumb.

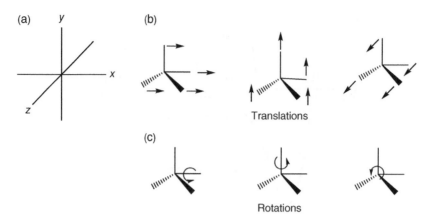

FIGURE 4.13 Translational and rotational motion of molecules on an (a) x-, y-, and z-coordinate system, (b) shows translation of methane on the x-, y-, and z-axes from left to right, and (c) depicts rotations about the x-, y-, and z-axes from left to right.

The $3N - 6$ rule arises from the fact that *locating* an atom in space requires three coordinates (x, y, z). So each atom in a molecule is free to have its own set of three coordinates, giving rise to $3N$ degrees of freedom.

Describing how a molecule *moves* requires considering (1) its translation through space, (2) its rotation, and (3) the movement of the atoms relative to one another (i.e., the vibrations of interest to us in IR spectra). Translation through space requires knowing three coordinates: motion in the x, y, and z directions. Because all of the atoms in a molecule are bonded together in a fixed arrangement, knowing the translation of one atom means knowing the translation of all of the atoms (see Figure 4.13b). As an analogy, think about moving a car from one place on a street to another. Knowing how far the front bumper moved in space also means knowing how far the back bumper (and all of the other parts of the car for that matter) moved through space because of the fixed relationship of all the parts. So only three degrees of freedom are required to specify how a whole molecule moves.

Likewise, three more degrees of freedom are required to specify a rotation of a molecule (see Figure 4.13c), namely, rotation about the x-, y-, and z-axes. So of the total $3N$ degrees of freedom, six are lost in specifying the translation and rotation of a molecule. This leaves $3N-6$ degrees of freedom for vibrational motion and, therefore, the same number of possible peaks in an IR spectrum.

For linear molecules like CO_2, rotation about the long axis does not change the molecule (see Figure 4.14b), so rotational motion requires only 2 degrees of freedom, leaving $3N-5$ degrees of freedom for linear molecules.

The $3N-6$ (or $3N-5$) vibrations are known as normal modes. Each has a frequency associated with it and thus should produce a peak in an IR spectrum. Frequently, however, not all of the peaks predicted by the $3N - 6$ rule actually appear. Reasons for this include:

1. Peaks are too weak to be observed.
2. Violation of the requirement that the dipole moment change as a result of the vibration, arising from the symmetry of the molecule (see Section 4.1.2).

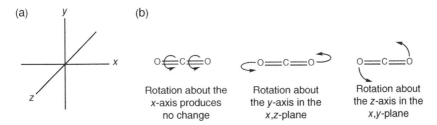

FIGURE 4.14   Rotation of linear molecules like CO_2. (a) Coordinate system. (b) Left: rotation about the long axis of the molecule (x-axis in the figure) produces no change in the molecule. Middle and right: rotation about the other axis changes the orientation of the molecule.

3. Frequencies are above or below the range scanned by the instrument.
4. Overlap of peaks arising from different vibrations but which coincidentally have nearly identical frequencies.
5. Overtone peaks that violate the selection rule of $\Delta v = \pm 1$ can appear in spectra.
6. Combination bands that result when molecules absorb photons with enough energy to excite two vibrations at the same time can also appear.

The first four items in the list above provide reasons that fewer peaks than predicted are present in a spectrum. In contrast, the potential for overtone bands and combination bands (numbers 5 and 6 in the list) can create more peaks than predicted. In the case of overtone bands, peaks at nearly 2 and/or 3 times the fundamental frequency arising from transitions of $\Delta v = \pm 2$ or ± 3 can be observed, although they are generally quite weak. Combination bands can also lead to more peaks than expected. For example, vibrations at 950 and 1050 cm^{-1} could be simultaneously excited by a single photon of 2000 cm^{-1}.

4.2.   FTIR INSTRUMENTS

As we have seen in previous chapters, instruments are designed to measure the fundamental behavior of molecules and thereby gain chemical information. Infrared spectrometers are no different. In this section, we explain how modern IR instruments work and describe the components – sources, detectors, etc. – they contain. We focus our attention initially on the mid-IR region of the spectrum. This region is frequently discussed in other courses, such as organic chemistry, because it provides valuable information regarding the functional groups that are, or are not, present in a sample.

We need to distinguish between two different types of IR instruments: dispersive spectrometers and Fourier transform infrared (FTIR) spectrometers. Dispersive instruments are so named because they disperse the electromagnetic radiation emitted by a source according to its wavelengths, commonly using a grating. In this way, dispersive infrared spectrometers work in the same way as scanning UV-visible spectrophotometers described in Chapter 2. However, dispersive infrared instruments have largely been replaced by FTIR instruments, so we only describe FTIR instruments in detail.

The reason FTIR instruments have replaced dispersive ones in the mid-IR region is that FTIR instruments provide considerably better signal-to-noise ratios than dispersive instruments in the same amount of analysis time or allow much faster analyses with the same S/N. The reasons for this require understanding how FTIR instruments generate spectra, which is the topic of the next section.

4.2.1. The Michelson Interferometer and Fourier Transform

FTIR instruments are based on an optical system known as an interferometer, first designed in 1880 by Albert Abraham Michelson (see Figure 4.15) as part of an experiment to detect an "aether" through which light waves were thought to propagate. The details of the scientific history of the "aether theory" and its ultimate refutation are quite interesting, and we encourage readers to explore it in more depth [16, 17]. However, for the purpose of this book, we only note here that Michelson ultimately was awarded the Nobel Prize in Physics in 1907 for developing the interferometer and using it to accurately measure wavelengths of electromagnetic radiation. He was the first American to win the Nobel Prize (the first prizes were awarded in 1901).

The basic components of an interferometer are shown in Figure 4.16. The source continuously emits electromagnetic radiation (EMR), which is split into two beams by the beam splitter. One beam is reflected at a 90° angle to a stationary mirror (M_1). The other beam continues through the beam splitter and strikes a mirror that can move (M_2). The EMR then reflects off the mirrors and returns to the beam splitter. Part of the EMR reflected by M_1 is reflected by the beam splitter, and the other part continues to the detector. Similarly, EMR from M_2 either continues through the beam splitter back toward the source or is reflected 90° to the detector.

The important aspect for understanding how an interferometer works is understanding how the light from M_1 that goes to the detector compares with the EMR from M_2 that also goes to the detector. Consider what happens when the source emits only one wavelength of EMR (i.e., monochromatic radiation). This, of course, is not ultimately

FIGURE 4.15 Albert Abraham Michelson, used with permission from AstroLab; the original uploader was Bunzil at English Wikipedia (Public domain), via Wikimedia Commons.

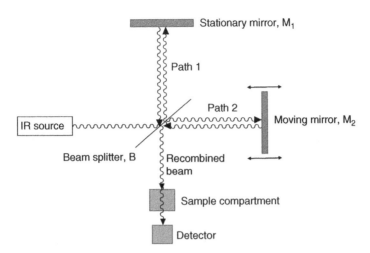

FIGURE 4.16 Bird's-eye view of the main components of an interferometer. Infrared radiation emitted by the source strikes the beam splitter, B. Some of the radiation is transmitted through the splitter, strikes the movable mirror, M_2, and returns to the beam splitter (Path 2). The other part of the radiation is reflected over to the stationary mirror, M_1, and is reflected back to the beam splitter (Path 1). The waves from M_1 and M_2 recombine at the beam splitter and are directed through the sample compartment and into the detector.

useful for obtaining an entire spectrum, and actual sources emit a wide range of infrared radiation, but temporarily adopting this simplification helps to understand how an interferometer works.

When the distances from the beam splitter to M_1 and M_2 are the same, the electromagnetic wave on each path travels the same distance in going from the beam splitter to the mirror and back again. In other words, the distance $BM_1B = BM_2B$. In this case, the two different beams arrived at the beam splitter in phase, meaning their crests and troughs overlap as shown in Figure 4.17.

When this occurs, the waves constructively interfere and appear at the detector as a single wave with the same frequency that was originally emitted by the monochromatic source. It may help to consider an analogy of two people marching side by side, in step with one another. Imagine they start together, split at some point, but march the same distance at the same speed with the same stride length and then return to the point where they split. When they meet, they will still be in step.

The condition in which the distance between the beam splitter and the two mirrors is equal is known as the zero path difference (ZPD) condition.

Now consider what happens when M_2 moves away from the beam splitter. The distance it moves is measured relative to the ZPD condition. The mirror displacement is given the symbol Δ. Now, light that travels along the path from the beam splitter to M_2 must travel an extra distance compared to that on the path to M_1. In fact, the extra distance that it must travel in going from the beam splitter to the mirror and then back again is $2 \times \Delta$. The factor of two arises because of the extra distance *going to* the mirror and then *back* again. The total extra distance is known as the retardation factor because it now takes longer for radiation on Path 2 to reach the detector compared to that on Path 1; hence, it has been retarded. The retardation is given the symbol δ such that

$$\delta = 2\Delta \tag{4.12}$$

Consider the special case in which Δ is equal to half the wavelength of the monochromatic radiation coming from the source. The total extra distance traveled on Path 2 is thus equal to one full wavelength, meaning that when the beam arrives back at the beam splitter, its crests meet the crests from the light on Path 1. In other words, the two beams are in phase and constructively interfere (see Figure 4.17). So constructive interference occurs when $\delta = \lambda$ (i.e., when $\Delta = \frac{1}{2}\lambda$).

Again, the analogy to marchers may help. In this special circumstance, the extra path length amounts to one additional step away from the beam splitter (i.e., $\frac{1}{2}\lambda$) and one additional step toward it (i.e., an extra "left/right" sequence). When the person on Path 2 rejoins the person on Path 1, they will be in step.

In fact, the result is generalizable: Constructive interference occurs whenever

$$\delta = n\lambda \quad n = 0,1,2,3,\ldots \tag{4.13}$$

where λ is the wavelength of electromagnetic radiation coming from the source.

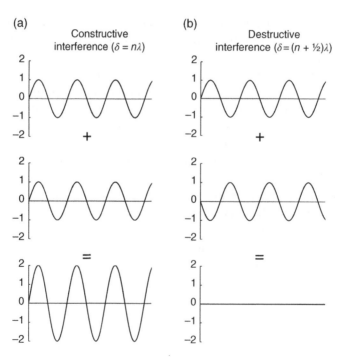

FIGURE 4.17 Constructive and total destructive interference of waves. Constructive interference occurs when waves have the same frequency and are in phase with each other, resulting in a single wave of the same frequency but of greater amplitude. Total destructive interference occurs when waves of the same frequency are 180° out of phase with one another, resulting in the complete cancelation of the waves.

Total destructive interference (see Figure 4.17) occurs when $\delta = (n + \frac{1}{2})\lambda$. In this case, rather than traveling an extra wavelength or multiple thereof, the light travels a total of an extra $\frac{1}{2}\lambda$ when $n = 0$ (i.e., an extra $\Delta = \frac{1}{4}\lambda$ and $\delta = \frac{1}{2}\lambda$). As a result, troughs on Path 2 meet crests on Path 1 and cause complete cancelation of the waves. In the marching analogy, one person is planting their left foot, while the other plants their right foot. Partial interference occurs in between the two conditions of $\delta = n\lambda$ (perfect constructive interference) and $\delta = (n + \frac{1}{2})\lambda$ (total destructive interference).

Because the wavelengths that are typically examined in the mid-IR region extend from 2.5 to 25 μm (4000–400 cm^{-1}), a mirror displacement of just a few centimeters means that thousands of oscillations for all wavelengths will be observed at the detector. A plot of the intensity, or power, of electromagnetic radiation as a function of the retardation is called an interferogram. In the case of monochromatic radiation, the plot looks like that in Figure 4.18.

There are several important points to stress here before considering what happens when multiple wavelengths are present, as occurs in practice when measuring spectra:

1. An interferogram results from the fact that one beam of EMR travels an extra distance compared to a second beam.
2. Constructive and destructive interference occur as a function of the mirror displacement and also depend on the wavelength (frequency) of the EMR.
3. The frequency of the signal observed at the detector is <u>not</u> the frequency of the original electromagnetic radiation, but rather depends on the speed with which the mirror is moved.

Point 3 requires more attention as it underpins interferometry. The frequency of the electromagnetic oscillation of EMR in the IR region is roughly 5.4×10^{13} oscillations per second (a heck of a lot of oscillations every second). Just like UV-visible radiation, these oscillations are far too fast to be tracked by any instrument. But the oscillation intensity shown in Figure 4.18 that results from the displacement of a mirror can, in fact, be tracked by IR-sensitive detectors. To understand this, simply consider moving the mirror backward very slowly. The more slowly the mirror is moved, the more slowly the

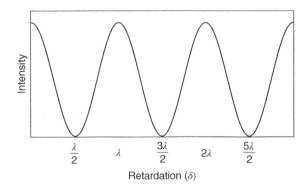

FIGURE 4.18 The interferogram that results from monochromatic radiation passing through an interferometer.

conditions of $\delta = n\lambda$ and $\delta = (n + \frac{1}{2})\lambda$ are experienced. By controlling the speed of the mirror, the oscillations in intensity that are observed at the detector occur much more slowly (i.e., lower frequency) than the oscillations of the electromagnetic fields of the radiation itself. In other words, the frequency of the oscillations at the detector is dependent on the mirror speed, which we can control.

To be more specific, let the velocity of the mirror be v_M and let the time, t, be the time it takes to move the mirror a distance of $\frac{1}{2}\lambda$ – the displacement required for $\delta = \lambda$ (i.e., a full oscillation) – such that

$$v_M t = \frac{1}{2}\lambda \tag{4.14}$$

Given this, the time required for one full oscillation (i.e., constructive to destructive and back to constructive interference) to be observed at the detector is

$$t = \frac{\lambda}{2v_M} \tag{4.15}$$

If this is the time for one cycle (i.e., units of time/cycle), then the frequency of the signal, f_{signal}, is given by the reciprocal (i.e., cycles per unit time).

$$f_{signal} = \frac{1}{t} = \frac{2v_M}{\lambda} \tag{4.16}$$

To relate the frequency of the observed signal (f) to the frequency of the electromagnetic radiation, v_{IR}, we can use the relationship $c/v_{IR} = \lambda$ such that

$$f_{signal} = \left(\frac{2v_M}{c}\right)v_{IR} \tag{4.17}$$

where c is the speed of light.

Notice that the speed of light is in the denominator. The speed of light is a very large number, and dividing by it usually results in a very small number. In this case, we can plug in a mirror velocity of 2 cm/s, which is a typical mirror velocity in commercial instruments, and obtain $f_{signal} \approx 1.3 \times 10^{-10} v_{IR}$. This shows that the frequency of the signal is 10^{-10} times *slower* than the frequency of the IR radiation being used. Another way to look at it is that the fast oscillation of the IR EMR has been modulated down *ten orders of magnitude* to a frequency that is possible to measure in real time.

EXAMPLE 4.3

If a mirror is moved at a speed of 0.100 cm/s, and 1250 cm^{-1} monochromatic radiation (approximately that absorbed by C–N stretching vibrations) is passing through it, what is the frequency of the signal observed by the detector?

Answer:

Recall $E = h\upsilon = hc\bar{v}$, so

$$f_{signal} = \left(\frac{2v_M}{c}\right)\upsilon_{EMR} = \left(\frac{2v_M}{c}\right)c\bar{v}_{EMR} = 2v_M\bar{v}_{EMR} = 2\left(\frac{0.100\text{ cm}}{s}\right)\left(\frac{1250}{cm}\right) = 250\text{ s}^{-1} = 250\text{ Hz}$$

Another question:
   (a) What is the frequency (in Hz) of photons that have wavenumber = 3000 cm^{-1}?
   (b) At what speed would a mirror have to move to modulate this frequency of electromagnetic radiation (typical of EMR absorbed by C–H bond vibrations) down to 9000 Hz?

Answers:
   (a) 8.994×10^{13} s^{-1}
   (b) 1.5 cm/s

A little foreshadowing is in order here. The signal ultimately measured by the instrument is the power of the radiation vs. *time* (i.e., time domain data). In FTIR, the time domain signal is analyzed using Fourier transform (FT) methods to calculate the intensity of each frequency present in order to produce a spectrum.

4.2.1.1. Interferometry with Multiple Frequencies. Sources in FTIR instruments emit radiation with frequencies spanning the region of interest in typical mid-IR studies (1.2×10^{14}–1.2×10^{13} s^{-1} corresponding to 4000–400 cm^{-1}). To understand how an interferogram is produced when an entire range of frequencies is present, we first look at the interferogram that results when just two wavelengths are present. We will let the wavelengths be λ and 4λ (e.g., in the IR region, this could correspond to frequencies of 8.60×10^{13} and 2.15×10^{13} s^{-1}). The interferograms that would be observed for λ and 4λ individually are shown in Figure 4.19, top and middle, respectively. When both wavelengths are present simultaneously, the detector simply detects the total signal (modulated down in frequency as described above) or the sum of all of the individual components. This is shown at the bottom of Figure 4.19. One can imagine that as more wavelengths are added, the total signal becomes more complex. Such an interferogram for a broadband IR source is shown in Figure 4.20. The big spike in the center coincides with the zero displacement point. When the two mirrors are equidistant from the beam splitter, all of the EMR travels the same distance. Hence, all frequencies experience constructive interference, resulting in the highest observed intensity. As the mirror moves away from or toward the beam splitter, some frequencies begin to experience destructive interference, and the signal diminishes, creating the "wings" on each side of the center burst shown in the figure (see Figure 4.20).

4.2.1.2. Converting Interferograms into Spectra: The Fourier Transform.
The interferogram shown in Figure 4.20 is a plot of signal intensity versus δ, but δ is just a function of time because the mirror moves at a fixed rate. So the plot is really just signal versus time (i.e., time domain data). It is clear that it is difficult to discern any chemical information from such a plot.

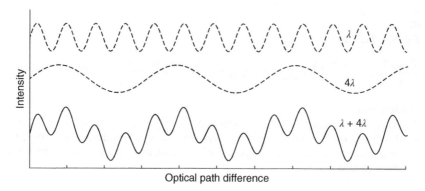

FIGURE 4.19 The superposition of two waves [16]. Top: a wave with wavelength λ. Middle: a wave with wavelength 4λ. Bottom: the single wave that results from adding the top and middle waves together. Source: Adapted and republished with permission of Taylor & Francis Group LLC Books, from Smith [16], permission conveyed through Copyright Clearance Center, Inc.

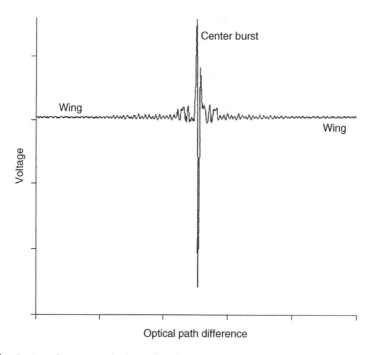

FIGURE 4.20 An interferogram of a broadband source. The center burst occurs at the point of zero path difference (ZPD) [16]. The y-axis is voltage because detectors convert the electromagnetic radiation into an electric signal. Source: Republished with permission of Taylor & Francis Group LLC Books, from Smith [16], permission conveyed through Copyright Clearance Center, Inc.

Fourier transform mathematics is used to convert the data from the time domain into the frequency domain, yielding a plot of power versus frequency. The Fourier transformation process is depicted in Figure 4.21. For historical reasons, it is common to convert frequencies to wavenumbers when plotting infrared spectra. This conversion is based on the relationships $E = h\upsilon = hc\bar{\nu}$, such that $\upsilon = c\bar{\nu}$, showing that frequencies are proportional to

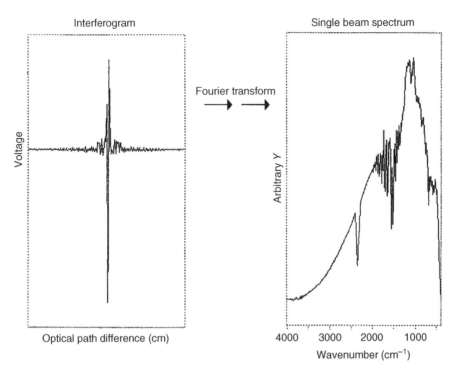

FIGURE 4.21   Depiction of the Fourier transformation of an interferogram (time domain) into an IR spectrum (frequency domain) [16]. Source: Republished with permission of Taylor & Francis Group LLC Books, from Smith [16], permission conveyed through Copyright Clearance Center, Inc.

wavenumbers, with the speed of light being the proportionality constant. It is also common to plot the percent transmittance, rather than absorbance, in IR spectra. As a result, many IR spectra are a plot of percent transmittance versus wavenumber, whereas UV-visible spectra are conventionally plots of absorbance versus wavelength.

Figure 4.22 shows a background spectrum collected in the absence of any sample. It is clear that some specific spectral features are present, and there is considerable variation in intensity across the spectrum. When collecting a background or blank spectrum, CO_2 and water vapor are always present unless the sample compartment is purged with nitrogen or another gas. The prominent peaks seen in Figure 4.22 around 2350 and 667 cm^{-1} are due to CO_2, and the spectral features around 3500 and 1630 cm^{-1} arise from water.

After collecting the blank signal, the interferogram of the sample is recorded. As an example, a spectrum based on an interferogram collected with a film of polystyrene (see Figure 4.23 inset) as the sample is shown in Figure 4.23. Because polystyrene has a number of IR-active vibrations, a number of spectral features appear compared to those in the background spectrum, and the CO_2 and water features can also be observed.

By collecting both a background spectrum (P_0 vs. wavenumber) and a sample spectrum (P vs. wavenumber), the transmittance at each wavenumber can be calculated ($T = P/P_0$). The spectrum for polystyrene after taking the ratio is shown in Figure 4.24. Notice that the CO_2 and water features have ratioed out, leaving just the features arising from the bond vibrations present in polystyrene.

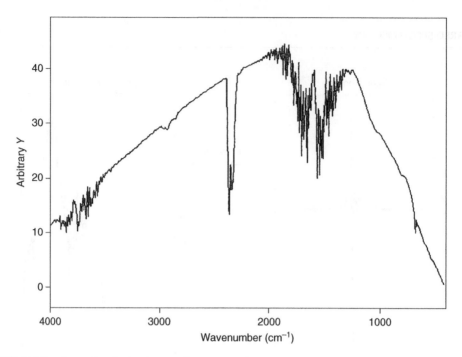

FIGURE 4.22 Example of a background spectrum that ultimately gets used as P_o to determine the absorbance created by a sample ($A = \log P_o/P$) [16]. Signals in the blank mainly arise from CO_2 and water. Source: Republished with permission of Taylor & Francis Group LLC Books, from Smith [16], permission conveyed through Copyright Clearance Center, Inc.

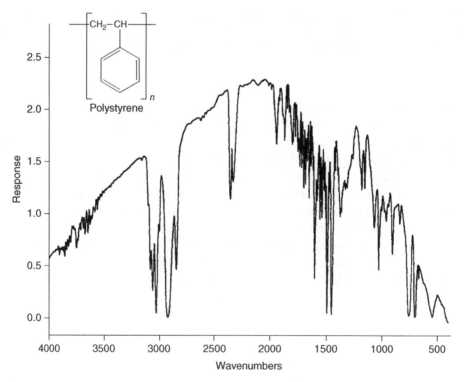

FIGURE 4.23 Spectrum with a film of polystyrene in the sample compartment [16]. The ratio of this spectrum to the background spectrum has not yet been taken in this image, but if you visually compare the two, you will see that several signals that are not in the background are present here. Source: Republished with permission of Taylor & Francis Group LLC Books, from Smith [16], permission conveyed through Copyright Clearance Center, Inc.

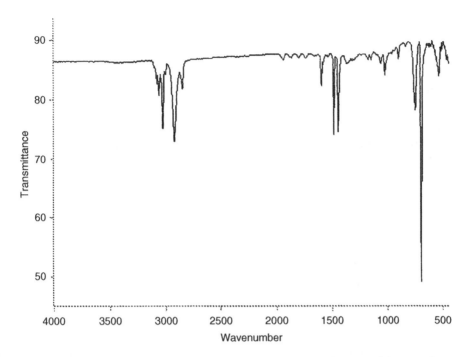

FIGURE 4.24 A spectrum of polystyrene [16] obtained by taking the ratio of the sample spectrum (*P*) to the background spectrum (*P$_o$*), which were shown in the previous figures, to obtain the transmittance of the polystyrene film. Source: Republished with permission of Taylor & Francis Group LLC Books, from Smith [16], permission conveyed through Copyright Clearance Center, Inc.

Thus, just as in UV-visible spectroscopy, two measurements – a blank and a sample – are required to create a single spectrum. It is possible to present IR spectra as absorbance (*A* = –log*T*) vs. wavelength as is done with UV-visible spectra, but for historical reasons, this is not typically done in the mid-IR region.

4.2.1.3. Advantages of FTIR over Dispersive Instruments: The Fellgett Advantage.

Given that traditional dispersive instruments were in use well before FTIR instruments were developed, designing an entirely different optical system that also required complex mathematical transformation of the resulting data may seem unnecessarily complicated. It is therefore important to point out that FTIR instruments offer significant advantages over dispersive ones. There are two primary advantages – known as the Fellgett (or multiplex) advantage [18] and the Jacquinot (or throughput) advantage [19] – named after those who elucidated these advantages.

The Fellgett advantage of FTIR derives from the fact that an entire spectrum is collected in just a few seconds. In a scanning (dispersive) instrument, it can take several minutes to collect a spectrum. The speed of FTIR instruments derives from the fact that the mirror is displaced only a short distance – on the scale of a few centimeters – which can be done quite rapidly. Because spectra only take a few seconds to collect, dozens or hundreds of them can be collected and averaged together in the time it would take a dispersive instrument to measure a single spectrum. The ability to collect, and then average, multiple spectra is significant because the signal to noise ratio (S/N) improves with the square root of the number of measurements. Therefore, a spectrum that is the result of averaging 100 scans has an S/N ratio that is 10 times (i.e., $\sqrt{100}$) better than a spectrum from a single scan. This is critical in IR spectroscopy because many spectral features are weak, and improved S/N ratios help discern chemical information from noise. It is possible to add several hundred spectra together to achieve even greater improvements in the S/N ratio. However, at very large numbers of scans (e.g., hundreds to thousands), various effects can ultimately limit the improvement in S/N. These limitations, coupled with the greatly increased total analysis time required to collect hundreds or thousands of spectra, ultimately limit the desirability of collecting large numbers of spectra on a single sample. For many samples, 16–100 spectra are sufficient.

It is important to not only understand that there *is* a significant improvement in signal-to-noise ratios with FTIR instruments over dispersive ones but also to understand *why*. It fundamentally relates to the *time* spent measuring the signal at each wavelength (i.e., wavenumber) [16]. Ultimately, the more time, t, spent measuring the noise, the more likely it averages out, leading to higher S/N ratios – more specifically

$$\frac{S}{N} \propto \sqrt{t} \qquad\qquad (4.18)$$

Consider the typical IR spectrum that extends from 4000 to 400 cm^{-1}. If the resolution is set at 4 cm^{-1}, which is typical for IR spectra, then 900 data points ([4000 – 400 cm^{-1}]/4 cm^{-1}) are recorded. Suppose it takes two minutes to record these data points with a dispersive instrument. This means each point is observed for an average of 0.133 seconds.

With an FTIR instrument, essentially all of the data points are recorded all of the time because the source EMR is not separated according to wavelength (this statement is admittedly a bit of an oversimplification as not all wavelengths can be recorded simultaneously) [20]. So to a first approximation, we can say that each data point is observed for a full two minutes, giving an average of 120 seconds/point.

Based on Eq. (4.18), the advantage of an FT instrument over a dispersive one is calculated as

$$\left(\frac{120 \text{ seconds}}{0.133 \text{ seconds}}\right)^{1/2} = 30 \qquad\qquad (4.19)$$

indicating that for the same resolution and the same analysis time, the FTIR instrument offers a 30-fold improvement in the signal to noise ratio [16].

4.2.1.4. Advantages of FTIR over Dispersive Instruments: The Jacquinot Advantage.
The basis for the throughput advantage (i.e., the Jacquinot advantage) of FTIR over dispersive instruments derives from the fact that no slits are present in FTIR instruments. As we saw in the description of UV-visible scanning instruments, slits are used to select a narrow range of wavelengths to pass through the sample. In doing so, they limit the power of the EMR reaching the detector, thus ultimately limiting the signal. Additionally, higher resolution requires narrower slits, further limiting the fraction of EMR emitted by the source that is detected.

Because FTIR instruments are based on a fundamentally different method of discriminating the wavelengths of light – using interferometry as opposed to dispersion – all of the wavelengths emitted by the source are continuously passed through the instrument, instead of just a fraction as occurs in dispersive instruments. The greater intensity of electromagnetic radiation, along with the multiplex advantage described above, serves to significantly improve the S/N ratios obtained with FTIR compared to dispersive instruments.

4.2.1.5. Resolution. In a dispersive instrument, resolution is determined by the slit width, as described in Chapter 2. In FTIR instruments, the resolution is dictated by the retardation, δ, which is ultimately related to the maximum mirror displacement away from the zero path difference point. Defining resolution as the difference between the two nearest wavenumbers ($\bar{v}_1$ and $\bar{v}_2$) that can be separated by an instrument ultimately yields the result that

$$R = \Delta\bar{v} = \bar{v}_1 - \bar{v}_2 = \frac{1}{\delta} \tag{4.20}$$

The last relationship in Eq. (4.20) shows that larger mirror displacements (bigger δ) result in smaller differences between wavenumbers that can be distinguished. Instruments with very small values of $\Delta\bar{v}$ are required to resolve fine spectral details that nearly overlap and therefore require greater mirror displacement.

EXAMPLE 4.4

Commercial instruments can obtain a resolution of 0.5 cm⁻¹. What is the maximum mirror displacement (i.e., Δ – see Eq. 4.12) required to achieve this resolution?

Answer:

$\frac{1}{\delta} = R = 0.5 \text{ cm}^{-1}$

$\delta = 2 \text{ cm}$

$\delta = 2\Delta = 2 \text{ cm}$

$\Delta = 1 \text{ cm}$

Another question:
    (a) What mirror displacement is required for a resolution of 4 cm⁻¹?
    (b) If a mirror moves at 0.1 cm/s, how long will this displacement take?

Answers:
    (a) 0.125 cm
    (b) 1.25 seconds

Experimentally, choosing a higher resolution setting means that the mirror must be displaced a greater distance for each spectrum collected. Given a constant mirror velocity, this means that higher resolution spectra take longer to collect. You can experience this on commercial instruments by recording the same number of spectra, for example, 64 replicates, at two different resolution settings and timing how long it takes to collect the entire data set.

While resolution is limited in fundamentally different ways in FTIR and dispersive instruments (i.e., mirror displacement versus slit widths), the tradeoff between resolution and S/N is the same. Higher resolution spectra are inherently noisier than lower resolution spectra. There are a number of factors that contribute to this tradeoff when using interferometers. The practical consequence, however, is that for each sample analyzed, we must consider the fact that high resolution spectra may contain more chemical information by discriminating nearly overlapping spectral features, but they will be noisier than low resolution spectra. So the appropriate resolution setting depends on the sample and the type of information sought. A resolution of 4 cm^{-1} is common for spectra of typical organic compounds.

4.2.1.6. *Note About FT UV-Visible Spectroscopy.*

Before moving on to discuss the requirements and components for FTIR instruments, followed by practical considerations for using IR spectroscopy, we pause here to answer a question that may have arisen; namely, if FTIR instruments have so many advantages over dispersive ones in terms of signal-to-noise ratios and throughput, then why are so many UV-visible instruments dispersive (i.e., scanning) rather than FTUV-vis instruments?

FTUV-vis instruments have indeed been built, but they do not show the same gains in performance as do instruments operating in the IR region. The reason for this relates to the fact that IR instruments use entirely different detectors and sources than UV-vis instruments. IR radiation is not strong enough to cause ejection of an electron from a photoemissive surface as happens in the photomultiplier tubes used to detect the higher energy UV-vis EMR. Instead, IR detectors respond to thermal fluctuations caused by variations in the intensity of IR radiation striking them. As such, IR detectors are much more prone to thermal noise (e.g., temperature variations in the environment of the detector unrelated to the incoming IR radiation). In contrast, UV-visible detectors are shot noise limited and inherently less noisy than IR detectors. Because of this difference, the advantages of signal averaging that are made possible by an interferometer system are not nearly as significant in the UV-visible region of the spectrum as they are in the IR region [21, 22].

4.2.1.7. *Interferometer Requirements.*

Several conditions are required to achieve the benefits associated with interferometry. First, the mirror must move at a constant velocity to ensure the accurate measurement of the retardation and, ultimately, the frequencies (wavenumbers) of the electromagnetic radiation. Additionally, the face of the mirror must remain at 90° to the incoming source radiation. Imagine what happens if the mirror is tilted relative to the incoming radiation. In this case, light striking the top of the mirror travels a different distance than light striking the bottom, so different wavelengths of radiation experience constructive interference for the same mirror displacement, degrading the resolution. These two considerations (mirror displacement and planarity) are established by the quality of the instrument and are generally out of the control of the analyst.

Additional requirements for FTIR instruments are introduced because of the desirability of averaging multiple spectra and because of Fourier transformation of the raw data. Regarding the requirements imposed by the averaging of spectra, imagine what would result if each spectrum is slightly shifted from the others in terms of the wavenumbers at which spectral features occur, as depicted in Figure 4.25. In this case, averaging all of the spectra actually decreases the signal and degrades the resolution. To counter this, the instrument must have a way of "knowing" when the collection of each new spectrum begins. This is achieved by using the highest point in the interferogram. This point coincides with the zero retardation point because only when the two paths are equal do *all* wavelengths constructively interfere, producing the greatest signal. Using this point allows all spectra to be aligned in terms of wavenumbers to preserve signal intensity and resolution during the averaging process.

In addition to knowing where the zero point displacement occurs, the timing of the digitization of the data must be highly reproducible and in sync with the mirror movement from one spectrum to the next. If not, averaging multiple spectra again leads to a loss of signal intensity. To achieve reproducible digitization, FTIR instruments contain a laser source in addition to the polychromatic IR source. The monochromatic radiation from the laser also goes through the interferometer and is detected by a sensor that is different than the main signal detector. As we saw earlier in this chapter, monochromatic radiation creates a signal with a single frequency, as shown in Figure 4.18. This signal is converted electronically to a square wave, and the peaks and valleys are used to trigger the sampling of data (i.e., the instrument records a data point from the *sample detector* every time a maximum signal in the *laser detector* is observed). Because the frequency of the laser output is highly stable and because the oscillating frequency observed as a result is tied to

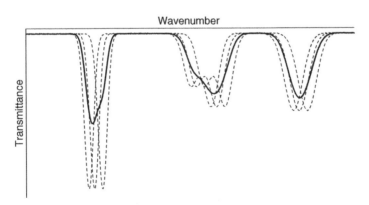

FIGURE 4.25 The effect of small differences in the starting time of each spectrum. Here, three simulated spectra (dashes) are slightly offset from one another. The solid line shows the average of the three spectra. Notice that in this case, because the spectral features do not perfectly overlap, taking the average distorts and diminishes the actual signal. This emphasizes the importance of having a mechanism for the instrument to start the collection of a new spectrum at exactly the same point in the mirror displacement. Note that the shifts in the spectra have been exaggerated so that the effect is easily seen.

the movement of the mirror, such a system allows for highly reproducible digitization of the signal arising from the main interferometer.

4.2.2. Components of FTIR Instruments: Sources

The main components of interest in FTIR instruments are the sources and detectors as these dictate the wavenumber range that can be examined, the intensity of the signal, and the sensitivity of the measurements.

There are three main sources of IR radiation: (1) Nernst glowers, (2) coiled metal wires, and (3) globars.

4.2.2.1. Nernst Glower. Nernst glowers are rods composed of rare earth oxides such as zirconium, yttrium, and thorium oxides that have wires sealed into their ends [12]. Current is passed through a rod via the wires, causing the rod to heat up. As it does, it emits infrared radiation in a manner that is similar to that of blackbody radiation. The rod is maintained at a specified temperature (usually 1200–2200 K). Examples of the output of two Nernst glowers are shown in Figure 4.26.

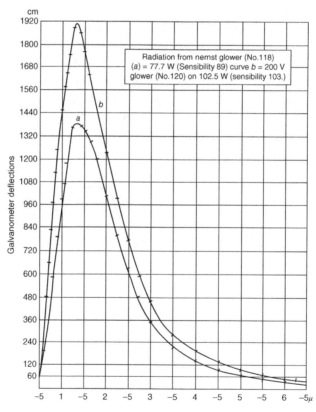

FIGURE 4.26 Spectral output of two Nernst glowers operated under different conditions at approximately 2100 °C. (a) 77.7 W (sensibility 89). (b) 102.5 W (sensibility 108). These curves approximate that of blackbody radiation at a comparable temperature. Source: Reprinted from Coblentz [23].

4.2.2.2. Wire Sources. Like Nernst glowers, coils of metals such as rhodium or nichrome emit IR radiation when heated by having a current passed through them. This is similar in operation to the visible light emitted by incandescent light bulbs containing a tungsten wire. They are generally operated around 1100 K.

Incandescent sources and Nernst glowers are air cooled, in contrast to the globars discussed below that require circulating water to carry away excess heat. They are also generally cheaper and easier to operate than globars, but may not provide sufficient intensity for certain applications. However, some commercial instruments include an optical design that uses a reflector to reflect emitted radiation that would otherwise be transmitted away from the sample. Such systems can produce power outputs that are similar to that of globars without the need for water cooling.

4.2.2.3. Globars. As hinted at above, globars, which are rods of silicon carbide (SiC), emit a higher intensity of IR radiation than Nernst glowers at shorter wavelengths (see Figure 4.27) [25]. They are also operated at higher temperatures to increase their output. Older globar sources required water cooling to keep from overheating. Recent advances in ceramic metal alloys, however, have led to sources that do not need to be water cooled. Globars are also more expensive than air-cooled sources.

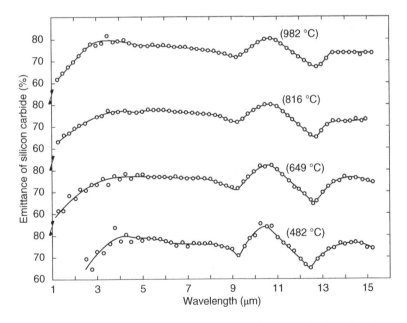

FIGURE 4.27 Globar output at four different temperatures. Emittance is the ratio of the output of the source relative to that of a blackbody. In this plot, the ratio is converted to a percent. So in this plot, for example, silicon carbide emits 80% of the radiation at 3 μm and 982 °C that a blackbody would emit. Source: Reprinted from Stewart and Richmond [24].

4.2.3. Components of FTIR Instruments: DTGS and MCT Detectors

As mentioned earlier, detectors for IR radiation are necessarily different than for UV-visible radiation because IR radiation is not energetic enough to cause the ejection of electrons from photoemissive materials. Therefore, photomultiplier tubes cannot be used in IR instruments. Instead, two different detectors are commonly employed: deuterated triglycine sulfate (DTGS) and mercury cadmium telluride (MCT) detectors.

The DTGS detector – a pyroelectric bolometer – is most commonly used. "Bolometer" is a general term for a device that measures small amounts of radiant heat [26]. The word has its roots in the Greek words bolē, meaning something thrown, and ballein, meaning to throw, both referring to the rays of radiation as the thing being "thrown," and in the Greek word "metron," meaning "measure."

The term "pyroelectric" refers to the phenomenon of a crystalline substance becoming electrically polar (i.e., developing a voltage across the crystal) as a result of heating. The root "pyro" refers to fire or heat, and the word "electricity" has its root in the Greek word for amber (ἤλεκτρον or ēlektron), which develops a charge when rubbed with silk.

Consistent with these definitions, DTGS crystals are in fact sensitive to temperature changes. The formula for triglycine sulfate is $(NH_2CH_2COOH)_3 \cdot H_2SO_4$, and the deuterated version replaces all of the hydrogen atoms with deuterium atoms. By applying an initial electric field across a DTGS crystal, it becomes polarized and remains so after the field is removed. The polarization (i.e., voltage) across the crystal surface varies with the temperature of the crystal, decreasing in magnitude as the temperature increases. The change in the polarization results from changes in the orientation of the molecules within the crystal as it warms and cools. A circuit is used to measure the voltage across the crystal. The temperature of the crystal depends on the power of IR radiation striking it. Recall that as the interferometer moves back and forth, different wavelengths of light constructively and destructively interfere and then pass through the sample. When the sample absorbs specific wavelengths, less radiant energy reaches the crystal, and it cools relative to times when more radiation is striking it. As a result, the crystal gets more polarized and the voltage across it increases. When more radiation passes through the sample, the crystal warms up, and its polarization decreases. The changes in the potential across the crystal are recorded by a circuit and ultimately converted into IR spectra.

The non-deuterated form (i.e., TGS) loses its polarization above 39 °C (~102 °F), meaning it no longer functions as a detector. By using the deuterated analog, this temperature is raised to 49 °C. Furthermore, DTGS crystals are sometimes doped with L-alanine to improve their performance characteristics [27]. In this case, the abbreviation DLaTGS is used.

The advantages of DTGS and DLaTGS detectors include the fact that they can be operated at room temperature (although some commercial instruments provide Peltier cooling), their response is essentially flat across the wavelengths in the near-IR and far-IR, and they can respond to variations in signals that occur at several thousand Hz. However, this speed does limit the speed with which the interferometer can be moved and, consequently, the time required to collect a spectrum.

The main disadvantage of DTGS and DLaTGS detectors is that they are not as sensitive as the other common detector, mercury cadmium telluride (HgCdTe), or MCT [16]. MCT detectors are metal alloys that are semiconductors. When they absorb photons, the electrons are promoted from the valence (i.e., nonconducting) band to the conduction band. Electrons in the conduction band respond to an externally applied voltage, producing an electric current, which is measured to provide the signal. The current is directly proportional to the number of IR photons striking the detector.

The main advantage of MCT detectors is their sensitivity, being ten times more sensitive than DTGS detectors. They are therefore frequently found in applications in which the signal is expected to be weak. MCT detectors also respond faster. Therefore, spectra can be recorded more quickly, which is useful in applications such as following reactions with fast kinetics or when used as a detector for gas chromatography [16].

Regarding disadvantages, MCT detectors that operate from 4000 down to 400 cm^{-1} are five to ten times noisier than those that operate in a more limited range of 4000–700 cm^{-1}. Increased noise for such detectors nearly offsets the increased signal as far as signal-to-noise ratios are concerned [16]. Also, MCT detectors must be cooled with liquid nitrogen, adding expense and hassle to operating them relative to DTGS detectors. Lastly, MCT detectors can saturate more easily than DTGS detectors, meaning that their response becomes nonlinear when the IR radiation striking it is too intense. This sometimes requires attenuating the IR source radiation.

4.2.4. Sample Handling

IR instruments contain a sample compartment that allows the IR electromagnetic radiation coming from the interferometer to pass through the sample (see Figure 4.16). Depending on the nature of the sample (i.e., solid, liquid, or gas), it is mounted in the sample compartment in one of several ways, which are described below. Sample compartments are designed to accommodate a wide variety of interchangeable sample holders, ranging from large (10 cm) cells for gas samples down to very thin (μm) solid films.

In UV-visible spectroscopy, the sample is often dissolved in a solvent that does not absorb UV-visible radiation, and a 1 cm cuvette is used as the sample holder. Unfortunately, this cannot be done in the IR region because most solvents absorb IR radiation and would therefore dominate the spectrum and swamp out signals from the analyte. For this and other reasons, sample preparation and sample holders used in FTIR instruments are different than those used in UV-visible spectrophotometers. In this section, we examine sample holders and sample preparation techniques used for analyzing gases, liquids, and solids by FTIR.

4.2.4.1. Gases. Gases, by definition, are dilute (few molecules per unit volume). Because IR spectroscopy follows Beer's law ($A = \varepsilon bc$), the low concentration of molecules in the gas phase can lead to low absorbance values. Sample cells with long path lengths (i.e., large b) are used to offset the low concentrations. A 10 cm cell for gas analysis is

FIGURE 4.28   A 10 cm gas cell for FTIR analysis. Input and outlet spouts with rotating valves allow gases to be pumped into the cell. Once the cell is full, the valves are close and a spectrum recorded. The gas can then be displaced by another gas. Continuous monitoring of the gas is also possible by continuing to pump the gas of interest through the cell and recording spectra as a function of time. Source: Reproduced with permission of Specac, Ltd.

shown in Figure 4.28. Path lengths up to several meters are obtained with cells that have mirrors inside that reflect the light beam back and forth through the sample multiple times before it exits the cell. Gases are pumped into the cell that is sealed at both ends with windows made of materials such as KBr, CaF_2, or ZnSe that are IR transparent. Cells such as that shown in Figure 4.28 are mounted inside the FTIR instrument in the sample compartment and positioned so the infrared radiation emerging from the beam splitter passes through the sample and proceeds to the detector.

4.2.4.2.   Liquids. Liquids are frequently measured in a cell that can be removed from its mounting in the sample compartment. For qualitative analyses where the path length need not be known, a cell such as that shown in Figure 4.29 can be used. A spacer of approximately 0.01 mm is placed on top of a cell window, which is also called a salt plate. A drop of the liquid sample is added, and a second salt plate is placed on top of the spacer to sandwich the sample between the salt plates. In some instances, a spacer is not even used, and the liquid is just spread out in a thin film between the salt plates. This assembly is then clamped in the sample holder and mounted in the sample compartment.

The salt plates are an important consideration here. They must be transparent to IR radiation. The data provided in Table 4.1 show the approximate wavenumber above which the materials transmit radiation [12]. Below these values, the material absorbs significant amounts of IR radiation. Thus, materials such as glass and quartz cannot be used because they absorb in the region often studied in IR spectroscopy.

Salt plates for routine IR spectroscopy are frequently made of NaCl. KBr is also used. This is an important consideration because these salts dissolve in water. Therefore,

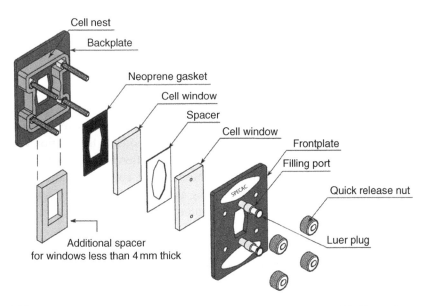

FIGURE 4.29 A demountable liquid sample cell for FTIR spectroscopy [12]. The cell is assembled by placing the gasket and one cell window between the four screws on the backplate. The spacer is then put in place and a drop of the liquid sample added to it. The second cell window is placed on top of the spacer to sandwich the sample between the two windows. Lastly, the final metal housing piece is slipped over the four screws. The screws are gently tightened to hold the entire assembly in place when it is mounted inside the instrument. Source: Reproduced with permission of Specac, Ltd.

TABLE 4.1 Cutoff Wavenumbers for Optical Materials

Material	Wavenumber (cm^{-1})
Glass	3000
Quartz	2500
LiF	1500
ZnS	750
NaCl	600
KCl	500
ZnSe	450
KBr	350
AgCl	350
CsI	250

Source: From Colthup et al. [12]. Reproduced with permission of Academic Press.

aqueous samples cannot be analyzed using them. In fact, proper handling of the plates is critical as even moisture from hands or from hygroscopic solvents can cause the salt plates to cloud up, thus reducing their transparency. Salt plates made from salts that are insoluble in water, such as AgCl, must be used for aqueous samples.

Sample cells with known thickness are also available and used when reproducibility of absorbance values is desired, primarily for quantitative analyses. Such a cell is shown in Figure 4.30. The sample is flushed through the cell via the input port and fills the cell. It is then mounted in the sample compartment and analyzed. Fixed path lengths from 0.01 mm to several millimeters are commercially available.

4.2.4.3. Solids. Multiple methods for analyzing solids exist. These include (1) dissolution, (2) films, (3) mulls, (4) pellets, and (5) reflectance methods. In this section we examine the first four as they do not require any significant modification to the sample compartment and do not represent dramatically different methods of collecting signal. Reflectance methods are conceptually different than passing the beam directly through the sample. Therefore, we discuss them in a separate, dedicated section below.

4.2.4.3.1. Dissolution. Some solid samples can be dissolved in appropriate solvents and analyzed via the liquid methods described above. However, very few solvents are transparent across the entire IR spectrum. Carbon tetrachloride, chloroform, and carbon disulfide are the most generally transparent, but health issues and solubility limitations exist with these solvents, so alternative methods may be more desirable.

4.2.4.3.2. Films. A thin film of a solid sample can be made by dissolving it in a solvent and applying it to a cell window or salt plate, followed by evaporating the solvent. This leaves behind a thin film of the sample, making this method of sample preparation particularly useful for polymer analysis.

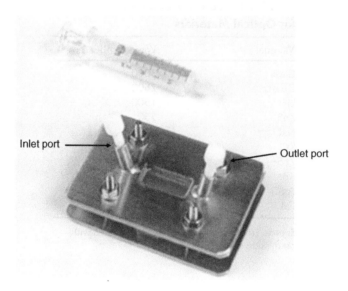

Inlet port ⟶   ⟵ Outlet port

FIGURE 4.30 A fixed path length IR cell [28]. Such cells, unlike those in the previous figure, are not made to be disassembled. They have a spacer of fixed, known thickness that provides the fixed path length. The sample solution is added to the cell by injecting it into the inlet port using a syringe. Once the sample space is full, excess sample comes out from the outlet side. The ports are then plugged and the assembly is mounted in the instrument sample compartment. Source: Courtesy of Shimadzu Corporation, Kyoto, Japan.

Another approach, which is somewhat similar but which does not involve a solvent, is to melt the solid on a salt plate, spread it out to create a thin film, and then cool it. This approach does not work well for crystalline materials, but is applicable to amorphous materials and waxy substances.

4.2.4.3.3. Mulls. Mulls are created by grinding the solid sample into a fine powder that is added to a supporting matrix such as mineral oil. Nujol, a specific brand of mineral oil composed of a mixture of heavy hydrocarbons, is commonly used for this purpose. The powder does not dissolve in the oil, but rather is suspended in it. The resulting viscous material is then spread between salt plates as described above for the analysis of liquids.

4.2.4.3.4. Pellets. Solids are also analyzed by mixing some of the solid sample with a salt, commonly KBr, and using high pressure to squeeze the KBr/sample mixture into a thin, translucent disk, called a pellet. The pellet is then mounted in a sample holder in the sample compartment and its spectrum recorded. KBr is generally transparent in the mid-IR region, which is why it is usually chosen as the salt to mix with the sample. CsI has a cutoff of 200 cm^{-1}, so it is used if spectral information below 400 cm^{-1} (the cutoff for KBr) is required.

4.2.5. Reflectance Techniques

In all of the analysis methods considered so far, IR radiation is passed, or transmitted, *through* the sample. In other words, they are "transmission" techniques. In reflectance techniques – as the name implies – radiation is *reflected off* the surface of the sample. If the sample is perfectly smooth, the angle of incidence equals the angle of reflection (see Figure 4.31), and the technique is known as specular reflectance. When the surface is rough or the sample is a powder, radiation is reflected in any and all directions – a process known as diffuse reflectance (see Figure 4.31).

IR radiation penetrates a short distance (a few microns) into a solid material. Thus, the reflected light contains information about which wavelengths are absorbed by the surface of the sample. Collecting the reflected light, however, requires special accessories designed for this purpose, which adds expense while increasing the flexibility and applicability of IR analyses. The techniques and special accessories are discussed in greater detail in the following sections.

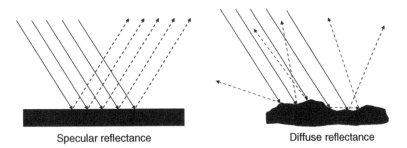

Specular reflectance   Diffuse reflectance

FIGURE 4.31 Reflection of radiation off a perfectly smooth surface (specular reflectance) and a rough surface (diffuse reflectance). Solid lines indicate incoming radiation, and dashed lines represent reflected radiation.

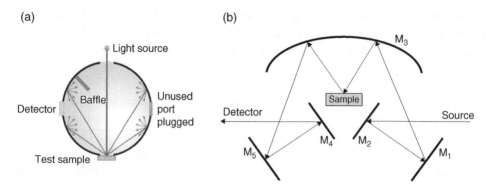

FIGURE 4.32 Examples of an (a) integrating sphere and (b) ellipsoidal mirror [27, 29, 30]. Sources: (a) Image of the integrating sphere created by Cmglee CC BY-SA 3.0 via Wikimedia Commons. (b) Adapted from Griffiths and De Haseth [27]. Reproduced with permission of John Wiley & Sons, Inc.

4.2.5.1. *Diffuse Reflectance (DRIFTS).*

We focus here on diffuse reflectance rather than on specular reflectance because samples are seldom perfectly smooth. Therefore, DRIFTS (*diffuse reflectance infrared Fourier transform spectroscopy*) is more commonly encountered. The challenge with diffuse reflectance is to capture the reflected light. Unlike transmission methods, in which the beam propagates straight through the sample and is therefore easily focused on a detector, in DRIFTS, the radiation is scattered in all directions. A special accessory collects the scattered radiation. The component directs the beam emerging from the interferometer onto the sample and uses a large ellipsoidal mirror or integrating sphere to collect the scattered radiation and direct it onto the detector (see Figure 4.32).

In DRIFTS, solid samples are usually ground into a fine powder and mixed with KBr or KCl to dilute the sample to 1–10%. The intensity of the signal is then recorded. Signal is also recorded with just pure KBr or KCl present as a reference. The ratio, R_∞, known as the reflectance and defined as

$$R_\infty = \frac{R_\infty\,(\text{sample})}{R_\infty\,(\text{reference})} \tag{4.21}$$

is measured.

The reflectance spectrum is converted into a spectrum that resembles an absorbance spectrum by calculating the function, $f(R_\infty)$, defined by Kubelka and Munk [27, 31, 32]:

$$f(R_\infty) = \frac{(1-R_\infty)^2}{2R_\infty} = \frac{k}{s} \tag{4.22}$$

where k is the molar absorption coefficient and s is the scattering coefficient (which varies with particle size and packing density). For dilute samples, the molar absorption coefficient is related to concentration, c, and molar absorptivity, ε, via [33–35]

$$k = 2.303\,\varepsilon c \tag{4.23}$$

Reflectance spectra, then, are ultimately plots of $f(R_\infty)$ vs. wavenumber.

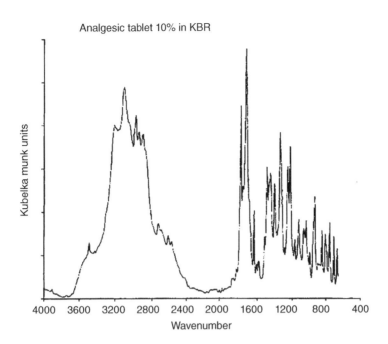

Analgesic tablet 10% in KBR

FIGURE 4.33 A diffuse reflectance spectrum of an analgesic tablet ground up and diluted to 10% with KBr [16]. Source: Republished with permission of Taylor & Francis Group LLC Books, from Smith [16], permission conveyed through Copyright Clearance Center, Inc.

Figure 4.33 shows a diffuse reflectance spectrum of an analgesic tablet. Because $f(R_\infty)$ is related to absorbance, the spectra appear more like UV-visible absorbance spectra than IR spectra.

4.2.5.2. Attenuated Total Reflection (ATR).

Attenuated total reflection (ATR) is another common reflectance technique. It is used not just for solids but also liquids and thin films with little sample preparation. As with DRIFTS, a special ATR accessory is required. IR radiation from the interferometer enters the ATR accessory and is directed at an angle into a crystal (see Figure 4.34) made of a material with a higher refractive index than that of the sample. The sample is placed on the surface of the crystal, and pressure is applied to press it against the crystal. The angle at which the radiation enters the crystal is chosen so that total internal reflection occurs at each interface of the crystal. At each point of reflection, however, the beam penetrates a short distance into the sample prior to reflection. The portion of the beam that penetrates into the sample is called an "evanescent wave." The key is that this portion of the IR beam can be absorbed by the sample if its frequency matches vibrational frequencies present in the sample. Absorption decreases, or "attenuates" the reflected beam, leading to the name "attenuated total reflection."

In some systems, only a single reflection occurs (see Figure 4.34a), but in other systems, the light reflects within the crystal multiple times (see Figure 4.34b). Reflections at multiple interfaces increase the opportunities for absorption to occur and thereby improve the S/N ratio.

The ATR spectrum of a sample is often different than a direct transmittance spectrum of the same sample because of the characteristics of the evanescent wave and how its

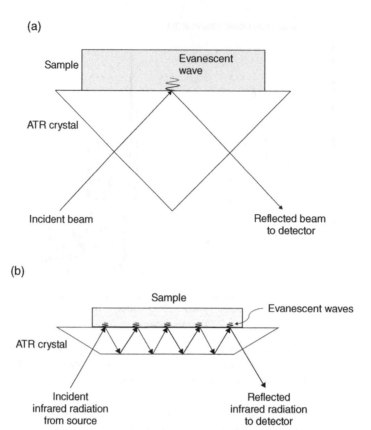

FIGURE 4.34 ATR accessories. (a) A single reflection ATR and (b) multiple reflection ATR cell [36]. Note that the depth of penetration of the evanescent waves, which is generally on the order of a few microns, is greatly exaggerated relative to the thickness of the samples in these images. Sources: (a) and (b) Adapted from Cerruti [36]. Reproduced with permission of The Royal Society.

depth of penetration varies with wavelength, sample properties, and angle of interaction. ATR is valuable for analyzing surfaces, though, because the beam only penetrates a short distance. Furthermore, as long as conditions are held constant, ATR provides a highly reproducible method for obtaining chemical information about samples. Another important advantage of ATR is the fact that samples that are difficult to analyze by other methods, such as solids, pastes, thin films, and fabrics, can be analyzed using ATR. This is of significant benefit in areas such as polymer analysis, forensic investigations, art conservation, and pharmaceutical studies. It also essentially eliminates sample preparation time compared to making mulls and pellets.

Because of the simplicity of ATR, it has rapidly become one of the most popular choices for IR analyses.

4.3. APPLICATIONS OF IR SPECTROSCOPY, INCLUDING NEAR-IR AND FAR-IR

The fact that IR spectra can be collected in just a few seconds makes it attractive in a number of practical applications, particularly in what is known as process analytical chemistry (PAT) in industry. A significant number of such applications ranging from the analysis of wine, food, microbes, bodily fluid, gasoline, and pharmaceuticals have been compiled by Cozzolino [37]. Students planning careers in industrial chemistry are likely to use IR spectroscopy in the types of applications discussed in that book and in the sections below.

Our focus on applications in this section requires broadening our considerations beyond the mid-IR region to include near-IR (NIR) and far-IR (FIR) spectroscopy. First, though, we must point out the application that you are already likely familiar with, namely, the use of mid-IR spectroscopy to assist in structure determination.

4.3.1. Structure Determination with Mid-IR Spectroscopy

As discussed at the start of the chapter, the frequency of a bond vibration depends on the masses of the atoms and the strength of the bond. Therefore, C—H bonds vibrate at different frequencies than N—H, C—Cl, O—H, and C=O bonds. Knowing the frequencies at which these bonds appear in the IR spectra helps determine if such bonds are present or absent in a sample. For example, if you think you have synthesized a compound containing a carbonyl group but a strong signal in the $1690–1760 \text{ cm}^{-1}$ range – the known frequency range for C=O bonds – does not appear in the IR spectra, you can begin to suspect that the synthesis did not produce the expected compound. Correlation charts that show a wide variety of bonds and their characteristic frequencies are widely available in organic chemistry textbooks and on the Internet.

IR spectra are commonly used to complement the information obtained from nuclear magnetic resonance (NMR) and mass spectrometry when determining molecular structures. The combination of all three techniques, each mutually supporting the other, can lead to definitive structure determination of molecules. This application of IR spectra, most often performed using FT instruments, is common in academic and industrial laboratories focused on synthesis.

4.3.2. Gas Analysis

IR spectroscopy is also widely used in monitoring gases, particularly in the workplace. This is important because exposure to some gases can be harmful. For this reason, OSHA (Occupational Health and Safety Administration) sets limits on the allowable concentration of some gases to which workers can be exposed. It is easy to imagine that volatile organic chemicals (e.g., hydrocarbons and halogenated compounds) are in the air in chemical factories and manufacturing plants.

The simplest designs of gas sensors based on IR spectroscopy consist of an IR source, a sample compartment, a filter to select the desired frequency, and a detector. Such a setup is designed to monitor only a single wavelength or a narrow range of wavelengths, rather than the entire range of IR frequencies. The specific wavelength is selected based on the analyte of interest. Two different designs are shown in Figure 4.35. In the first example, air is pumped through the sample compartment. If it contains the chemical(s) being monitored, the power of the IR radiation reaching the detector decreases because some of it is absorbed by the chemical of interest. The decreased power is registered by the detector. Calibrating the sensor allows for actual concentrations to be monitored and allows precautions to be taken if levels of harmful gases rise above threshold values. In the second

(a)

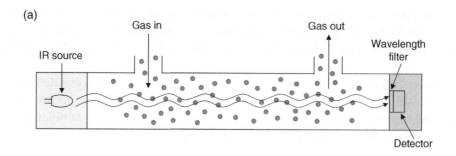

(b)

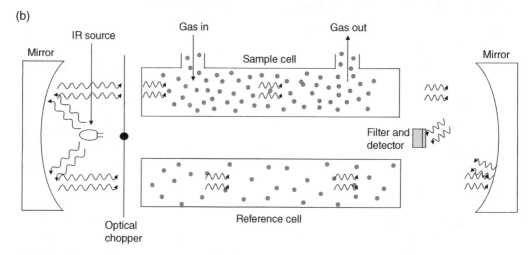

FIGURE 4.35 Gas sensors based on IR detection [38]. (a) single beam sensor. (b) Double beam in time gas sensor. A rotating chopper is used to create the two beams. Source: Adapted from Chou [39]. Reproduced with permission from Tom Chou and International Sensor Technology (IST).

image shown, the signal generated by the sample compartment is compared to signal arising from a reference detector. The reference side of the detector is usually sealed and filled with an inert gas such as nitrogen. Having a dual-beam system allows compensation for changes in the power of the signal arising from artifacts other than the presence of the chemical of interest, such as decreasing signal intensity simply due to the IR source aging over time.

4.3.3. Near-Infrared Spectroscopy (NIR)

While mid-IR analysis for structure elucidation purposes has been conducted for many decades, the analytical use of a greater span of the IR region has recently increased dramatically. More specifically, the near-IR portion of the spectrum has gained considerable popularity for use in fast on-site *quantitative* applications.

The near-IR region is so named because it is "near" the visible region – generally taken as the region extending from approximately 800–2500 nm, or 12 500–4000 cm^{-1} [40] (note that the ISO – International Organization for Standardization – definition specifies a more limited range of 760–1400 nm) [41]. 4000 cm^{-1} (2500 nm) is the

TABLE 4.2   Approximate Wavenumber and Wavelength Ranges of UV, Visible, Near-IR, Mid-IR, and Far-IR Electromagnetic Radiation

	UV	Visible	Near-IR	Mid-IR	Far-IR
Wavenumbers (cm^{-1})	50 000–28 500	28 500–12 500	12 500–4 000	4000–200	200–10
Wavelengths (nm)	200–350	350–800	800–2500	2 500–50 000	50 000–1 000 000

FIGURE 4.36   A depiction of the relative ranges in wavelengths covered by various spectroscopic techniques. As stated in the text, the specific wavelengths associated with each region vary slightly from source to source. Specific definitions are available from governing bodies such as the ISO.

wavenumber commonly associated with the high end of the mid-IR region, so the near IR extends from the edge of the visible (800 nm) to the edge of the mid-IR (2500 nm) (see Table 4.2 and Figure 4.36).

It should be stressed that these ranges are not strictly defined, and different sources frequently cite slightly different ranges. The ISO provides strict definitions of the ranges, but sometimes the ranges specified overlap with our colloquial definitions of different types of spectroscopy.

Photons in the near-IR region of the spectrum are of higher energy than those in the mid-IR. The effect that they have on molecules still relates to vibrational energies. However, in the NIR, quantum transitions typically involve $\Delta v = \pm 2$ or ± 3. For example, the primary vibration (i.e., $\Delta v = \pm 1$) for C—H bonds in alkanes occurs around 2900 cm^{-1} in the mid-IR. In the NIR region, signals for C—H bonds are detected around 5800 cm^{-1}, or at approximately twice the fundamental vibration wavenumber [42]. Recall that earlier in the chapter we noted that such transitions are called overtone bands and that they are forbidden (i.e., of low probability). Just as the C—H overtone band is found at roughly twice the fundamental frequency, overtone bands for O—H (~6450 cm^{-1}, twice the fundamental vibration), N—H (1475 cm^{-1}), and C=O (5100–5200 cm^{-1} – from the second overtone) are found in the near IR region. In Table 4.3, we have converted these wavenumbers into wavelengths because NIR spectra are often plotted as a function of wavelength rather than wavenumbers. As an example of a near-IR spectrum, an annotated spectrum of biscuit dough with the overtone bands of C—H and O—H bonds labeled is shown in Figure 4.37.

As Figure 4.37 shows, spectral features arising from combination bands are also observed in the NIR. Combination bands result from the absorption of a photon that has an energy that is equal to the sum of two different vibrational frequencies, activating both vibrational modes simultaneously. Because they are combinations of two fundamental frequencies, they generally have higher energy than photons used in mid-IR spectroscopy.

TABLE 4.3 Approximate Wavenumbers and Wavelengths of the Overtone Bands of C—H, O—H, N—H, and C=O Vibrations

Bond	Wavenumber (cm⁻¹)	Wavelength (nm)
C—H	5800	1725
O—H	6450	1550
N—H	1475	6780
C=O	5100–5200	1960–1920

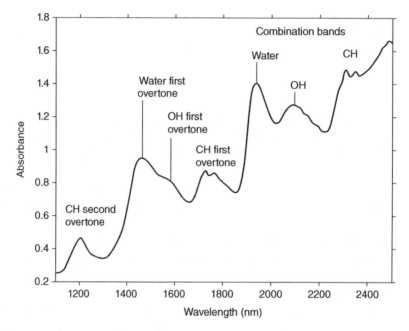

FIGURE 4.37 NIR spectrum of biscuit dough. Source: From Davies [43]. Reproduced with permission of SAGE publishing.

The NIR spectrum in Figure 4.37 shows that such spectra generally have broad, over-lapping features and therefore do not lend themselves well to the qualitative structural analyses that are common with mid-IR spectra. Instead, NIR spectra are routinely used for quantitative analyses. Because the peaks in NIR spectra overlap, quantitative applications are frequently based on chemometric analyses [37, 44–47]. In such analyses, spectra are collected spanning a wide range of the variables expected to be present for the target sample (for example, with pharmaceutical tablets, these would include excipients, binders, the drug itself, water, known impurities, etc.). Multivariate calibrations are built using these spectra of *known* concentrations of components and analyzed with mathematical techniques such as principal component analysis, partial least squares, or artificial neural networks. The spectra collected for this purpose are collectively known as the "training set." Spectra of unknown samples are then collected, and the concentrations of the species of interest are determined using the multivariate calibration. There is great skill and knowledge required to build robust calibration methods when relying on chemometric techniques, and small changes in components that are not accounted for when building the training set can have significant negative consequences. Thus, the development of the training set and its validation are critical when employing chemometric methods for quantitative purposes.

EXAMPLE 4.5

(a) Calculate the energy of photons in the near-IR portion of the spectrum having $\lambda = 800.0$ nm and $\lambda = 2500$ nm.

(b) As a reference, calculate the energy of a photon in the middle of the mid-IR region, for example, at 2100 cm^{-1}.

Answers:

(a) $E = \dfrac{hc}{\lambda} = \dfrac{\left(6.626 \times 10^{-34} \text{ Js}\right)\left(2.998 \times 10^{8} \text{ m/s}\right)}{800.0 \times 10^{-9} \text{ m}} = 2.483 \times 10^{-19} \text{ J}$

$E = \dfrac{hc}{\lambda} = \dfrac{\left(6.626 \times 10^{-34} \text{ Js}\right)\left(2.998 \times 10^{8} \text{ m/s}\right)}{2500 \times 10^{-9} \text{ m}}$

$= 7.946 \times 10^{-20} \text{ J}$ or 7.9×10^{-20} J to two significant figures.

(b) $E = hc\bar{v} = \left(6.626 \times 10^{-34} \text{ Js}\right)\left(\dfrac{2.998 \times 10^{8} \text{ m}}{\text{s}}\right)\left(\dfrac{2100}{\text{cm}}\right)\left(\dfrac{100 \text{ cm}}{\text{m}}\right)$

$= 4.172 \times 10^{-20} \text{ J}$ or 4.2×10^{-20} J to two significant figures.

Another question:

(a) Calculate the ratio of the energy of a photon of $\lambda = 800$ nm relative to that of one at 2100 cm^{-1}.

(b) Calculate the ratio of the energy of a photon of $\lambda = 2500$ nm relative to that of one at 2100 cm^{-1}. Doing these calculations will show roughly how much more energetic near-IR photons are than mid-IR photons because 2100 cm^{-1} is approximately in the middle of near-IR spectra and 800–2500 nm is an approximate range of near-IR photons.

Answer:

Photons in the near-IR portion of the spectrum are approximately 1.90–5.95 times greater in energy than those in the middle of the mid-IR (or more roughly, about 2–6 times greater).

Because NIR spectroscopy is fast and nondestructive, its applications are quite varied and rapidly increasing. In the pharmaceutical industry, for instance, it is used to determine the concentration of the active ingredients (i.e., drugs) in tablets and suspensions [48]. To analyze intact tablets, instruments employing the reflectance techniques discussed above can be used to bounce near-IR radiation off the surface of the tablet and collect the signal from the reflected radiation. Such analyses are fast compared to, for example, dissolving the tablet and analyzing it by chromatography. Plus, the tablet is still intact after the analysis. Additionally, because spectrometers are getting increasingly smaller (some are even handheld), rapid analyses can be made at the point of production or at various points along the production process (i.e., process analytical technology). This is advantageous because if something begins to go wrong, it can be detected and corrected quickly rather than waiting for an entire batch of medication to be produced only to find out later that it is unusable.

NIR spectroscopy is also quite popular for polymer analysis [49, 50]. Figure 4.38 shows IR spectra for polymers made with different blends of polyethylene and polypropylene [37, 51]. The peak at 1.21 μm is from the C—H overtone band. Its intensity increases as the percentage of polyethylene increases. There is also a small peak around 1.19 μm that looks like a shoulder on the peak at 1.21 μm. This signal arises from the CH_3 groups on polypropylene. Notice that its intensity increases as the percent of polypropylene increases. The two peaks merge as polyethylene and polypropylene are blended in different proportions.

Just by looking at the spectra, we can see that it would be difficult to quantify the proportion of polyethylene and polypropylene based simply on peak heights because the peaks are broad and overlapping. However, with careful calibration using materials of known composition combined with chemometric techniques, the composition of unknown samples can be determined. Based on this example and the spectra shown in Figure 4.38, one can further imagine that it is possible to follow the kinetics of polymerization reactions in real time by periodically measuring spectra during a reaction and determining the relative concentrations of different components. This allows industrial operators to stop the reaction at desired compositions. It could also be used to indicate whether or not a reaction has come to completion. Such uses of NIR are not limited to polymerization reactions and are found in pharmaceutical, petrochemical, and biochemical manufacturing processes to monitor reactions.

Many applications of NIR involve measuring the moisture content of different products because several peaks related to the O—H overtones and combination bands arising from water appear in NIR spectra. Moisture is important in things like grain, seed, food products, and pharmaceutical tablets as it affects the shelf life and acceptable storage time of such materials. NIR spectroscopy is also used to measure the protein content in commodities such as grain, milk, seeds, infant formula, and protein powders [52–54]. The analyses are based on spectral features in the NIR region that arise from the N—H bonds in proteins, just as moisture content is related to the spectral features arising from O—H bonds. The price of these substances and their dietary health effects on animals and humans are directly related to their protein content. NIR provides a rapid, nondestructive method for making such measurements.

4.3.3.1. Clinical Uses of NIR. In addition to the above agricultural and industrial uses of NIR, clinical applications of NIR spectroscopy are remarkably important [55–57]. One of the most common examples of NIR in medicine is its use in pulse oximeters, which measure the oxygen saturation of blood [58]. You may have seen, or even used, one. They are the device that is clipped onto the fingertip of patients in emergency rooms and hospitals (see Figure 4.39). They transmit two different wavelengths of light through the

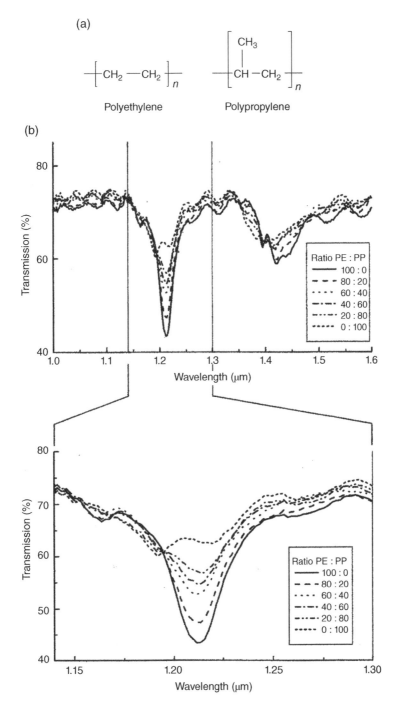

FIGURE 4.38 (a) The structures of polyethylene (PE) and polypropylene (PP) [37, 51]. (b) The NIR spectra of different blends of PE and PP. Examining the 100% PE (solid black line) and 100% PP (dotted line) shows that the two materials have different spectra, but determining the percentage of each in the blends is difficult without chemometric methods due to the high degree of overlap [37, 51]. Sources: (a) From Cozzolino [37]. Reproduced with permission of Nova Science Publishers, Inc. (b) Reprinted from Rohe et al. [51]. Reproduced with permission of Elsevier.

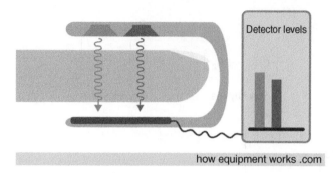

FIGURE 4.39 Figure of a finger oximeter showing that two wavelengths are used to measure the percent saturation of hemoglobin in blood [59]. Source: Reproduced with permission of the image creator, Pras Tila.

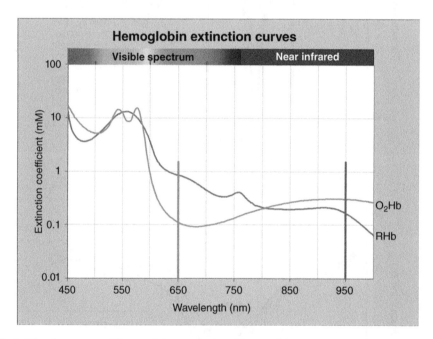

FIGURE 4.40 Spectrum of hemoglobin and deoxyhemoglobin in a portion of the visible and near-infrared regions of the spectrum [60]. Source: Reproduced with permission from Jonas Pologe, Kestrel Labs, Inc., Boulder, CO.

fingertip. One wavelength – usually around 630–660 nm, which is in the visible portion of the spectrum – is chosen as a wavelength at which deoxygenated hemoglobin (Hb) absorbs. The other wavelength – usually around 910–950 nm, which is in the near-infrared region – is chosen because oxygenated hemoglobin (HbO_2) absorbs in this region (see Figure 4.40).

With increasing oxygenation of the blood (or more specifically, oxygenation of the hemoglobin in the blood), less 940 nm radiation is transmitted through the fingertip, increasing the ratio of the amount of 650 to 940 nm light transmitted (Figure 4.41).

Additional signal manipulations and calibration are required to account for other phenomena such as movement by the patient, ambient light, and absorption by other

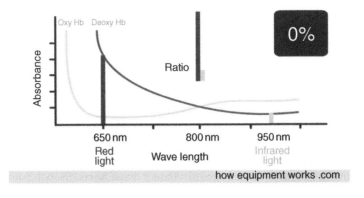

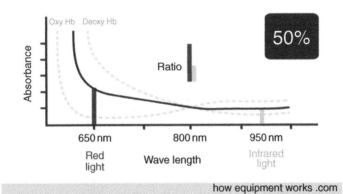

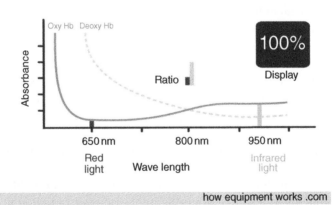

FIGURE 4.41 A depiction of how the ratio of absorbance at 650–950 nm changes as the percent oxygenation of hemoglobin in blood changes [59]. Source: Reproduced with permission of the image creator, Pras Tila.

components such as skin and fat. Ultimately, however, the percent of the hemoglobin that is bound to O_2 in the arteries of a patient can be measured using such a device. Like many applications of IR spectroscopy, pulse oximeters rely on the fact that IR radiation is generally transmitted quite well through most materials (like fingertips), which allows specific wavelengths that are selectively absorbed by target molecules to be monitored.

(a)   (b)   (c)

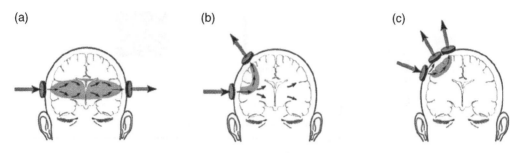

FIGURE 4.42   Depiction of (a) transmission, (b) reflectance, and (c) multidistance modes of NIR spectroscopy to measure oxygenation changes in different tissue layers [63, 64]. Source: From Bakker et al. [63], © 2012, 1999 under CC BY 3.0 license. Available from: http://dx.doi.org/10.5772/32493.

Devices based on reflectance techniques that detect scattered radiation have been developed to measure blood flow and oxygenation in the chest or on the forehead where transmission is not possible [61, 62]. Figure 4.42 shows the difference between transmission and reflectance measurements.

While pulse oximeters have been used for decades, newer devices for monitoring other regions of the body for blood flow and oxygen levels are in development or have already been commercialized. For example, NIR has been used to measure the oxygenation of blood in the brain in neonates. Transmission of NIR radiation through their heads is possible because of their small size and because IR radiation generally is not absorbed by the skull [65]. Numerous other clinical applications have been developed, and it is likely that medical applications of NIR spectroscopy will expand for many years [66–69].

It is evident from all of the above examples, which included applications in medicine, pharmaceuticals, polymer science, food safety, coatings, art analysis, and many other areas, that the use of NIR spectroscopy is exciting, widespread, and growing [53, 70–74]. While we must constrain the discussion within this text, we encourage interested readers to pursue the topic via the Internet and books regarding NIR and its many applications. Students who pursue professions in medicine or in the chemical industry will likely encounter applications of NIR spectroscopy throughout their careers.

4.3.3.2. NIR Instrumentation. There are many configurations and components used for NIR instruments. In fact, many UV-visible instruments are capable of making measurements well into the NIR region. Tungsten-halogen lamps like those found in UV-visible instruments as a source of visible light also emit radiation in the NIR region. Scanning instruments using gratings and instruments based on interferometers are both used in NIR applications. For specific, well-defined applications, filter instruments that lack the capability of scanning are also used. Detectors are frequently semiconductors such as PbS and PbSe [75]. Photodiodes and photodiode arrays are also used.

NIR instruments, like their mid-IR counterparts, operate via transmission of radiation through the sample or via reflection of radiation off of the sample (such as in DRIFTS and ATR discussed above). The selection of scanning instruments vs. interferometers and of transmission vs. reflection, as well as the selection of specific sources and detectors, is ultimately dictated by the requirements of the application.

4.3.4 Far-Infrared Spectroscopy (FIR)

Radiation in the far infrared portion of the spectrum, like that in the near- and mid-IR regions, is nonionizing. Furthermore, far-IR radiation can penetrate nonconducting materials like clothing, paper, wood, masonry, and dust. These qualities make it attractive for a number of applications, particularly those related to safety and security, such as screening passengers and luggage for explosives at airports. While the use of far-IR spectroscopy is not nearly as common today as that of near- or mid-IR spectroscopy, advances in sources and detectors make FIR a rapidly growing field.

The term "far-IR" is, for all intents and purposes, synonymous with "terahertz spectroscopy." While there is no general consensus on the precise wavelength range covered by these terms, most scientists accept 0.3–3 THz (10–100 cm^{-1}) as the working definition (T = tera = 10^{12}) [76]. By the definitions established by the ISO, the range is 3000–1000000 nm (1000000 nm = 1 mm) [41].

EXAMPLE 4.6

Calculate the wavelength associated with 0.300 and 3.00 THz. Looking at your answers, does it make sense that spectroscopy in this region is also referred to as submillimeter spectroscopy?

Answer:

$$\frac{hc}{\lambda} = h\upsilon \quad \text{so} \quad \frac{c}{\lambda} = \upsilon \quad \text{and} \quad \frac{c}{\upsilon} = \lambda$$

$$\frac{c}{\upsilon} = \frac{2.998 \times 10^8 \text{ m/s}}{0.300 \times 10^{12} \text{ s}^{-1}} = 9.99 \times 10^{-4} \text{ m} \quad \text{or} \quad 0.999 \text{ mm}$$

$$\frac{c}{\upsilon} = \frac{2.998 \times 10^8 \text{ m/s}}{3.00 \times 10^{12} \text{ s}^{-1}} = 9.99 \times 10^{-5} \text{ m} \quad \text{or} \quad 0.0999 \text{ mm}$$

Both values are at or under 1 mm, so it makes sense that THz spectroscopy is also referred to as submillimeter spectroscopy.

Given this range, energy of electromagnetic radiation follows the order UV > visible > near-IR > mid-IR > far-IR > microwave. The far-IR region, like the mid-IR, probes vibrational modes of molecules. However, the far-IR probes vibrations with low frequencies, including those associated with metal–metal and metal–ligand bonds, as well as large-scale conformational changes of polypeptides and polynucleotides. Metal–metal and metal–organic ligand bonds tend to appear at low frequency because the high masses of many metals lead to high reduced masses, which, according to Hooke's law discussed early in this chapter, lead to low-frequency vibrations. The position of bands in the THz region is also sensitive to hydrogen bonding and other intermolecular interaction effects, making terahertz spectroscopy particularly useful for studying the binding of water to large biomolecules [76].

Another important application of THz spectroscopy is the analysis of polymorphic forms of drug crystals and their hydration states. The crystalline form of a drug and its hydration state affect its solubility and dissolution rate. Many drugs can crystallize into different crystal forms (known as polymorphs) or change from one crystalline form to another while on the shelf. Information about the degree of crystallinity or the polymorphic composition in a tablet can be obtained using THz spectroscopy because it is sensitive to large-scale interactions, unlike mid-IR spectroscopy which only provides information at the molecular level.

Examples of spectra obtained from two different crystalline forms of acetazolamide dispersed in polyethylene as a diluent are shown in Figure 4.43 [77]. Acetazolamide is used to treat the symptoms of altitude sickness, glaucoma, and some types of seizures [79]. Note that both forms have distinct spectral features. For example, Form I has a band at ~40 cm^{-1} and Form II has a different ratio of intensities for the peaks at 63 and 73 cm^{-1} than Form I. Such characteristics, when observed in tablet formulations, help identify the crystalline forms present in the tablet.

THz spectroscopy can be used to monitor tablets in the production line in real time. This continuous monitoring has a number of advantages over batch monitoring and is possible because THz spectroscopy is fast and nondestructive.

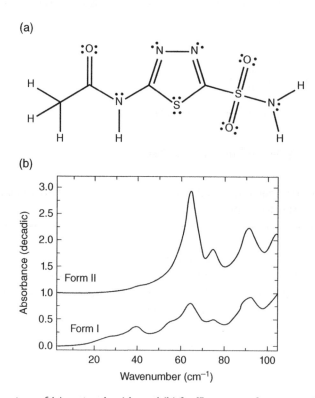

FIGURE 4.43 Structure of (a) acetazolamide and (b) far-IR spectra of two crystalline forms of it [77]. Source: Republished with permission of Taylor & Francis Group LLC Books, from Zeitler et al. [78], permission conveyed through Copyright Clearance Center, Inc.

4.3.4.1. Security Applications of FIR Spectroscopy. Security applications of far-IR spectroscopy include using THz radiation to detect explosives and narcotics based on their distinct spectral characteristics and the ability of THz radiation to penetrate through packaging in which such substances might be enclosed [78, 80, 81]. Spectra of two explosives – pentaerythritol tetranitrate (PETN) and 1,3,5-trinitroperhydro-1,3,5-triazine (commonly known as research department explosive (RDX)) – are shown in Figure 4.44 as an example of the ability of THz spectroscopy to probe explosives.

The full-body scanners in airports are sometimes associated with the terahertz frequency range, although they operate at even lower energies (i.e., longer wavelengths) than usually considered in chemical applications of THz spectroscopy. Furthermore, the scanners are imagining instruments, meaning that they provide a picture rather than a spectrum, and are therefore not detecting the specific chemical composition of items.

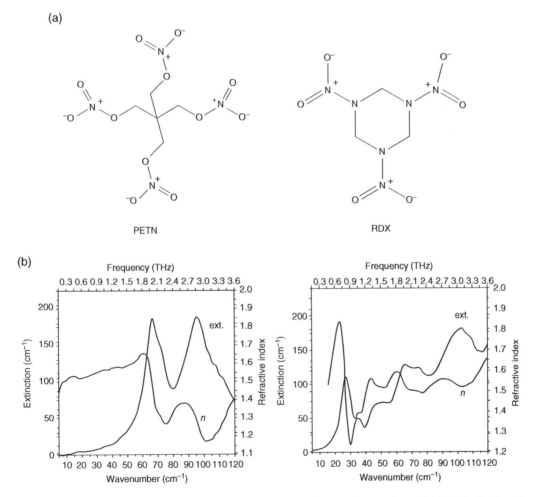

FIGURE 4.44 Structures of (a) PETN and RDX and (b) their terahertz absorption (extinction) and refractive index spectra. PETN spectra are on the left and RDX spectra are on the right [78]. Source: Republished with permission of Taylor & Francis Group LLC Books, from Zeitler et al. [78], permission conveyed through Copyright Clearance Center, Inc.

4.3.4.2. Instruments for FIR. Advances in sources and detectors have allowed for increased exploration of the far-IR region of the spectrum. Quantum cascade lasers and solid-state sources, coupled with increasingly sensitive semiconductor detectors, all aid in obtaining THz spectra with acceptable S/N ratios. Experimentally, spectra are collected either in transmission or reflectance modes, depending on the application. Some applications, like the analysis of pharmaceutical tablets, have been conducted in both modes with the source radiation either propagating through the entire tablet (transmission) or being scattered off its surface (reflectance).

While we have only scratched the surface of the methods and applications of FIR, we encourage students to remain aware of advances in this area as it has significant potential to become a widely used instrumental technique.

4.4. SUMMARY

The bonds in molecules vibrate with characteristic frequencies. The frequencies of these vibrations are determined primarily by bond strength and the masses of the atoms, with informative perturbations dictated by the immediate chemical environment of the bonds. Most of the vibrations in common organic molecules fall into the mid-IR region of the spectrum, while the near-IR also contains overtones and combination bands. Far-IR, or terahertz, spectroscopy also probes vibrational modes, but of lower frequency than those in the near- and mid-IR region.

Instrumental advances, particularly the development of interferometers, allow for the rapid acquisition of spectra in the time domain, with subsequent Fourier transformation to the frequency domain. Because a spectrum can be collected in a second or less, hundreds of spectra on the same sample can be collected and averaged together, significantly increasing the signal-to-noise ratio.

Applications of the entire infrared range (near-, mid-, and far-IR) include, but are certainly not limited to, aiding structure determinations and following reaction kinetics (primarily in the mid-IR); quantifying moisture, cellulose, and protein content in samples (near-IR); and detecting crystalline structures of pharmaceutical products and the presence of explosives and narcotics in packages (far-IR). Biomedical imaging applications are also emerging. Given all of these applications, it is highly likely that students entering the medical, chemical, or biochemical professions will use infrared spectroscopy or imaging in their work.

PROBLEMS

4.1 UV-visible spectra are normally plotted using absorbance, whereas IR spectra are plotted using transmittance.

(a) If a peak in an IR spectrum has a 10.0% transmittance (i.e., only 10% of the photons get through), what is the absorbance?

(b) Repeat for a 1.00% transmittance.

4.2 Would a resolution setting of 4 cm^{-1} be able to resolve peaks arising from ^{1}H—^{35}Cl and ^{1}H—^{37}Cl, assuming a force constant of 481.0 N/m for both bonds? Hint: Calculate the expected wavenumber for each. Find masses for the isotopes using the Internet.

4.3 The force constant for a C—C single bond in an alkane is approximately 450 N/m².

    (a) Based on this, estimate the force constant for the C—C bond in benzene (think carefully about the nature of the C—C bonds in benzene).

    (b) In benzene and toluene, a prominent peak arising from a C—C stretch appears near 1480 cm⁻¹. Estimate the force constant for this vibration using Hooke's law. The reduced mass for a C—C bond is 9.963×10^{-27} kg.

4.4 Consider Hooke's law and C—C, C=C, and C≡C bonds. Which bond has the highest stretching frequency and which has the lowest?

4.5 (a) How many normal vibrational modes are expected for methyl salicylate based on the $3N - 6$ rule? How many peaks, approximately, are observed in the IR spectrum shown in Figure 4.1?

    (b) List two reasons why fewer peaks than expected may be observed in an IR spectrum.

    (c) List two reasons why more peaks than expected may be observed in an IR spectrum.

4.6 Why does it take longer to collect the same number of spectra using a better resolution setting (e.g., 1 cm⁻¹ compared to 4 cm⁻¹) assuming the same mirror velocity for both settings?

4.7 (a) Based on the $3N - 6$ rule, how many peaks are expected in the IR spectrum of SO_2?

    (b) Depict the motions and decide which are IR-active and which are IR-inactive.

    (c) Compare the SO_2 predictions with those of CO_2 in Figure 4.7.

    Hint: Searching Internet sites may be useful for visualizing the stretching and bending motions.

4.8 Why is an IR cell with a known path length sometimes used?

4.9 (a) Using the Boltzmann equation (see Chapter 1), calculate the population of the $v = 0$ vibrational level relative to that of the $v = 1$ vibrational level for carbon monoxide, which has a fundamental vibration at 2143 cm⁻¹. Assume that P^* and $P°$ in the equation are both equal to 1. Perform the calculation at 25.0 °C.

    (b) Will this ratio increase or decrease if the temperature is increased? Explain.

    (c) For electronic transitions at 25 °C, the ratio of ground to excited state molecules is around 10^{52}, much greater than that for vibrational transitions. Why?

4.10 What advantages do FTIR instruments provide over dispersive ones, and how do they help improve spectra?

4.11 (a) If 10 spectra yield a signal-to-noise ratio of 2.7 for a particular spectral feature, how many are required to increase the signal-to-noise ratio to 5?

    (b) If the 10 scans take one minute to collect, how long will it take to collect the new set of scans?

4.12 The C≡N bond in acetonitrile (CH_3CN) has a fundamental stretching vibration ($\Delta v = \pm 1$) at 2267 cm⁻¹.

    (a) At what wavenumbers would we expect to the first ($\Delta v = \pm 2$) and second ($\Delta v = \pm 3$) overtone bands for this stretch?

    (b) What region of the spectrum are these wavenumbers associated with?

4.13 The spectra below were measured on the same sample on the same FTIR spectrometer using a series of different mirror retardations, ranging from 0.0625 to approximately 1.50 cm^{-1}.

(a) Which spectrum in the figure has the best resolution (i.e., smallest R)?

(b) Which spectrum (very top or very bottom) is associated with the smallest retardation and which with the largest retardation?

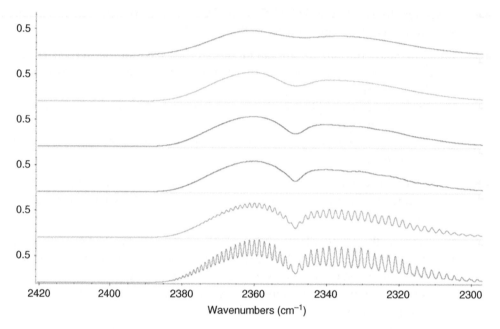

Effect of retardation on the resolution of peaks in carbon dioxide spectra measured on a Thermo Scientific™ Nicolet™ iS5 FTIR spectrometer. Source: Reprinted with permission from Thermo Fisher Scientific.

4.14 NIR spectra generally contain broad peaks, whereas mid-IR spectra are characterized mainly by narrow peaks. Based on this, which is expected to be better for producing linear calibration curves?

4.15 What advantage is there to using a multiple reflection ATR cell versus a single reflection arrangement?

4.16 Why are chemometric techniques frequently needed in order to gain quantitative information from near-IR spectra?

4.17 Most cell windows in IR instruments are made of various salts. Why don't plastic or glass windows work in IR spectroscopy like they do in visible spectroscopy?

4.18 (a) What are two advantages of MCT detectors over DTGS detectors? Given these advantages, what kinds of applications are they best for?

(b) Despite their advantages, MCT detectors are not the most common detectors in commercial instruments. Why not?

REFERENCES

1. Stein, S.E. (2017). *Evaluated Infrared Reference Spectra in NIST Chemistry WebBook, NIST Standard Reference Database Number 69* (ed. P.J. Linstrom and W.G. Mallard), 20899. Gaithersburg, MD: National Institute of Standards and Technology. www://webbook.nist.gov (accessed 14 January 2015).

2. Avram, M. and Matteescu, G.H. (1972). *Infrared Spectroscopy*, 70. New York: Wiley-Interscience.

3. http://www.chem.purdue.edu/gchelp/vibs/index.html (accessed 22 April 2015).

4. http://www.chemtube3d.com/vibrationsC6H6.htm (accessed 22 April 2015).

5. http://www.chemeddl.org/resources/models360/models.php?pubchem=1119 (accessed 22 April 2015).

6. http://csi.chemie.tu-darmstadt.de/ak/immel/misc/oc-scripts/vibrations.html (accessed 22 April 2015).

7. https://www.youtube.com/watch?v=3RqEIr8NtMI (accessed 22 April 2015).

8. http://www.rsc.org/learn-chemistry/collections/spectroscopy/introduction#IRSpectroscopy (accessed 22 April 2015).

9. https://www.youtube.com/watch?v=K-AwK2jsEzk&list=PLmt1fgYjGn-Dkbvm6WH_MS5YhhxIQeUt4&index=14 (accessed 22 April 2015).

10. https://www.youtube.com/watch?v=3szR-xQF4CA&index=15&list=PLmt1fgYjGn-Dkbvm6WH_MS5YhhxIQeUt4 (accessed 22 April 2015).

11. Pavia, D.L., Lampman, G.M., and Kriz, G.S. (2001). *Introduction to Spectroscopy: A Guide for Students of Organic Chemistry*, 3e, 16. Victoria: Thomson Learning.

12. Colthup, N.B., Daly, L.H., and Wiberley, S.E. (1990). *Introduction to Infrared and Raman Spectroscopy*, 3e, 10, 75, 83, 86. San Diego: Academic Press.

13. Avram, M. and Matteescu, G.H. (1972). *Infrared Spectroscopy*, 40. Wiley-Interscience: New York.

14. http://www2.ups.edu/faculty/hanson/Spectroscopy/IR/IRfrequencies.html (accessed 15 January 2015).

15. Stuart, B. (2007). *Infrared Spectroscopy: Fundamentals and Applications*, 12. New York: Wiley.

16. Smith, B.C. (1996). *Fundamentals of Fourier Transform Infrared Spectroscopy*, 5, 31–41, 44–45, 112. Boca Rotan: CRC Press.

17. Swenson, L.S. (2012). *The Ethereal Aether: A History of the Michelson-Morley-Miller Aether-drift Experiments, 1880–1930*. Austin: University of Texas Press.

18. Fellgett, P.B. (1949). *J. Opt. Soc. Am.* 39: 970–976.

19. Jacquinot, P. (1954). *J. Opt. Soc. Am.* 44: 761–765.

20. Rubinson, K.A. and Rubinson, J.F. (2000). *Contemporary Instrumental Analysis*, 768. Upper Saddle River: Prentice Hall.

21. Glick, M. R. (1989). Fourier transform spectrometry in the ultraviolet-visible region. PhD thesis. University of Florida, p. 12.

22. Glick, M.R., Jones, B.T., Smith, B.W., and Winefordner, J.D. (1989). *Appl. Spectrosc.* 43: 342–344.

23. Coblentz, W.W. (1908). *Bull. Bur. Stand.* 4: 533–551.

24. Stewart, J.E. and Richmond, J.C. (1957). *J. Res. Natl. Bur. Stand.* 59: 407.

25. http://chemwiki.ucdavis.edu/Analytical_Chemistry/Instrumental_Analysis/Spectrometer/How_an_FTIR_instrument_works/FTIR%3A_Hardware (accessed 22 April 2015).

26. www.oed.com (accessed 22 April 2015).

27. Griffiths, P.R. and De Haseth, J.A. (2007). *Fourier Transform Infrared Spectrometry*, 147, 350. New York: Wiley.

28. http://www.shimadzu.com/an/ftir/support/ftirtalk/talk9/intro.html (accessed 22 April 2015).

29. https://www.newport.com/t/light-collection-and-systems-throughput (accessed 31 August 2016).

30. http://www.wikiwand.com/en/Integrating_sphere (accessed 22 April 2015).

31. Kubelka, P. and Munk, E. (1931). *Z. Tech. Phys.* 12: 593–601.

32. Kubelka, P. (1948). *J. Opt. Soc. Am.* 38: 448–457.

33. Culler, S.R. (1993). *Practical Sampling Techniques for Infrared Analysis* (ed. P.B. Coleman), 94. Boca Raton: CRC Press.

34. Kortum, P., Braun, W., and Herzog, G. (1963). *Angew. Chem. Int. Ed. Engl.* 2: 333–341.

35. Fuller, M.P. and Griffiths, P.R. (1978). *Anal. Chem.* 50: 1906–1910.

36. Cerruti, M. (2012). *Phil. Trans. R. Soc. A 370*: 1281–1312.

37. Cozzolino, D. (2014). *Infrared Spectroscopy: Theory, Developments, and Applications*. New York: Nova Science Publishers, Inc.

38. http://www.intlsensor.com/pdf/infrared.pdf (accessed 22 April 2015).

39. Chou, J. (1999). *Hazardous Gas Monitors: A Practical Guide to Selection, Operation, and Applications*. New York: McGraw Hill and SciTech Publishing, Inc.

40. Hof, M. and Macháň, R. (2014). *Handbook of Spectroscopy* (ed. G. Gauglitz and D.S. Moore), 31. Hoboken: Wiley.

41. http://www.spacewx.com/pdf/SET_21348_2004.pdf (accessed 22 April 2015).

42. Pelletier, M.J. and Pelletier, C.C. (2010). *Raman, Infrared, and Near-Infrared Chemical Imaging* (ed. S. Slobodan and Y. Ozaki), 9. Hoboken: Wiley.

43. Davies, A.M.C. (2005). An introduction to near infrared (NIR) spectroscopy. *NIR News* (1 November). http://www.impublications.com/content/introduction-near-infrared-nir-spectroscopy.

44. Mark, H. (1989). *Anal. Chem. Acta 223*: 75–93.

45. Mark, H. and Workman, J. Jr. (2001). *Spectroscopy 22*: 20–26.

46. Mark, H. (2001). *Pract. Spectrosc. 25*: 291–321.

47. Heise, H.M. and Winzen, R. (2008). *Near-Infrared Spectroscopy: Principles, Instruments, and Applications* (ed. H.W. Siesler, Y. Ozaki, S. Kawata and H.M. Heise), 125–162. Hoboken: Wiley.

48. Fortunato De Carvalho Rocha, W., Nogueira, R., Fontes, G.N. et al. (2014). *Infrared Spectroscopy: Theory, Developments, and Applications* (ed. D. Cozzolino). New York: Nova Science Publishers, Inc.

49. Kradjel, C. and Lee, K.A. (2008). *Handbook of Near-Infrared Analysis*, 3e (ed. D.A. Burns and E.W. Ciurczak). Boca Raton: CRC Press.

50. Workman, J. Jr. and Weyer, L. (2008). *Practical Guide to Interpretive Near-Infrared Spectroscopy*. Boca Raton: CRC Press.

51. Rohe, T., Becker, W., Kolle, S. et al. (1999). *Talanta 50*: 283–290.

52. Roberts, C.A., Workman, J. Jr., and Reeves, J.B. III (2004). *Near-Infrared Spectroscopy in Agriculture*. Madison, WI: American Society of Agronomy, Crop Science Society of America, Inc., Soil Science Society of America, Inc.

53. Burns, D.A. and Ciurczak, E.W. (2008). *Handbook of Near-Infrared Analysis*, 3e. Boca Raton: CRC Press.

54. Palmer, S. (February/March 2014). Food quality and safety. http://www.foodquality.com/details/print/5862751/FebruaryMarch_2014.html (accessed 2 February 2015).

55. DeBlasi, R.A., Conti, G., and Gasparetto, A. (1995). *1995 Yearbook of Intensive Care and Emergency Medicine* (ed. J.-L. Vincent), 652–660. Berlin: Springer-Verlag.

56. Simonson, S.G. and Piantadosi, C.A. (1996). *Crit. Care Clin. 12*: 1019–1029.

57. Smith, M. (2011). *Phil. Trans. R. Soc. A 369*: 4452–4469.

58. Nitzan, M., Romem, A., and Koppel, R. (2014). *Med. Devices 7*: 231–239.

59. http://www.howequipmentworks.com/physics/respi_measurements/oxygen/oximeter/pulse_oximeter.html#introduction (accessed 22 April 2015).

60. http://www.oximetry.org/pulseox/principles.htm (accessed 22 April 2015).

61. Mendelson, T., Kent, J.C., Yocum, B.L., and Birle, M.J. (1998). *Med. Instrum. 22*: 167–173.

62. Nogawa, M., Kaiwa, T., and Takatani, S. (1998). *IEEE Eng. Med. Biol. 4*: 1858–1861.

63. Bakker, A., Smith, B., Ainslie, P., and Smith, K. (2012). *Applied Aspects of Ultrasonography in Humans* (ed. P. Ainslie), 70. London: InTech.

64. Wahr, J.A., Tremper, K.K., Samra, S.K., and Delpy, D.T. (1996). *J. Cardiothorac. Vasc. Anesth. 10*: 406–418.

65. Takami, T. (2014). *Infrared Spectroscopy: Theory, Developments, and Applications* (ed. D. Cozzolino). New York: Nova Science Publishers, Inc.

66. Mesquida, J., Gruartmoner, G., and Espinal, C. (2013). *BioMed Res. Int. 2013*: 502194.

67. Murkin, J.M. and Arango, M. (2009). *Br. J. Anaesth. Suppl. 1*: i3–i13.

68. Ferrari, M., Muthalib, M., and Quaresima, V. (2011). *Philos. Trans. A Math. Phys. Eng. Sci. 369*: 4577–4590.

69. http://cedars-sinai.edu/Patients/Programs-and-Services/Pediatrics/NIRS---Near-Infrared-Spectroscopy.aspx (accessed 22 April 2015).

70. Salzer, R. and Siesler, H.W. (eds.) (2014). *Infrared and Raman Spectroscopic Imaging*, 2e. Weinheim: Wiley-VCH.

71. Ciurczak, E.W. and Igne, B. (2015). *Pharmaceutical and Medical Applications of Near-Infrared Spectroscopy*, 2e. Boca Raton: CRC Press, Taylor & Francis Group.

72. Jue, T. and Masuda, K. (2013). *Application of Near Infrared Spectroscopy in Biomedicine*, vol. 4. New York: Springer.

73. Sasic, S. and Ozaki, Y. (eds.) (2010). *Raman, Infrared, and Near-Infrared Chemical Imaging*. Hoboken: Wiley.

74. Siesler, H., Ozaki, Y., Kawata, S., and Heise, H.M. (2002). *Near-Infrared Spectroscopy: Principles, Instruments, and Applications*. Weinheim: Wiley-VCH.

75. McCarthy, W.J. and Kemeny, G.J. (2008). *Handbook of Near-Infrared Analysis*, 3e (ed. D.A. Burns and E.W. Ciurczak), 82. Boca Raton: CRC Press.

76. Mantsch, H.H. and Naumann, D. (2010). *J. Mol. Struct. 964*: 1–4.

77. Zeitler, J.A., Rades, T., and Taday, P.F. (2008). *Terahertz Spectroscopy: Principles and Applications* (ed. S. Dexheimer). Boca Raton: CRC Press.

78. Zeitler, J.A., Rades, T., and Taday, P.F. (2008). *Terahertz Spectroscopy: Principles and Applications* (ed. S. Dexheimer). Boca Raton: CRC Press.

79. http://www.webmd.com/drugs/2/drug-6755/acetazolamide+oral/details (accessed 24 July 2015).

80. Davies, A.G., Burnett, A.D., Fan, W. et al. (2008). *Mater. Today 11*: 18–26.

81. Konek, C., Wilkinson, J., Esenturk, O. et al. (2009). *Terahertz Physics, Devices, and Systems III: Advanced Applications in Industry and Defense* (ed. M. Anwar, N.K. Dhar and T.W. Crowe). Bellingham: Society of Photo-Optical Instrumentation Engineers (SPIE).

FURTHER READING

1. Burns, D.A. and Ciurczak, E.W. (2008). *Handbook of Near-Infrared Analysis*, 3e. Boca Raton: CRC Press.

2. Chalmers, J.M. (2010). *Biomedical Applications of Synchrotron Infrared Microspectroscopy: A Practical Approach (RSC Analytical Spectroscopy Series)* (ed. D. Moss). Cambridge: RSC Publishing.

3. Christy, A.A., Ozaki, Y., and Gregoriou, V.G. (2001). *Wilson and Wilson's Comprehensive Analytical Chemistry, Volume XXXV, Modern Fourier Transform Infrared Spectroscopy* (ed. D. Barceló). Amsterdam: Elsevier.

4. Coleman, P.B. (1993). *Practical Sampling Techniques for Infrared Analysis*. Boca Raton: CRC Press.

5. Colthup, N.B., Daly, L.H., and Wiberley, S.E. (1990). *Introduction to Infrared and Raman Spectroscopy*, 3e, 10, 75, 83, 86. San Diego: Academic Press.

6. Cozzolino, D. (2014). *Infrared Spectroscopy: Theory, Developments, and Applications*. New York: Nova Science Publishers, Inc.

7. Griffiths, P.R. and De Haseth, J.A. (2007). *Fourier Transform Infrared Spectrometry*. New York: Wiley.

8. Ingle, J.D. Jr. and Crouch, S.R. (1988). *Spectrochemical Analysis*. Englewood Cliffs: Prentice Hall, Inc.

9. Mantsch, H.H. and Naumann, D. (2010). *J. Mol. Struct.* 964: 1–4.

10. Salzer, R. and Siesler, H.W. (2014). *Infrared and Raman Spectroscopic Imaging*, 2e. Weinheim: Wiley-VCH.

11. Slobodan, S. and Ozaki, Y. (2010). *Raman, Infrared, and Near-Infrared Chemical Imaging*. Hoboken: Wiley.

12. Smith, B.C. (1996). *Fundamentals of Fourier Transform Infrared Spectroscopy*. Boca Rotan: CRC Press.

13. Ozaki, Y. (2012). *Anal. Sci.* 28: 545–563.

RAMAN SPECTROSCOPY 5

Jason R. Maher

University of Rochester, Rochester, NY, USA

As we have seen in the previous chapters, many forms of molecular spectroscopy are used to study transitions between rotational, vibrational, and electronic energy states by evaluating the absorption or emission of electromagnetic radiation. This chapter presents an alternative method for probing energy states by measuring inelastic scattering. The inelastic scattering of light is called Raman scattering, named after Sir Chandrasekhara Venkata Raman (see Figure 5.1) who first observed the effect in 1928 [1] and was awarded the Nobel Prize in Physics for the discovery two years later. Raman spectroscopy is primarily used to detect changes in molecular vibrations and, like infrared (IR) spectroscopy, can provide information about chemical structure, identify a substance based on its characteristic spectrum, or quantify analyte concentrations (see Table 5.1).

This chapter presents mathematical descriptions of Raman scattering, a survey of relevant instrumentation and applications, and a comparison with IR absorption, which provides complementary information about vibrational structure. The focus is primarily conventional, spontaneous Raman spectroscopy; however, variants including surface-enhanced Raman spectroscopy, resonance Raman spectroscopy, and nonlinear Raman spectroscopy are also discussed.

5.1. ENERGY-LEVEL DESCRIPTION

When light scatters from a molecule, the majority of the photons are elastically scattered, meaning that the scattered photons have the same energy as the incident light. A small fraction of light, generally around 1 in 10 million scattered photons [2], is *inelastically* scattered, and the energy difference between the incident and scattered photons is transferred to the molecule. This process is known as Raman scattering. Because Raman spectroscopy is most commonly used to measure vibrational transitions, this chapter uses the term Raman scattering (without an additional prefix) to refer to *vibrational* Raman scattering only.

Figure 5.2 is an energy-level diagram depicting Raman scattering and related interactions involving vibrational transitions. Raman scattering occurs after an incident photon promotes the molecule to a *temporary* high-energy state called a virtual level. The virtual

Spectroscopy: Principles and Instrumentation, First Edition. Mark F. Vitha.
© 2019 John Wiley & Sons, Inc. Published 2019 by John Wiley & Sons, Inc.

FIGURE 5.1   Sir Chandrasekhara Venkata Raman. Source: Image reproduced from https://en.wikipedia. org/wiki/C._V._Raman, uploaded by Pieter Kuiper, https://www.nobelprize.org/nobel_prizes/physics/ laureates/1930/raman-bio.html.

TABLE 5.1   Qualitative Relationships Between Molecular Properties and Raman Spectral Features

Spectral feature	Molecular property
Intensity	Concentration
Vibrational frequency	Bond strengths and atomic masses
Peak width	Degree of environmental heterogeneity
Polarization dependence	Molecular symmetry

level is not a quantum mechanically allowed energy state, so the molecule can remain there for only a short time governed by Heisenberg's uncertainty principle.

After promotion to the virtual level, a second photon is scattered as the molecule returns to one of its allowed vibrational states. Elastic scattering occurs when the initial and final states are the same and no energy is transferred between the molecule and photon. Raman scattering occurs when there is an energy transfer that causes the molecule to return to a *different* vibrational state than the one it was in before the interaction. The result is that there is also a difference in the energy of the incident photon and the scattered photon.

Unlike Raman scattering, fluorescence occurs after absorption of an incident photon that causes an electronic transition. As described in Chapter 4, vibrational relaxation in the excited state prior to fluorescence causes the photons emitted as fluorescence to have lower energy, and therefore longer wavelengths, than those initially absorbed by the molecule. The average time required for the absorption, relaxation, and fluorescence emission process is called the fluorescence lifetime and depending upon the molecule can range from picoseconds to hundreds of nanoseconds. The fluorescence process is generally much slower than Raman scattering.

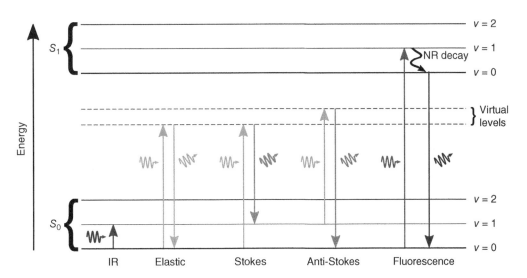

FIGURE 5.2   An energy-level diagram depicting infrared absorption (IR), elastic scattering, Stokes and anti-Stokes Raman scattering, and fluorescence. Upward straight arrows represent increases in the energy of a molecule caused by incident photons, which are symbolized by horizontal squiggles to the left of upward arrows. Downward straight arrows represent decreases in the energy of a molecule caused by scattering or emission of photons, which are symbolized by the squiggles going off at an angle to the right of the downward lines. The downward squiggly arrow between $v = 1$ and $v = 0$ in the excited state, S_1, represents nonradiative (NR) decay. S_0 and S_1 refer to the ground and first excited electronic states, while the vibrational states are labeled by their corresponding vibrational quantum numbers, v. For elastic scattering, the incident and scattered photons have the same energy. Scattering in which the scattered photon has less energy than the incident photon is known as Stokes scattering. Anti-Stokes scattered photons have more energy than the incident photons.

Figure 5.2 also illustrates a fundamental difference between Raman scattering and IR absorption. Infrared absorption involves a direct transition to an excited vibrational state. This occurs by absorption of a photon whose energy matches the energy difference between the states. Conversely, Raman scattering is induced by photons with energies greater than those required to cause direct vibrational excitation. Further, instead of measuring absorbance, *in Raman spectroscopy, the energy differences between incident and Raman-scattered photons are measured.*

Raman scattering has two possible outcomes: The scattered photon can have either less or more energy than the incident photon due to the molecule either gaining or losing energy. When the scattered photon has less energy than the incident photon, the process is known as Stokes Raman scattering, and when the scattered photon has more energy, the process is known as anti-Stokes Raman scattering. The relative intensities of Stokes and anti-Stokes scattering are dependent upon the initial populations of the states. At room temperature, most molecules are present in the ground vibrational state, with a small number present in excited vibrational states due to thermal energy. Anti-Stokes scattering is generally significantly weaker due to the lower population of molecules in excited states. Raman instruments are therefore typically configured to measure Stokes Raman scattering.

5.2.   VISUALIZATION OF RAMAN DATA

The energy shifts of the Raman-scattered photons can be measured with an optical spectrometer producing a Raman spectrum. Such spectra are often referred to as "molecular fingerprints." A representative Raman spectrum of N-acetyl-p-aminophenol, better known as Tylenol, is shown in Figure 5.3. The spectrum exhibits a number of characteristic peaks corresponding to some of the allowed vibrational states of the molecule [3]. The vertical axis gives the magnitude of the detected Raman-scattered energy, typically called Raman intensity. The horizontal axis is the relative energy shift, in wavenumbers, between the incident and Raman-scattered photons. This wavenumber shift, usually plotted in units of inverse centimeters, is called the Raman shift ($\Delta \bar{v}$) and can be calculated as

$$\Delta \bar{v} = \frac{1}{\lambda_0} - \frac{1}{\lambda} \tag{5.1}$$

where λ_0 is the excitation or illumination wavelength and λ is the detection wavelength.

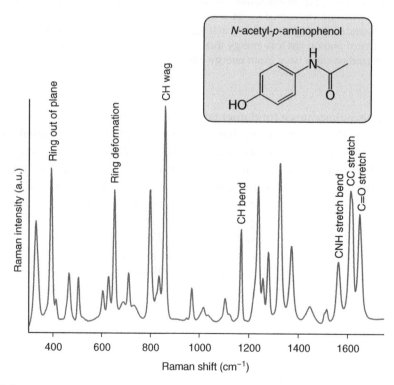

FIGURE 5.3   Representative Raman spectrum of N-acetyl-p-aminophenol (Tylenol) with peak assignments highlighting some of the Raman-active vibrational modes. The structural formula of the molecule is shown in the inset.

EXAMPLE 5.1

Cyclohexane has a Raman-active C—C stretching mode near 800 cm⁻¹. Using an excitation source with a wavelength of 532 nm, what is the wavelength of the Raman-scattered light associated with this vibration?

Answer:
The wavelength can be determined by solving Eq. (5.1) for λ.

$$\Delta\bar{v} = \frac{1}{\lambda_0} - \frac{1}{\lambda}$$

$$\frac{1}{\lambda} = \frac{1}{\lambda_0} - \Delta\bar{v}$$

$$\lambda = \left[\frac{1}{\lambda_0} - \Delta\bar{v}\right]^{-1}$$

$$\lambda = \left[\frac{1}{532\,\text{nm}} - 800 \times 10^{-7}\,\text{nm}^{-1}\right]^{-1} \approx 555\,\text{nm}$$

Another question:
After measuring the Raman spectrum of cyclohexane, you notice an additional Raman peak near 563 nm. What is the Raman shift of this peak in inverse centimeters?

Answer:
1035 cm⁻¹

5.3. MOLECULAR POLARIZABILITY

As shown in Figure 5.3, multiple vibrations can produce Raman scattering. In order for Raman scattering to occur, however, there must be a change in the polarizability of the molecule (α) during the vibration. Polarizability describes the ease with which electrons rearrange in an applied electric field creating an *induced* dipole moment. I–I molecules, for example, are more polarizable than F–F molecules. This is because the outer electrons in the I–I molecule are significantly farther from the nucleus and can therefore be distorted more easily by an applied electric field. Light-induced distortion of the electron cloud surrounding a molecule is shown conceptually in Figure 5.4.

Vibrational modes that can produce Raman scattering are called *Raman active*. Symmetric vibrations usually exhibit the largest changes in polarizability and consequently the most intense Raman scattering. Conversely, IR absorption occurs if the vibration causes a change in the *permanent* molecular dipole moment (see Chapter 4), which typically occurs for antisymmetric vibrations. Infrared absorption and Raman scattering therefore provide complementary molecular information (see Figure 5.5). In fact, symmetric vibrations are Raman active but not IR active, and antisymmetric vibrations are IR active but not Raman active for any molecule having a center of symmetry. This condition is known as the *mutual exclusion principle*.

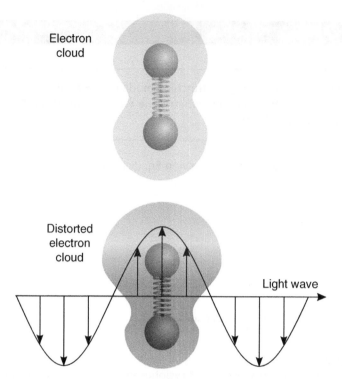

FIGURE 5.4 An electromagnetic field or light wave can distort the electron cloud surrounding a molecule.

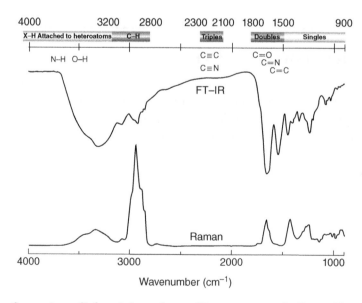

FIGURE 5.5 Comparison of infrared absorption and Raman spectra of collagen. The absorption spectrum was acquired with a Fourier transform infrared (FT-IR) spectrometer. Notice the similar but unique and complementary spectral features present in the Raman and FT-IR spectra. Source: From Lin et al. [30]. Licenced under CC BY 3.0.

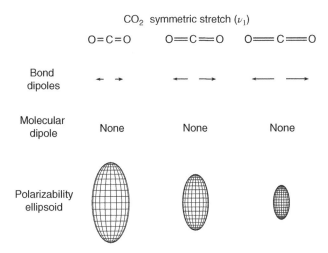

FIGURE 5.6 Dipole moments and polarizability ellipsoids for the CO_2 symmetric stretch vibration. There is no molecular dipole moment because the two bond dipoles cancel. The vibration is therefore not IR active. The polarizability ellipsoid depicts the magnitude of the polarizability in three dimensions. As explained in the text, the ellipsoids reflect the *inverse* square root of the polarizability, so as the polarizability of the CO_2 molecule decreases as the C=O bonds lengthen, the polarizability ellipsoid increases in size. As the C=O bonds gets shorter during the symmetric stretch, the polarizability increases, which is reflected in a smaller polarizability ellipsoid. As seen above, the size of the ellipsoid changes during vibration around the equilibrium position and the vibration is therefore Raman active.

The mutual exclusion principle is illustrated for the CO_2 symmetric stretch vibration in Figure 5.6. The figure depicts the bond dipoles, total molecular dipole, and molecular polarizability during vibration. The electron cloud surrounding CO_2 has an elongated, watermelon-like shape, and the electrons are more polarizable (larger α) along the chemical bond than perpendicular to it. Plotting α in all directions from the center of the molecule yields a three-dimensional surface. Conventionally, $1/\sqrt{\alpha}$ is plotted, and the resulting surface is called a *polarizability ellipsoid* [4]. The changes in the CO_2 polarizability ellipsoid during the symmetric stretch vibration are shown conceptually in Figure 5.6. It is important to note that because the surface reflects the *inverse* square of the polarizability, the size of the ellipsoid increases as the polarizibility decreases. The polarizability increases as the bond lengths increase because the electrons are less tightly held by the nuclei.

5.4.   BRIEF REVIEW OF MOLECULAR VIBRATIONS

Various types of molecular vibrations and the models used to describe them are discussed in detail in Chapter 4. As a reminder, vibrational frequencies of diatomic molecules are estimated using Hooke's law:

$$v = \frac{1}{2\pi}\sqrt{\frac{k(m_1+m_2)}{m_1 m_2}} \tag{5.2}$$

where k is the force constant of the bond and m_1 and m_2 are the masses of the two atoms. Note that this vibrational frequency can be converted to energy in wavenumbers, $\bar{v}$, by dividing by the speed of light ($\bar{v} = v/c$). Equation (5.2) can be used to understand the approximate energy of specific vibrations. All else being equal, stronger bonds vibrate at higher frequencies; for example, C—C bonds vibrate near $\bar{v} = 1650$ cm^{-1}, while C≡C bonds vibrate around 2150 cm^{-1}. Lighter atoms also vibrate at higher frequencies; C—H vibrations occur near 3000 cm^{-1}, while C—I bonds vibrate at less than 500 cm^{-1}.

Just as in infrared spectroscopy, the potential number of normal vibrational modes observed in a Raman spectrum can be determined by considering the degrees of freedom needed to describe the motion. Three degrees of freedom or coordinates are needed to completely describe the motion of a single atom: x, y, and z. For a molecule consisting of N atoms, there are $3N$ degrees of freedom, corresponding to 3 degrees of freedom per atom. These degrees of freedom correspond to translational, rotational, and vibrational motion. Three of the total $3N$ degrees of freedom describe molecular translation along the x, y, and z axes, while an additional three describe rotation about these axes. Subtracting the translational and rotational degrees of freedom leaves $3N - 6$ degrees of freedom for vibrational motion. This applies for all molecules except linear ones, which only have 2 degrees of rotational freedom because rotation about the bond axis leaves the molecule unchanged. Linear molecules therefore have $3N - 5$ vibrational degrees of freedom.

5.5. CLASSICAL THEORY OF RAMAN SCATTERING

A classical description of Raman scattering can be developed by examining the effect of an oscillating electric field on a molecule. Complicated vibrational motion is often expressed as a superposition of *normal vibrations* that are completely independent. Each of these normal vibrations can be described as

$$Q = Q_0 \cos(2\pi v_m t) \tag{5.3}$$

where Q is the "progress" along the oscillation, Q_0 is the amplitude of the oscillation, v_m is the frequency of the oscillation, and t is time. Assuming monochromatic illumination at frequency v, the time-varying electric field intensity at the molecule is given by

$$E = E_0 \cos(2\pi v t). \tag{5.4}$$

The dipole moment induced by this field is given by

$$\mu = \alpha E. \tag{5.5}$$

where α is the polarizability of the molecule. Remember that the polarizability describes the tendency of the molecule's electron cloud to be distorted by an applied electric field (see Section 5.3 and Figure 5.4).

Expanding the polarizability in a Taylor series about the molecule's equilibrium position yields

$$\alpha = \alpha_0 + \left(\frac{\partial \alpha}{\partial Q}\right) Q + \cdots \approx \alpha_0 + \left(\frac{\partial \alpha}{\partial Q}\right) Q_0 \cos(2\pi v_m t) \tag{5.6}$$

where α_0 is the polarizability at the equilibrium position. Using the trigonometric identity $\cos(A)\cos(B) = \cos(A + B)/2 + \cos(A - B)/2$, Eq. (5.5) can be rewritten as

$$
\begin{aligned}
\mu = \alpha E \approx{} & \alpha_0 E_0 \cos(2\pi\upsilon t) + \left(\frac{\partial\alpha}{\partial Q}\right) Q_0 \cos(2\pi\upsilon_m t) E_0 \cos(2\pi\upsilon t) \\
={} & \alpha_0 E_0 \cos(2\pi\upsilon t) + \left(\frac{\partial\alpha}{\partial Q}\right)\frac{E_0 Q_0}{2} \\
& \times \left\{ \cos\left[2\pi(\upsilon - \upsilon_m)t\right] + \cos\left[2\pi(\upsilon + \upsilon_m)t\right] \right\}.
\end{aligned}
\tag{5.7}
$$

This equation has three terms oscillating at υ, $\upsilon - \upsilon_m$, and $\upsilon + \upsilon_m$. Assuming the vibration causes a change in polarizability near the equilibrium position (i.e., $\partial\alpha/\partial Q \neq 0$), the induced dipole emits light at frequencies corresponding to elastic (υ), Stokes Raman ($\upsilon - \upsilon_m$), and anti-Stokes Raman scattering ($\upsilon + \upsilon_m$). The optical and vibrational frequencies involved in the interaction are summarized in Figure 5.7.

The intensity of the Raman-scattered light (I_{Raman}) can be determined by calculating the energy radiated from the induced oscillating dipole and is given by

$$
I_{\text{Raman}} \propto I_0 \upsilon^4 N f(\alpha^2)
\tag{5.8}
$$

where I_0 is the incident intensity, υ is the optical frequency, N is the molecular number density, and $f(\alpha^2)$ is a function of the squared polarizability. Of crucial importance is the linear dependence on N as it allows for accurate determination of chemical concentrations based on the Raman intensity. The equation also shows that the Raman-scattered intensity is enhanced by increasing either the incident intensity or excitation frequency. Lasers are frequently used for excitation because they are generally more intense than other light sources. In other words, they provide greater I_0 values and therefore produce more intense Raman scattering, I_{Raman}. High-frequency excitation sources can also provide significantly stronger Raman scattering because of the υ^4 dependence. In practice, however, intense high-frequency radiation can cause sample photodegradation. Thus, there are often limits on the intensity and frequency that can be used for excitation.

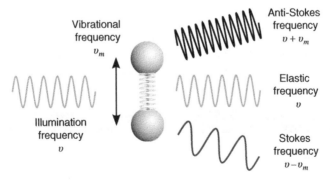

FIGURE 5.7 Raman-active modes vibrating at frequency υ_m and excited with monochromatic illumination at frequency υ will emit light corresponding to elastic (υ), Stokes Raman ($\upsilon - \upsilon_m$), and anti-Stokes Raman scattering ($\upsilon - \upsilon_m$).

EXAMPLE 5.2

Raman spectroscopy has been used extensively to study biological tissues. Unfortunately, tissue exhibits significant fluorescence that can overwhelm the Raman signal. Near-infrared radiation is therefore frequently used in Raman spectroscopy because it causes significantly less fluorescence to be produced by the sample than do visible and ultraviolet light. Calculate the ratio of Raman signal excited by a near-infrared diode laser ($\lambda_1 = 830$ nm) to that produced by a visible helium–neon laser ($\lambda_2 = 633$ nm). Assume that both sources have the same intensity.

Answer:
The ratio of Raman-scattered intensity can be calculated with Eq. (5.8).

$$\frac{I_0 \upsilon_1^4 Nf\left(\alpha^2\right)}{I_0 \upsilon_2^4 Nf\left(\alpha^2\right)} = \frac{\upsilon_1^4}{\upsilon_2^4} = \frac{(c/\lambda_1)^4}{(c/\lambda_2)^4} = \left(\frac{\lambda_2}{\lambda_1}\right)^4 = \left(\frac{633\,nm}{830\,nm}\right)^4 \approx \frac{1}{3}$$

Another question:
What excitation wavelength is required to produce a signal 100 times more intense than that generated by the 830 nm laser? Where does this wavelength fall in the electromagnetic spectrum, and is there a risk of generating significant fluorescence? Assume no change in incident intensity.

Answer:
$\lambda = 262$ nm would produce a 100-fold increase in sample intensity. However, because this falls in the ultraviolet range, it would likely generate significant fluorescence that could overlap and mask the weak Raman emission signals.

Equation (5.7) suggests that the Stokes and anti-Stokes scattering intensities are equal, but the derivation ignored the initial populations of the vibrational states. The ratio of the populations in excited (N_e) and ground (N_0) vibrational states is determined by Boltzmann's law:

$$\frac{N_e}{N_0} = e^{-(E_e - E_0)/kT} \tag{5.9}$$

where E_e and E_0 are the energies of the excited and ground vibrational states, respectively, k is the Boltzmann constant (0.695 cm^{-1}), and T is the temperature of the system in Kelvin. This ratio is also equivalent to the ratio of anti-Stokes to Stokes Raman signal.

EXAMPLE 5.3

A typical Raman transition involves an energy transfer of about 1000 cm^{-1} between the photon and the molecule. Calculate the ratio of anti-Stokes to Stokes Raman signal for this transition at room temperature (20 °C).

Answer:
The population ratio is calculated with Boltzmann's law.

$$\frac{N_e}{N_0} = \exp\left[-(E_e - E_0)/kT\right]$$

$$= \exp\left[-1000\text{cm}^{-1}\big/\left(0.695\,\text{cm}^{-1}/\text{K} \times 293\text{K}\right)\right]$$

$$\approx \frac{1}{135}.$$

There are approximately 135 times fewer molecules in the excited vibrational state causing the anti-Stokes signal to be about 135 times weaker than the Stokes signal.

Another question:
What is the energy of this transition (1000 cm^{-1}) in joules?

Answer:
1.986×10^{-20} J

5.6. POLARIZATION OF RAMAN SCATTERING

As discussed in Section 5.3, polarizability describes the ease with which electrons surrounding a molecule rearrange in an applied electric field. So far, we have primarily considered scalar representations of polarizability, but polarizability is not a single scalar number; we've seen that it can be different along different axes of the molecule and it can also change during molecular vibration. Remember that a change in polarizability during vibration is necessary for Raman scattering to occur (more precisely $\partial\alpha/\partial Q \neq 0$ near the molecule's equilibrium position). The three-dimensional nature of polarizability and its dependence on vibrational position are often illustrated with polarizability ellipsoids (see Figure 5.6).

In addition to the dependence on molecular orientation and vibrational position, polarizability (and therefore the magnitude and direction of an induced dipole moment) depends on the polarization of the incident light. This means that spectral differences can be observed between Raman-scattered light polarized parallel and perpendicular to the incident light. Taking polarization into account, the full expression for the dipole moment induced by the electric field (at a single vibrational position) is

$$\begin{bmatrix} \mu_x \\ \mu_y \\ \mu_z \end{bmatrix} = \begin{bmatrix} \alpha_{xx} & \alpha_{xy} & \alpha_{xz} \\ \alpha_{yx} & \alpha_{yy} & \alpha_{yz} \\ \alpha_{zx} & \alpha_{zy} & \alpha_{zz} \end{bmatrix} \begin{bmatrix} E_x \\ E_y \\ E_z \end{bmatrix}. \tag{5.10}$$

The dipole moment, μ, and the electric field, E, are both vector quantities each having a magnitude and a direction. The polarizability, α, is a 3×3 matrix where α_{xx} describes the contribution of the x-component of the applied electric field, E_x, to the x-component of

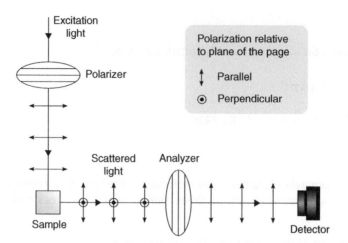

FIGURE 5.8 Bird's-eye view of a polarization-sensitive Raman measurement. The excitation light is passed through a polarizer to deliver linearly polarized light to the sample (polarized parallel to the page). The Raman-scattered light in general has components polarized both parallel and perpendicular to the excitation light. A second polarizer called an analyzer is used to sequentially pass each of these components to the detector.

the induced dipole moment, μ_x. Similarly, α_{xy} describes the contribution of E_y to $\mu_{x'}$ α_{xz} describes the contribution of E_z to $\mu_{x'}$ α_{yx} describes the contribution of E_x to μ_y etc. If the off-diagonal terms of $\boldsymbol{\alpha}$ (i.e., every term except α_{xx}, α_{yy} and α_{zz}) are nonzero, the excitation and scattered light can have different polarizations.

5.6.1. Depolarization Ratio

Many Raman instruments use two polarizers to measure the polarization difference between the excitation and scattered light: One polarizer controls the excitation, while the other is used to analyze the scattered radiation. Rather than attempting to deduce the full polarizability, in practice the depolarization ratio is usually measured instead. The depolarization ratio, ρ, is calculated by dividing the perpendicular component of the scattered intensity by the parallel component ($\rho = I_{\text{perpendicular}}/I_{\text{parallel}}$). The analyzer can be used to sequentially measure these polarization components as shown in Figure 5.8.

The depolarization ratio depends on the symmetry of the vibration: Symmetric vibrations generally have low depolarization ratios (less than 0.75), while asymmetric vibrations exhibit high depolarization (greater than 0.75). Depolarization measurements can therefore be used to aid molecular identification or verify peak assignments.

5.7. INSTRUMENTATION AND ANALYSIS METHODS

This section describes the basic types of instruments used to detect Raman-scattered light. Because Raman scattering is a weak effect, intense laser light is typically used for excitation, and instruments are equipped with sensitive detectors to measure as many Raman-shifted photons as possible. The Raman signal can also be easily overwhelmed by other light such as elastically scattered light, fluorescence, and ambient light. Most instruments therefore use specialized filters, lens coatings, enclosures, and optical baffling to reduce interfering signals. Finally, good spectral resolution and range are often required to analyze multiple vibrational modes, and noninvasive and noncontact sampling geometries are desirable to measure samples in their native environments.

5.7.1. Filter Instruments

The most basic Raman instrument consists of a narrowband excitation source, a detector, and a filter placed between the sample and detector. The filter is used to block the excitation and elastically scattered light while passing the Raman-scattered light generated over a particular wavelength range of interest. The first observation of Raman scattering by C. V. Raman used an instrument of this type; filtered sunlight was used for excitation, a colored glass filter was used to reject elastically scattered light and pass the Raman-scattered light, and Raman used his eye as a detector [1]. Today, laser light is typically used for excitation, and technological developments have produced better filters and more sensitive detectors. Although filter instruments can provide high throughput in a small, inexpensive package, they are limited because they provide only a single measurement of Raman intensity over the bandwidth of the filter (i.e., measurement of only one Raman "peak").

5.7.1.1. *Multifilter Instruments.* A natural extension to the single-filter instrument involves the incorporation of multiple optical filters. Using an appropriate set of filters, multifilter instruments can simultaneously measure multiple vibrational modes for only a modest increase in cost. Alternatively, a variable-wavelength filter can be used to sequentially measure many more vibrations. These variable-wavelength filters include scanning monochromators, acousto-optic tunable filters, angle-tuned interference filters, and birefringent filters, which are described in further detail below.

5.7.1.2. *Scanning Monochromators.* Scanning monochromators transmit a selectable narrow wavelength range based on the mechanical movement of a diffraction grating or prism. The Czerny–Turner design, which was described in detail in Chapter 2, remains one of the most commonly used scanning monochromators. The instrument uses a curved mirror to collimate light from an input slit, a grating to diffract the light, and another curved mirror to refocus the dispersed light onto an exit slit, effectively rejecting light outside the selected bandwidth as shown in Figure 5.9.

5.7.1.3. *Acousto-optic Tunable Filters.* Acousto-optic tunable filters (AOTFs) are electrically tunable bandpass filters with transmission wavelengths that can be rapidly adjusted on the order of microseconds. Multiple frequencies can also be applied to simultaneously generate multiple passbands. Although the bandwidths are too large for many applications, these filters are often used for Raman imaging. Acoustic waves are generated in a birefringent crystal by applying a radio-frequency signal to transducers bonded to the crystal surface. The acoustic waves induce a periodic refractive index profile in the crystal that acts as a diffraction grating as shown in Figure 5.10. Changing the frequency of the transducer signal changes the period of the diffraction grating, which enables rapid wavelength tuning.

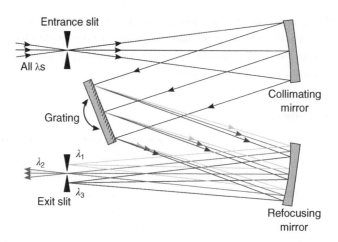

FIGURE 5.9 Bird's-eye view of the optical layout of a scanning Czerny–Turner monochromator. Source: Adapted with permission from Ref. [31]. https://commons.wikimedia.org/w/index.php?curid=40251125. Licenced under CC BY SA 4.0.

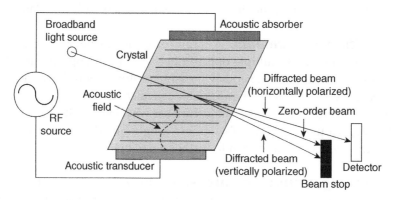

FIGURE 5.10 Major components of an AOTF. Source: From Xie et al. [32]. Used with permission.

5.7.1.4. Angle-Tuned Interference Filters.

Interference filters are commonly used for Raman spectroscopy and are engineered to provide a variety of transmission profiles including high-pass, low-pass, bandpass, or band rejection. Bandpass filters, for example, are designed to transmit a specific wavelength or wavelength range. However, they can be rotated relative to the incoming light to shift the transmission to a shorter wavelength. This behavior can be used to sequentially measure two Raman peaks: one at the design wavelength of the filter and one at a shorter wavelength that is transmitted after the filter is rotated.

Interference filters can generally be rotated 5–10° without degrading the performance. The corresponding wavelength shift is approximately

$$\Delta\lambda = \lambda_0 \left[1 - \sqrt{1 - \left(\frac{\sin\theta}{n_{\text{eff}}}\right)^2} \right] \tag{5.11}$$

where λ_0 is the center wavelength for normal incidence, θ is the angle of incidence relative to the surface normal, and n_{eff} is the effective index of refraction.

EXAMPLE 5.4

Although the light emitted from a laser is often described as having a single wavelength, the output is not perfectly monochromatic. The laser peak has a nonzero width, and there is often a small amount of power output in other longitudinal modes (i.e., wavelengths that can constructively interfere within the laser cavity). Bandpass filters are therefore often used to "clean up" the spectral output of a laser and deliver quasi-monochromatic light to the sample.

Suppose you find a narrow bandpass filter with a center wavelength of 835 nm and effective refractive index of 2 in the laboratory. Can you use this filter to "clean up" the output of an 830-nm diode laser?

Answer:
Plugging the design wavelength, effective refractive index, and maximum rotation angle ($\theta = 10°$) into Eq. (5.11) yields

$$\Delta\lambda = 835 \text{ nm}\left[1 - \sqrt{1 - \left(\sin 10°/2\right)^2}\right] = 3.2 \text{ nm}$$

It is not possible to use this filter to "clean up" the laser spectrum; angle tuning shifts the transmission wavelength at most 3.2 nm, and a 5 nm shift is required to transmit the 830 nm output of the laser.

Another question:
What is the maximum effective refractive index that would transmit 830 nm light? Assume the center wavelength is unchanged ($\lambda_0 = 835$ nm).

Answer:
1.59

5.7.1.5. Birefringent Filters. The final tunable filter type discussed here is a birefringent filter. These filters are made up of a birefringent crystal sandwiched between two polarizers. The refractive index of a birefringent crystal depends upon the polarization of the light; light polarized parallel to the optic axis of the crystal experiences a different refractive index than perpendicularly polarized light. The refractive index difference introduces a relative phase shift between parallel and perpendicular polarization components. This phase shift can be used to create a filter. For example, placing a birefringent crystal between crossed polarizers and orienting the optic axis of the crystal at 45° relative to the polarizers produces a transmission profile that is sinusoidal versus wavelength; maxima occur when the phase difference between the parallel and perpendicular polarization components is an odd multiple of π, while minima occur at even multiples. Several of these simple filters are combined in series to make a bandpass filter. Tunable filters can also be manufactured with a variable birefringent material such as a liquid crystal cell. These filters provide rapid wavelength tuning on the millisecond time scale.

5.7.2. Dispersive Spectrometers

The main limitation of filter-based instruments is that Raman-scattered light is detected only over the narrow wavelength range corresponding to the bandpass of the filter. Light scattered at other wavelengths is not detected because it is blocked by the filter.

In contrast, a dispersive spectrometer with an array of detectors can be used to separate and simultaneously measure Raman scattering at many wavelengths. The number of wavelengths measured is equal to the number of detectors or detector elements along the dispersion axis of the detector. Two of the most popular types of dispersive Raman spectrometers are the Czerny–Turner spectrometer and the axial transmissive spectrometer. These instruments probe many vibrational modes at the same time and gather significantly more information about the sample. This is analogous to how ultraviolet-visible absorption spectra provide more information than single absorption measurements made at one wavelength.

5.7.2.1. Czerny–Turner Spectrometer.
The Czerny–Turner spectrometer is similar to the monochromator shown in Figure 5.9 except that the exit slit is replaced with a detector array such as a charge-coupled device (CCD) camera. Instead of sequentially scanning through different wavelengths by rotating the grating, the grating is held at a fixed angle, and the full spectrum is recorded simultaneously with each element or pixel of the CCD camera acting as a separate exit "slit."

5.7.2.2. Axial Transmissive Spectrometer.
The axial transmissive spectrometer uses a volume holographic transmission diffraction grating to disperse the Raman-scattered light onto the detector. A set of lenses collimates light from the entrance slit or aperture and focuses the dispersed light onto the image plane as shown in Figure 5.11. Raman instruments built around these spectrometers often incorporate additional filters to ensure that only Raman-scattered photons are delivered to the spectrometer. For example, dichroic beam splitters that separate light by transmitting and reflecting different

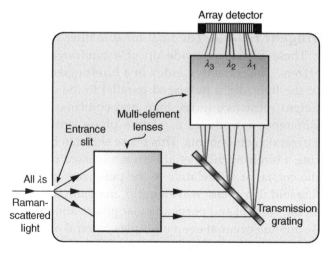

FIGURE 5.11   Bird's-eye view of the optical layout of an axial transmissive spectrometer.

wavelengths and notch filters that block only a narrow wavelength range are commonly used to attenuate stray laser light and elastically scattered light that could overwhelm weak Raman signals.

5.7.3. Fourier Transform Raman Spectrometers

Fourier transform spectrometers are also used to acquire Raman spectra. A Fourier transform Raman (FTR) instrument uses an interferometer to measure the interference between a sample beam and a reference beam as a function of pathlength difference between the beams. By taking the Fourier transform of the resulting interferogram, the Raman spectrum is recovered.

Although dispersive Raman instruments are more common, FTR spectrometers are especially useful for samples that exhibit large fluorescence backgrounds. This is because FTR instruments typically use near-infrared (NIR) excitation sources that produce little or no fluorescence background and are less likely to induce thermal damage.

5.7.4. Confocal Raman Instruments

Standard microscopes commonly found in biology and biochemistry labs form images of the sample on a detector, which is usually the observer's eye or a digital sensor. These microscopes are generally limited to imaging the surface of an object and are unable to detect subsurface structure or features. Conversely, a *confocal* microscope allows the observer to see light emitted from slices or planes below the sample surface.

Unlike a standard microscope, a confocal microscope forms images at one or more intermediate planes within the instrument. These planes are said to be *conjugate* to the others and are known as conjugate focal planes or confocal planes. By placing a pinhole in a confocal plane, light from a specific slice or plane in the sample is selected, while light from other sample regions is blocked as shown in Figure 5.12.

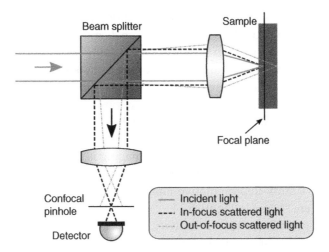

FIGURE 5.12 Basic principle of confocal microscopy. Backscattered light from the focus passes through the confocal pinhole, while light from out-of-focus planes is blocked by the pinhole.

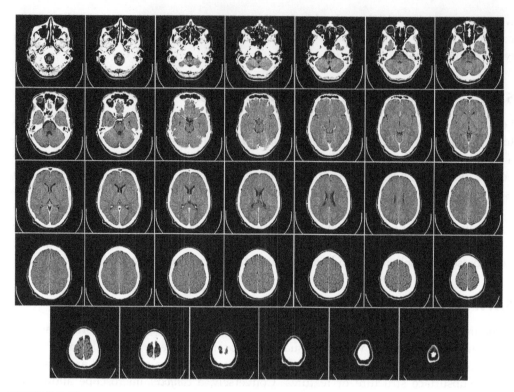

FIGURE 5.13 CT tomograms of a human brain from the base of the skull to the top. Source: Image used under Creative Commons CC0 1.0 Universal Public Domain Dedication, Author: Department of Radiology, Uppsala University Hospital. Uploaded by Mikael Hggstrm.

Confocal microscopy is a form of tomography, and each plane or slice imaged within the sample is called a tomogram. One of the most common forms of tomography is called computed tomography (CT). CT measures the absorption of X-rays as they pass through a sample at multiple angles in order to generate tomograms of the underlying structure. The technology is commonly used in medicine for both diagnostic and therapeutic purposes. CT tomograms of a human brain are shown in Figure 5.13.

Like CT, confocal Raman instruments can produce two- or even three-dimensional images derived from spectra acquired at many depths and lateral positions. Data from different depths are measured by adjusting the focal position within the sample, and different lateral positions can be selected by scanning the excitation beam across the sample with galvanometric mirrors. Confocal Raman microscopes often offer better spatial resolution and chemical specificity than CT, but the imaging is generally slow, and the penetration depth is limited (e.g., the maximum penetration depth in biological tissue is generally less than 1 mm). Representative images of live cells acquired with a confocal Raman instrument are shown in Figure 5.14.

In addition to providing depth selectivity, confocal microscopes generally provide better lateral (side-to-side) and axial (depth) resolution than standard microscopes; however, because some light is blocked by the confocal pinhole, the increased resolution and depth selectivity comes at the cost of decreased signal intensity.

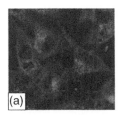 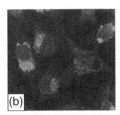 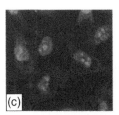 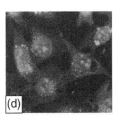

FIGURE 5.14 Confocal Raman images of LN-18 human malignant glioma cells. Each image shows different *spectral* information from the same spatial location (i.e., the same clustering of cells). Different cellular compartments are clearly visible in the panels including (a) the nucleus, (b) actin filaments, and (c) Golgi apparatus. Panel (d) shows a composite image of panels (a)–(c). Source: Reprinted from Klein et al. [33]. Reproduced with permission of Elsevier.

5.7.5. Light Sources

Lasers are ideal light sources for Raman spectroscopy because they can provide intense, narrowband, and polarized excitation that can be focused or reshaped with simple lens systems to illuminate and study a variety of macro- and microscopic samples. Prior to laser development in the 1960s, mercury lamps fitted with filters were frequently used instead. Complex systems were required to supply narrowband light from the lamp with enough intensity to produce significant Raman scattering. The lamps also required significant time to warm and ignite. Today, a variety of laser sources are used for Raman spectroscopy, a few of which are outlined below.

5.7.5.1. Ion Gas Lasers. Gas lasers that use argon or krypton gas can generate light over a wide range of visible wavelengths. The gas is contained within a water-cooled plasma tube. A large current is applied to ionize the gas and populate excited states. Depending upon the optics within the laser cavity, these lasers can output a number of discrete wavelengths. Argon-ion lasers, for example, can emit light between about 350 and 530 nm with 458, 488, and 514.5 nm being the most commonly used lines. The lasers can be configured to simultaneously output multiple wavelengths, but generally a prism is placed within the laser cavity to force operation at one specific wavelength.

5.7.5.2. Nd:YAG Lasers. Neodymium-doped yttrium aluminum garnet (Nd:YAG) lasers use a YAG crystal with a small fraction of the yttrium ions replaced with neodymium (the dopant). The neodymium ions provide the lasing activity after being optically pumped with flashtubes or laser diodes. The primary lasing wavelength is 1064 nm, which is ideal for FTR instruments. The laser can also be frequency doubled (i.e., wavelength divided by two) by placing a special crystal within the laser cavity. This generates intense light at 532 nm making the frequency-doubled Nd:YAG laser a good alternative to water-cooled gas lasers.

5.7.5.3. Diode Lasers. Diode lasers have been designed at specific wavelengths from the visible to the IR regions. These sources have been used in many consumer products including digital optical disk data storage (e.g., Blu-ray, DVD, CD), readers/writers, and fiber optic communications. The consumer applications have lowered the cost and accelerated

development of diode lasers. The sources are now inexpensive, compact, and highly efficient but generally require additional optics and careful temperature control to stabilize the output power and wavelength. By inserting a grating into the laser cavity and employing recent advancements that provide temperature control to within 0.01 °C, diode lasers become viable sources for many Raman applications.

5.8. QUANTITATIVE ANALYSIS METHODS

Many analytical methods have been developed to relate Raman spectra to quantitative sample properties, including composition and molecular orientation. These methods often take advantage of the linear relationship between concentration and Raman-scattered intensity (see Eq. 5.8). Please note that matrix notation is used to describe some of the methods in this section. These methods can significantly improve the accuracy of quantitative analyses, especially if the sample contains interfering spectral constituents; however, readers who are not familiar with matrix algebra may wish to skip Sections 5.8.3–5.8.5.

5.8.1. Calibration Curves

As is common in many forms of spectroscopy, a calibration curve can be used to quantify chemical concentrations. The curve is produced by measuring samples with known concentrations and plotting the height or area of a Raman band versus the associated analyte concentration. Light absorption and elastic scattering within the sample can complicate this procedure, so an internal standard, such as a different analyte with a constant concentration, is often used for normalization. After normalization, an equation that relates Raman intensity to analyte concentration is formulated by performing a linear least squares fit. This equation is then used to predict concentration in an unknown sample based upon the Raman spectrum alone.

5.8.2. Curve Fitting

In many samples, Raman peaks associated with the analyte of interest overlap with other interfering bands. If the interfering bands change intensity or shape between samples, the procedure outlined above may not generate a robust calibration curve. Curve fitting is commonly used to separate overlapping spectral components.

An example of curve fitting is displayed in Figure 5.15. The figure shows a representative Raman spectrum of bone as well as the best fit of the phosphate mineral peak. This peak is composed of at least two underlying components: one near 960 cm^{-1} arising from a phosphate vibration in carbonate-substituted apatite and another near 954 cm^{-1} due to a disordered phosphate vibration [5]. One can imagine curve fitting as the process of manually adjusting the heights of underlying components (the dash-dot and dotted lines in Figure 5.15) and adding them together until the sum accurately reproduces the larger, overall peak observed in the spectrum of the sample (the solid line in the inset in Figure 5.15). Then, the heights of the manually adjusted peaks can be used to quantify the

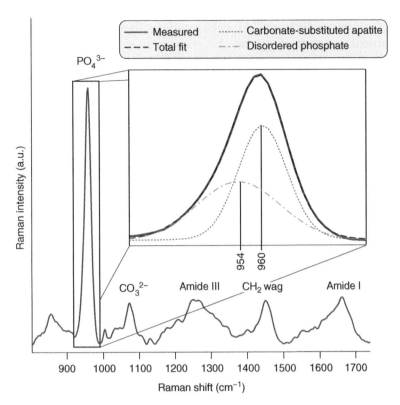

FIGURE 5.15 Representative Raman spectrum of bone including the best fit of the phosphate mineral band near 960 cm⁻¹.

individual components (e.g., the phosphate in the carbonate-substituted apatite and the disordered phosphate). Of course, doing so requires some prior knowledge of the shape of the underlying peaks that are contributing to the overall peak. While we have suggested that the analyst just manipulates the two peaks until the observed total peak is matched, in reality, mathematical methods are used to deconvolve the large peak into its individual components. Although curve fitting is useful for many applications, without prior knowledge of the approximate number of underlying bands as well as their approximate shapes and spectral locations, this method can produce variable results. This method can also require subjective user input, which makes the approach less reliable.

5.8.3. Ordinary Least Squares

Another method commonly used to study overlapping spectral features is ordinary least squares (OLS). OLS is an explicit linear model that requires knowledge of the spectrum associated with each varying chemical component of the sample. Compared to the methods outlined above, OLS uses all of the spectral data for calibration and generally produces superior results.

Consider the Raman spectrum $s(\Delta\bar{\nu})$ of a sample with multiple analytes. The concentration of a particular analyte of interest c is

$$c = \sum_{\Delta\bar{\nu}} s(\Delta\bar{\nu})b(\Delta\bar{\nu}) \tag{5.12}$$

where $b(\Delta\bar{\nu})$ is called the calibration vector. For the peak intensity calibration methods discussed above, $b(\Delta\bar{\nu})$ would be nonzero at the peak location and zero everywhere else. It is often convenient to express this equation in matrix form to simultaneously calculate concentrations of multiple analytes from multiple spectra. In matrix notation, the equation is

$$\begin{bmatrix} c_{11} & c_{12} & \cdots & c_{1n} \\ c_{21} & c_{22} & \cdots & c_{2n} \\ \vdots & \vdots & \ddots & \vdots \\ c_{m1} & c_{m2} & \cdots & c_{mn} \end{bmatrix} = \begin{bmatrix} - & s_1 & - \\ - & s_2 & - \\ & \vdots & \\ - & s_m & - \end{bmatrix} \cdot \begin{bmatrix} - & b_1 & - \\ - & b_2 & - \\ & \vdots & \\ - & b_n & - \end{bmatrix}^{T}, \tag{5.13}$$

which can be written more compactly as

$$\mathbf{C} = \mathbf{S} \cdot \mathbf{B}^{T}. \tag{5.14}$$

$\mathbf{C}$ is an $m \times n$ matrix where m is the total number of measured spectra and n is the number of chemical components. Similarly, $\mathbf{S}$ is an $m \times p$ matrix where p is the number of wavelengths per spectrum, $\mathbf{B}$ is the $n \times p$ calibration matrix with each row representing the calibration vector for a different analyte, and $\mathbf{B}^{T}$ is the transpose of $\mathbf{B}$ that is formed by interchanging the rows and columns of $\mathbf{B}$.

OLS, like other analytical methods, is used to calculate a calibration matrix $\mathbf{B}$ that will produce accurate predictions of analyte concentrations. If the Raman spectrum of each pure chemical component is known, the measured spectra can be expressed as

$$\mathbf{S} = \mathbf{C} \cdot \mathbf{P} + \mathbf{N} \tag{5.15}$$

where $\mathbf{P}$ is an $n \times p$ matrix of the pure spectral components and $\mathbf{N}$ an $m \times p$ matrix containing the noise at each wavelength. In order to predict the analyte concentrations $\mathbf{C}$, both sides of Eq. (5.15) are multiplied by $\mathbf{P}^{T}$ yielding

$$\mathbf{S} \cdot \mathbf{P}^{T} = \mathbf{C} \cdot \mathbf{P} \cdot \mathbf{P}^{T} + \mathbf{N} \cdot \mathbf{P}^{T}. \tag{5.16}$$

In general, the noise matrix cannot be modeled with $\mathbf{P}$ meaning that $\mathbf{N} \cdot \mathbf{P}^{T} \approx \mathbf{0}$ (i.e., the zero or null matrix). $\mathbf{C}$ can therefore be isolated by multiplying both sides by the inverse of $\mathbf{P}$ times its transpose. Note that $\mathbf{P} \cdot \mathbf{P}^{T}$ is invertible as long as the number of wavelengths p is greater than or equal to the number of pure spectral components n, which is usually true in practice. After performing these steps, Eq. (5.16) becomes

$$\mathbf{C} = \mathbf{S} \cdot \mathbf{P}^{T} \cdot (\mathbf{P} \cdot \mathbf{P}^{T})^{-1}. \tag{5.17}$$

Comparing to Eq. (5.14), we see that the transpose of the calibration matrix is

$$\mathbf{B}^T = \mathbf{P}^T \cdot (\mathbf{P} \cdot \mathbf{P}^T)^{-1}. \tag{5.18}$$

If the Raman spectrum of each pure component is known, OLS can be used to construct the calibration matrix $\mathbf{B}$ and predict component concentrations of an unknown sample using Eq. (5.14).

5.8.4. Classical Least Squares

Classical least squares (CLS) is a related technique that can be used if the concentrations of the chemical components in a calibration set are known, even if Raman spectra of the individual components have not been obtained. CLS is useful in biological applications where some biochemicals cannot be isolated for spectroscopic analysis while remaining biologically active. Although the biochemicals cannot be separated, other analytical techniques can sometimes be used to determine the component concentrations. If these concentrations are known, linear least squares fitting can be used to solve for $\mathbf{P}$, which can then be inserted into Eq. (5.18). This procedure is discussed in further detail in Ref. [6].

5.8.5. Implicit Analytical Methods

OLS and CLS are *explicit* analytical methods that require prior knowledge of either spectral properties or concentration levels for all components in a calibration sample. By performing the analysis described above, a calibration matrix is constructed and used to predict chemical concentrations in an unknown sample. *Implicit* analytical methods are best suited for cases when complete spectral or concentration information is not available. Principal component regression and partial least squares are two implicit methods that require concentration levels of only the analyte of interest to generate a calibration vector. Haaland and Thomas provide more detailed derivations and descriptions of these methods [6].

5.9. APPLICATIONS

Raman spectroscopy has been used in a variety of fields including art and archeology, polymers, electronics, pharmaceuticals, forensics, medicine, and biology. This section briefly covers some of these areas with an emphasis on biomedical applications.

5.9.1. Art and Archeology

Raman spectroscopy is an attractive analytical technique for studying art and archaeological objects because spectra can be obtained nondestructively from valuable samples. When coupled with confocal microscopy, multiple sample layers can be studied without requiring physical separation. This approach has been applied to identify specific dyes, pigments, resins, gemstones, and porcelains [2]. As an example, the identification of pigments used in a sixteenth-century German choir book is shown in Figure 5.16.

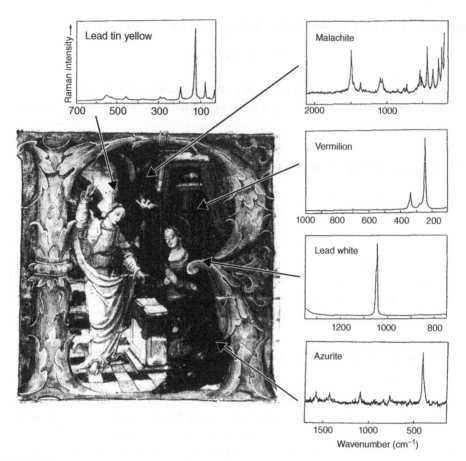

FIGURE 5.16 Sixteenth-century German choir book: historiated letter "R". Source: Reprinted from Clark [34]. Reproduced with permission of Elsevier.

5.9.2. Pharmaceuticals

The accurate characterization of pharmaceutical materials is important for manufacturing quality control. Unique Raman spectral profiles are obtained from a large variety of drugs. Quantitative analysis is used to measure the concentration distributions of the active pharmaceutical ingredients as well as other additives and binders. The spectra of a few drugs used to treat heart conditions and high blood pressure are shown in Figure 5.17. It is clear that the spectra are quite different and that each has spectral features that could be used to identify and quantify the drugs.

One of the main advantages of Raman spectroscopy for this application is the ability to acquire data nondestructively and *in situ* with little or no sample preparation. This reduces the risk of contamination and allows for simple and rapid characterization. Spectra of drugs can even be acquired after packaging depending upon the optical properties of the blister pack or bottle. There has been a significant increase in the use of Raman spectroscopy for pharmaceutical analysis because of these capabilities.

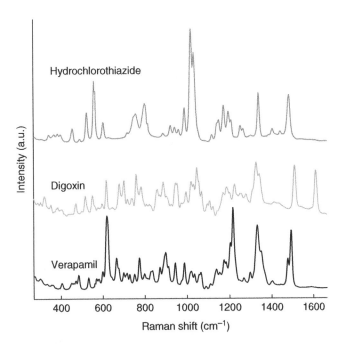

FIGURE 5.17 Representative Raman spectra of drug tablets used to treat hypertension (hydrochlorothiazide and verapamil), cardiac arrhythmia (digoxin and verapamil), and other related conditions.

5.9.3. Forensics

There are a number of forensic applications of Raman spectroscopy including the identification of plastic and liquid explosives and controlled drugs. Controlled drugs can even be identified in fingerprints. A recent study showed that Raman spectroscopy could distinguish five drugs of abuse (codeine phosphate, cocaine hydrochloride, amphetamine sulfate, barbital, and nitrazepam) from five noncontrolled substances with similar visual appearance (caffeine, aspirin, paracetamol, starch, and talc) in both sweat-rich and sebum-rich latent fingerprints [7]. (Sebum is an oily substance secreted by glands in the skin; it moisturizes, lubricates, and protects the skin from infection.) Drug identification is particularly challenging in this context because sweat and sebum produce interfering Raman signals. Raman spectra of a few controlled alkaloid drugs are shown in Figure 5.18.

5.9.4. Medicine and Biology

Disease and injury are often associated with biochemical changes that can be detected and characterized with Raman spectroscopy. Compared to other analytical techniques, Raman spectroscopy offers unique capabilities for measuring biological cells and tissue. For example, in some types of spectroscopy, molecules are chemically modified with tags that fluoresce, absorb, or scatter light to aid with identification. Sometimes these tags are toxic or alter the processes being studied. Raman measurements, however, are usually non- or minimally-invasive, require little or no sample preparation, and can often be performed without exogenous labels that could affect biological systems.

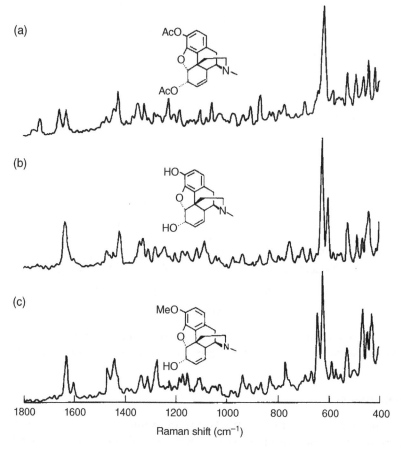

FIGURE 5.18 FTR spectra of the pure alkaloids: (a) heroin, (b) morphine, and (c) codeine. 50 scans, 6 cm^{-1} resolution, and incident laser power 200 mW. Scanning time 3 min. Source: Reprinted from Hodges and Akhavan [35]. Reproduced with permission of Elsevier.

Raman spectroscopy has been used to study a variety of tissues including the bone, skin, brain, eye, oral, artery, lung, breast, gastrointestinal, urinary tract, uterus, and cervical [8–15]. Many of these studies examined biochemical changes due to disease or injury. Raman spectroscopy has also been used to study atherosclerosis, breast cancer, Alzheimer's disease, blood analytes, topical drug delivery, single cells, and microorganisms [16–23].

In biological tissues, NIR radiation is often used for excitation instead of visible light because it produces less interfering fluorescence. An alternative strategy to reduce the fluorescence signal is to use ultraviolet (UV) excitation sources. In biological samples, the high-lying electronic states excited by UV light typically relax rapidly via non-radiative processes to the first electronic state, S_1. The fluorescence emission remains predominantly associated with transitions from S_1 to S_0 and therefore occurs at much longer wavelengths than the illumination wavelength. This effectively shifts the fluorescence emission outside the spectral range where Raman bands are observed. In addition, if the UV illumination is tuned to resonantly excite an electronic transition (a method known as UV resonance Raman spectroscopy), the probability of vibrational transitions, and therefore measured

Raman intensity, can be increased. Unfortunately, the penetration depth of UV light in tissue is on the order of microns, so this method is only suitable for superficial measurements, and overexposure to UV light can induce mutagenesis [16].

5.9.4.1. Bone Diagnostics.

Raman spectroscopy has been used extensively to study changes to bone biochemistry associated with growth, disease, fracture healing, and insertion of prosthetic implants. This section reviews a recent Raman spectroscopy study of bone disease [9] with an emphasis on the specific instrumentation and experimental choices that were made.

A diagram of the optical layout used in the study is shown in Figure 5.19. The system used a semiconductor laser with an output power of 500 mW and a wavelength of $\lambda = 830$ nm. An NIR source was chosen because bone produces significant fluorescence at shorter wavelengths. Illumination wavelengths longer than $\lambda = 830$ nm were avoided because Raman intensity decreases with increasing wavelength (Eq. 5.8), and the sensitivity of silicon CCD detectors is low in this spectral range.

The laser power was attenuated by optical elements in the system to deliver approximately 80 mW to the bone. The power delivered was chosen to maximize the Raman-scattered signal (remember Raman intensity is proportional to incident intensity as shown in Eq. 5.8) without inducing thermal damage that occurs at higher power [24].

A shutter in the illumination path blocked the laser light between measurements to avoid unnecessary exposure. Bandpass filters and dichroic beam splitters attenuated light outside the design wavelength of $\lambda = 830$ nm. The laser light was focused with a lens to a 1.5-mm diameter spot on the surface of the bone. The backscattered light generated at the laser focus was collected with the same lens and directed to the dichroic beam splitters. The first beam splitter (labeled BS1) reflected laser light at 830 nm but transmitted the Stokes-shifted Raman-scattered light. The second beam splitter (labeled BS2) reflected both the laser light and the Raman-scattered light but transmitted visible light to a CCD camera (labeled CCD2) used to acquire conventional images of the sample. A notch filter in the

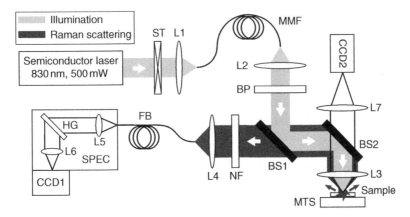

FIGURE 5.19 Raman spectroscopy system for bone diagnostics. Light gray corresponds to the laser illumination path, and dark gray corresponds to the Raman scattering collection path. Abbreviations: ST, shutter; L, lens; MMF, multimode fiber; BP, bandpass filter; BS, dichroic beam splitter; MTS, motorized translation stage; NF, notch filter; FB, fiber bundle; SPEC, spectrograph; HG, holographic grating; CCD, charge-coupled device. Source: From Maher et al. [13].

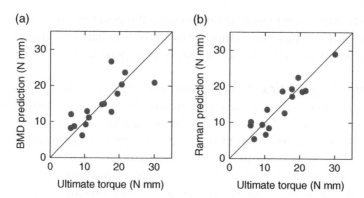

FIGURE 5.20 Scatterplots of the ultimate torque prediction for each bone sample versus its measured value for the (a) BMD and (b) Raman prediction models. The diagonal line is the perfect prediction line. The closer a data point lies to this line, the more accurate the prediction. Source: https://www.ncbi.nlm.nih.gov/pmc/articles/PMC5539739. Licenced under CC BY 4.0.

collection path further attenuated the laser light. After filtering, the Raman-scattered light was focused into an optical fiber bundle that delivered the light to a spectrometer.

An axial transmissive spectrometer with a holographic grating was coupled to a multichannel CCD camera for detection. This setup provided high throughput over a large spectral range enabling simultaneous measurement of multiple Raman peaks with high fidelity. The CCD camera was thermoelectrically cooled to $-60\ °C$ during acquisition to reduce electronic noise due to thermal fluctuations.

The study compared Raman spectra of bone samples extracted from normal mice and mice with osteoporosis, rheumatoid arthritis, or both. Significant differences in mineral content were observed in the diseased bones. The Raman data was also compared to the results of a destructive strength test that measured the maximum torque that could be applied prior to fracture (also known as the ultimate torque). Partial least squares regression, an advanced quantitative analysis method presented in Section 5.8, was used to predict the strength of each bone based on the nondestructive Raman measurements. The results show that the biochemical Raman information can accurately predict the corresponding strength of the bone. Each prediction versus the true measured strength is plotted in Figure 5.20. Predictions based on the clinical standard measurement of bone mineral density (BMD) measured with an X-ray-based technique are also shown for comparison. The results suggest that Raman spectroscopy is a valuable complementary tool for monitoring bone health and strength.

5.10. SIGNAL ENHANCEMENT TECHNIQUES

When photons scatter from a molecule, the majority are elastically scattered while only a small fraction (approximately one in $10^6 - 10^8$ [2]) are Raman scattered. There are a number of variants of Raman spectroscopy that enhance weak Raman signals. These include resonance Raman spectroscopy, surface-enhanced Raman spectroscopy, and nonlinear Raman spectroscopy. Although detailed descriptions are beyond the scope of this chapter, a discussion of general concepts and relevant applications is provided.

5.10.1. Resonance Raman Spectroscopy

Resonance Raman scattering occurs when the excitation photons promote the molecule to an excited electronic state rather than a virtual level lying between electronic states (see energy-level diagram in Figure 5.2). With this approach, the Raman intensities can be enhanced by a factor of 10^3–10^6.

Tunable laser sources are ideal for acquiring resonance Raman spectra because the frequency can be tuned to deliver the exact photon energy required to promote the molecule to an excited electronic state. Some enhancement still occurs, however, even if the photon energy does not exactly match the energy difference between electronic states. It is therefore often more practical to use a readily available, nontunable laser with a frequency that is close to the true resonance frequency.

The primary drawbacks of resonance Raman spectroscopy are increased risks of absorption, fluorescence, and photodegradation. Unfortunately, the ratio of scattering to absorption is a molecular property that is difficult to predict. Fluorescence interference, which can also be enhanced when using excitation frequencies near the resonance frequency, also limits the molecules that are suitable for study using resonance Raman spectroscopy. Moving the excitation frequency away from the resonance can mitigate some of these effects while retaining a degree of enhancement. Spinning sample holders or flow cells can also be used to reduce the chance of photodegradation as each sample region is excited for only a short period of time.

Despite these drawbacks, resonance Raman spectroscopy produces a large signal enhancement enabling a number of new applications. For example, water and proteins are often difficult to study because they generate particularly weak Raman signals. Resonance Raman spectroscopy is therefore especially useful for studying biological systems provided that significant absorption, fluorescence, and photodegradation are avoided. A thorough review of resonance Raman spectroscopy is provided in an article by Efremov et al. [25].

5.10.2. Surface-Enhanced Raman Spectroscopy

Another technique that significantly increases scattering efficiency is surface-enhanced Raman spectroscopy (SERS). This approach can provide enhancement factors of 10^6 up to as much as 10^{11} over normal Raman scattering. Despite the large improvement in signal intensity, the application of SERS has been somewhat limited because of the complexity of the technique and debate about the exact mechanism of the enhancement. Because SERS is effective with a wider range of molecules and can provide a larger sensitivity enhancement than resonance Raman spectroscopy, it is worth considering for some applications.

SERS enhancement occurs when molecules are adsorbed on rough metal surfaces or placed near metallic nanostructures. Laser excitation of these metallic structures causes resonant oscillations of conduction electrons at the surface, which are known as surface plasmons. This creates a highly localized electric field at the surface. Molecules that are adsorbed on or placed near the metallic surface are excited by this strong localized field, ultimately producing large enhancements in the Raman-scattered intensity. SERS has been used for a variety of applications from basic analytical chemistry, drug discovery, forensics, detection of trace chemical and biological agents, single-molecule detection, and

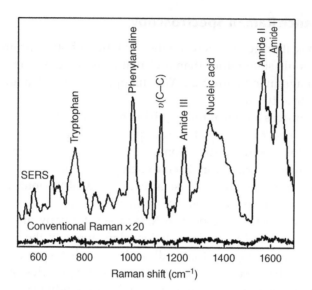

FIGURE 5.21   Comparison of SERS spectrum of blood plasma adsorbed on microwave-treated Au–PS substrates with conventional Raman spectrum of blood plasma. Note that different spectra have been multiplied by different factors as indicated in the figure for better visualization. The 785 nm laser excitation power is 2.5 mW for SERS measurements, while 80 mW for conventional Raman measurements. Source: Reprinted from Yuen et al. [26]. Reproduced with permission of Elsevier.

point-of-care medical diagnostics. A comparison of SERS and conventional Raman spectra of blood plasma is shown in Figure 5.21. The SERS substrate provided an intensity enhancement of 5.3×10^3 over conventional Raman measurements acquired on a glass substrate [26]. A comprehensive survey of chemical applications and a more extensive description of SERS theory are provided in a review article by Schlucker [27].

5.10.3.   Nonlinear Raman Spectroscopy

As discussed in Section 5.5, the applied electric field and the corresponding dipole moment induced in a molecule are linearly related by the equation

$$\mu = \alpha E \tag{5.19}$$

where α is the molecular polarizability or the ease with which the molecule's electron cloud can be distorted by the electric field. When the intensity of the incident light becomes sufficiently large, however, the relationship becomes nonlinear:

$$\mu = \alpha E + \frac{1}{2}\beta E^2 + \frac{1}{6}\gamma E^3 + \cdots \tag{5.20}$$

where β and γ are the first and second hyperpolarizabilities, respectively. In conventional Raman spectroscopy, the contributions of the β and γ terms are negligible because $\alpha \gg \beta \gg \gamma$. The contributions become significant when the sample is excited with extremely strong laser pulses as E^2 and E^3 become much larger. This leads to a variety of nonlinear Raman effects. Two of these effects are discussed here: coherent anti-Stokes Raman scattering (CARS) and stimulated Raman scattering (SRS).

Unlike conventional Raman measurements, CARS and SRS are generated by illuminating the sample with two high-energy laser pulses with frequencies v_1 and v_2. If the difference in frequencies $(v_1 - v_2)$ is tuned to the frequency of a Raman-active vibrational mode $(v_m = v_1 - v_2)$, the vibration is coherently driven, and a strong beam at a new frequency $v_1 + v_m$ is emitted. This multiphoton process is called CARS.

In addition to the CARS signal generated at $v_1 + v_m$, the two incident laser pulses exchange energy; the intensity of the pulse at frequency v_2 experiences a gain (stimulated Raman gain (SRG)), and the pulse at frequency v_1 experiences a loss (stimulated Raman loss (SRL)). If the frequency difference between the pulses does not match one of the molecule's vibrational frequencies, there is no energy transfer and no SRG or SRL signals. Unlike CARS, SRS does not occur at a new frequency. Instead, molecular vibrations are identified by measuring intensity changes (SRG or SRL) in the laser pulses after interaction with the sample. An SRS image of a brain tumor is shown in Figure 5.22. Although the tumor is not visible in the conventional bright-field microscope image, the SRS signal shows high contrast between tumor-infiltrated and normal brain tissue [28].

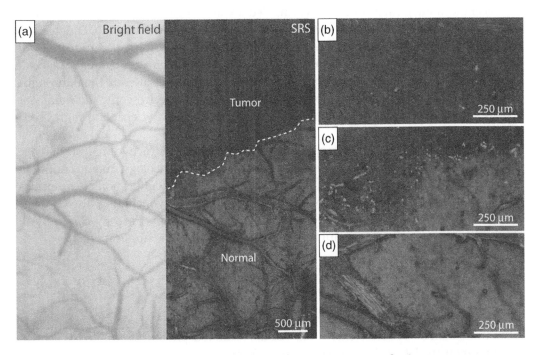

FIGURE 5.22 *In vivo* SRS microscopy images of human GBM xenografts. Images are representative of six mice. SRS imaging was carried out via acute cranial window preparation in mice 24 days after implantation of human GBM xenografts. (a) Bright-field microscopy appears grossly normal, whereas SRS microscopy within the same field of view demonstrates distinctions between tumor-infiltrated areas and noninfiltrated brain (normal), with a normal brain-tumor interface (dashed line). (b–d) High-magnification views (b) within the tumor, (c) at the brain surface, and (d) within normal brain. Source: From Ji et al. [28]. Reproduced with permission from AAAS.

The primary advantage of CARS and SRS is that the signal intensity is typically many times greater than conventional Raman spectroscopy. This has enabled Raman imaging at high speed by rapidly scanning the beam across a sample and measuring the coherent Raman signal at each position. Coherent Raman spectroscopy methods have not replaced conventional Raman spectroscopy, however, as they remain significantly more expensive and technically challenging to implement. A review of coherent Raman spectroscopy with an emphasis on biological applications is provided in an article by Camp and Cicerone [29].

5.11. SUMMARY

Raman spectroscopy probes the vibrational frequencies of a molecule providing information that can be used to determine chemical structure, identify analytes, or quantify chemical concentrations. Unlike other methods, Raman spectroscopy is based on inelastic light scattering rather than absorption or emission. Raman scattering occurs only if the polarizability of the molecule changes during vibration. Raman spectroscopy is therefore particularly sensitive to symmetric vibrations as these modes typically exhibit large polarizability changes during vibration.

Recent technological advances have enabled a wide range of applications for Raman spectroscopy. Stable laser diodes with narrow linewidths, multichannel CCD detectors, and holographic filters and gratings have all improved the performance and utility of Raman instruments. Although Raman scattering remains a weak effect, high-quality spectra can be acquired with modern instruments on the order of seconds. Recent breakthroughs in resonance Raman spectroscopy, surface-enhanced Raman spectroscopy, and nonlinear Raman spectroscopy offer additional increases in signal intensity enabling high-speed acquisition and Raman imaging.

Current applications of Raman spectroscopy include nondestructive analysis of dyes, pigments, resins, gemstones, and porcelains used in art and archeology, quality control in pharmaceutical manufacturing, forensic applications such as identification of explosives and controlled drugs, and applications in medicine and biology including disease diagnosis and treatment monitoring. Overall, the field of Raman spectroscopy continues to grow with many directions for future research, development, and application.

PROBLEMS

5.1 Benzene has a Raman-active ring stretching mode near 990 cm^{-1}. Using a near-infrared diode laser for excitation, Raman-scattered photons associated with this vibration are detected at a wavelength of 850 nm. What is the wavelength of the laser source?

5.2 The conducting polymer polyaniline has a C–H bending mode near 1192 cm^{-1}. Will the spectral position of this peak shift down or up if the molecule is fully deuterated by replacing hydrogen with deuterium? Estimate the Raman shift for the peak after deuteration using Hooke's law assuming no change in the force constant k.

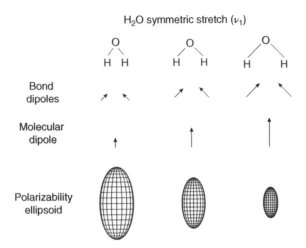

FIGURE 5.23   Dipole moments and polarizability ellipsoids for the H_2O symmetric stretching mode.

5.3   How many degrees of freedom does the linear molecule CO_2 have for vibrational motion? Given that CO_2 has a center of symmetry, how would you determine if these vibrational modes are Raman active, IR active, or both?

5.4   Figure 5.23 shows the dipole moments and polarizability ellipsoids for the H_2O symmetric stretching mode. Is this vibration Raman active, IR active, or both? Does the mutual exclusion principle apply to H_2O?

5.5   What incident power provided by a helium–neon laser ($\lambda = 633$ nm) produces the same Raman intensity as generated by 10 mW of ultraviolet excitation at 375 nm? Assume no difference in molecular polarizability.

5.6   At room temperature, what type of Raman scattering is more intense: Stokes or anti-Stokes? Does the molecule gain or lose energy during this process?

5.7   If a Stokes peak at 500 cm^{-1} is 10 times more intense than the corresponding anti-Stokes peak, what is the temperature of the sample according to Boltzmann's law?

5.8   If the depolarization ratio of a vibration is nonzero, what does this indicate about the polarizability matrix?

5.9   If the Raman intensity polarized parallel to the excitation polarization is 10 times greater than the perpendicular component, what is the depolarization ratio? Is this vibration symmetric or asymmetric?

5.10   Name four types of variable-wavelength filters that are used in filter-based Raman instruments. Which variable-wavelength filter uses a transducer?

5.11   What is the minimum angle of incidence that allows transmission of light from a helium–neon laser ($\lambda = 633$ nm) through a narrow dielectric bandpass filter with a center wavelength of 635 nm and an effective refractive index of 1.5?

5.12   Name two types of dispersive Raman spectrometers and an array detector commonly used with these instruments.

5.13   Confocal Raman instruments provide high spatial resolution by preventing out-of-focus light from reaching the detector. What element is used to reject this light and where is it placed?

5.14   (a) Quantitative analysis methods often rely on what relationship between the sample and the measured Raman spectrum? (b) Name two implicit analytical methods. How do these methods differ from explicit ones?

5.15   (a) What are the primary drawbacks of resonance Raman spectroscopy? (b) SERS enhancement is produced by a strong, localized electric field at a rough metal surface or nanostructure. What is the name of the resonant oscillation of conduction electrons that generates this localized electric field? (c) Unlike conventional Raman spectroscopy, the nonlinear CARS and SRS techniques use two high-energy laser pulses to coherently drive a specific vibration. What is the energy in relative wavenumbers that is probed by lasers with excitation wavelengths of $\lambda_1 = 850$ nm and $\lambda_2 = 1150$ nm? What is the wavelength of the CARS signal generated with these lasers?

REFERENCES

1. Raman, C.V. and Krishnan, K.S. (1928). *Nature 121*: 501–502.

2. Smith, E. and Dent, G. (2005). *Modern Raman Spectroscopy – A Practical Approach*. New York: Wiley.

3. Eckenrode, B.A., Bartick, E.G., Harvey, S. et al. (2001). *Forensic Sci. Commun.* 3.

4. Ferraro, J.R., Nakamoto, K., and Brown, C.W. (2003). *Introductory Raman Spectroscopy*, 2e. San Diego: Academic Press.

5. Tarnowski, C.P., Ignelzi, M.A. Jr., and Morris, M.D. (2002). *J. Bone Miner. Res. 17*: 1118–1126.

6. Haaland, D.M. and Thomas, E.V. (1988). *Anal. Chem. 60*: 1193–1202.

7. Day, J.S., Edwards, H.G., Dobrowski, S.A., and Voice, A.M. (2004). *Spectrochim. Acta A 60*: 563–568.

8. Krafft, C. and Sergo, V. (2006). *Spectroscopy 20*: 195–218.

9. Maher, J.R., Takahata, M., Awad, H.A., and Berger, A.J. (2011). *J. Biomed. Opt. 16*: 087012.

10. Takahata, M., Maher, J.R., Juneja, S.C. et al. (2012). *Arthritis Rheumatol. 64*: 3649–3659.

11. Beier, E.E., Maher, J.R., Sheu, T.-J. et al. (2013). *Environ. Health Perspect. 121*: 97–104.

12. Inzana, J.A., Maher, J.R., Takahata, M. et al. (2013). *J. Biomech. 46*: 723–730.

13. Maher, J.R., Inzana, J.A., Awad, H.A., and Berger, A.J. (2013). *J. Biomed. Opt. 18*: –077001.

14. Maher, J.R., Matthews, T.E., Reid, A.K. et al. (2014). *J. Biomed. Opt. 19*: 117001–117001.

15. Feng, G., Ochoa, M., Maher, J.R. et al. (2017). *J. Biophotonics 10*: 990–996.

16. Hanlon, E.B., Manoharan, R., Koo, T.-W. et al. (2000). *Phys. Med. Biol. 45*: R1–R59.

17. Swain, R. and Stevens, M. (2007). *Biochem. Soc. Trans. 35*: 544–549.

18. Krafft, C., Dietzek, B., and Popp, J. (2009). *Analyst 134*: 1046–1057.

19. Wax, A., Giacomelli, M.G., Matthews, T.E. et al. (2012). *Adv. Opt. Photon. 4*: 322–378.

20. Maher, J.R., Chuchuen, O., Henderson, M.H. et al. (2015). *Biomed. Opt. Express 6*: 2022–2035.

21. Chuchuen, O., Maher, J.R., Simons, M.G. et al. (2017). *J. Pharm. Sci. 106* (2): 639–644.

22. Chuchuen, O., Maher, J.R., Simons, M.G. et al. (2017). *J. Pharm. Sci. 106*: 639–644.

23. Chuchuen, O., Maher, J.R., Henderson, M.H. et al. (2017). *PLoS One 12*: e0185633.

24. Motz, J.T. (2003). Development of *in vivo* Raman spectroscopy of atherosclerosis. PhD thesis. MIT, p. 100.

25. Efremov, E.V., Ariese, F., and Gooijer, C. (2008). *Anal. Chim. Acta 606*: 119–134.

26. Yuen, C., Zheng, W., and Huang, Z. (2010). *Biosens. Bioelectron. 26*: 580–584.

27. Schlücker, S. (2014). *Angew. Chem. Int. Ed. 53*: 4756–4795.

28. Ji, M., Orringer, D.A., Freudiger, C.W. et al. (2013). *Sci. Transl. Med. 5*: 201ra119.

29. Camp, C.H. Jr. and Cicerone, M.T. (2015). *Nat. Photonics 9*: 295–305.

30. Lin, S.-Y., Li, M.-J., and Cheng, W.-T. (2007). *J. Spectrosc. 21*: 1–30.

31. https://commons.wikimedia.org/w/index.php?curid=35646782 (accessed 7 July 2016).

32. Xie, L., Wang, A., Xu, H. et al. (2016). *Trans. ASABE 59*: 399–419.

33. Klein, K., Gigler, A.M., Aschenbrenner, T. et al. (2012). *Biophys. J. 102*: 360–368.

34. Clark, R.J. (1995). *J. Mol. Struct. 347*: 417–427.

35. Hodges, C.M. and Akhavan, J. (1990). *Spectrochim. Acta A 46*: 303–307.

FURTHER READING

1. Ferraro, J.R., Nakamoto, K., and Brown, C.W. (2003). *Introductory Raman Spectroscopy*, 2e. San Diego: Academic Press.

2. Smith, E. and Dent, G. (2005). *Modern Raman Spectroscopy – A Practical Approach*. New York: Wiley.

3. Diem, M. (1993). *Introduction to Modern Vibrational Spectroscopy*. New York: Wiley.

4. Bright Wilson, E. Jr., Decius, J.C., and Cross, P.C. (1980). *Molecular Vibrations: The Theory of Infrared and Raman Vibrational Spectra*. New York: Dover Publications, Inc.

5. Larkin, P. (2017). *Infrared and Raman spectroscopy: Principles and Spectral Interpretation*, 2e. Cambridge: Elsevier.

6. Schrader, B. (ed.) (2008). *Infrared and Raman spectroscopy: Methods and Applications*. New York: Wiley.

7. Hanlon, E.B., Manoharan, R., Koo, T.-W. et al. (2000). *Phys. Med. Biol. 45*: R1–R59.

8. Krafft, C. and Sergo, V. (2006). *J. Spectrosc. 20*: 195–218.

9. Shipp, D., Sinjab, F., and Notingher, I. (2017). *Adv. Opt. Photon. 9*: 315–428.

SOLUTIONS

CHAPTER 1

1.1 (a) When electromagnetic radiation is propagating through space.
(b) When electromagnetic radiation interacts with matter.

1.2 (a) $E = h\upsilon = \left(6.626 \times 10^{-34} \text{ Js}\right)\left(1.73 \times 10^{17} \text{ s}^{-1}\right) = 1.15 \times 10^{-16} \text{ J}$

(b) $\upsilon = \dfrac{c}{\lambda} = 1.73 \times 10^{17} \text{ s}^{-1} = \dfrac{2.998 \times 10^{8} \text{ m/s}}{\lambda}$

$\lambda = \dfrac{2.998 \times 10^{8} \text{ m/s}}{1.73 \times 10^{17} \text{ s}^{-1}} = 1.73 \times 10^{-9} \text{ m}$

$\upsilon = c\bar{\nu}$ so $\dfrac{\upsilon}{c} = \bar{\nu} = \dfrac{1.73 \times 10^{17} \text{ s}^{-1}}{2.998 \times 10^{8} \text{ m/s}} = 5.77 \times 10^{8} \text{ m}^{-1}$

(c) This type of radiation falls in the X-ray region and therefore can cause core electrons to be excited or ejected from an atom. It can also cause bond breaking.

(d) The energy of one of these photons is 1.15×10^{-16} J, so the energy of a mole of them is given by multiplying the energy by Avogadro's number:

$E = \left(1.15 \times 10^{-16} \text{ J/photon}\right)\left(6.022 \times 10^{23} \text{ photons/mol}\right) = 6.93 \times 10^{7} \text{ J/mol}$

1.3 First convert kcal to joules because we know Planck's constant in terms of joules:

$$\dfrac{100.0 \text{ kcal}}{\text{mol}} = \dfrac{100 \times 10^{3} \text{ cal}}{\text{mol}} \times \dfrac{4.184 \text{ J}}{\text{cal}} = 418\,400 \text{ J/mol}$$

Next calculate the energy of a single photon using Avogadro's number:

$$\dfrac{418400 \text{ J}}{\text{mol}} \times \dfrac{1 \text{ mol}}{6.022 \times 10^{23} \text{ photons}} = 6.948 \times 10^{-19} \text{ J/photon}$$

Spectroscopy: Principles and Instrumentation, First Edition. Mark F. Vitha.
© 2019 John Wiley & Sons, Inc. Published 2019 by John Wiley & Sons, Inc.

Now use Planck's equation to calculate the wavelength

$$E = \frac{hc}{\lambda}, \text{ which rearranges to } \lambda = \frac{hc}{E} = \frac{\left(6.626 \times 10^{-34} \text{ Js}\right)\left(2.998 \times 10^{8} \text{ m/s}\right)}{6.948 \times 10^{-19} \text{ J}}$$
$$= 2.859 \times 10^{-7} \text{ m} \quad \text{or} \quad 285.9 \text{ nm}$$

This wavelength is in the UV region of the spectrum.

1.4 Set up a ratio. We know (and the question implies) that the red photon has higher energy than the microwave photon, so we put that in the numerator to get a value greater than 1.00:

$$\frac{E_{red}}{E_{microwave}} = \frac{hc/\lambda_{red}}{hc/\lambda_{microwave}}$$

The "hc" terms cancel because they are constants, leaving

$$\frac{E_{red}}{E_{microwave}} = \frac{1/\left(700.0 \times 10^{-9} \text{ m}\right)}{1/\left(2.000 \times 10^{-2} \text{ m}\right)} = 2.857 \times 10^{4}$$

So the red photon is 28 000 times more energetic than the microwave photon! That is a lot more energetic.

1.5 $\eta = \dfrac{c_{vacuum}}{c_{germanium}} = \dfrac{2.998 \times 10^{8} \text{ m/s}}{6.503 \times 10^{7} \text{ m/s}} = 4.610$

1.6 First convert the speed in miles per hour into meters per second, and then calculate the refractive index:

$$\frac{38 \text{ miles}}{\text{hour}} \times \frac{1 \text{hr}}{60 \text{min}} \times \frac{1 \text{min}}{60 \text{s}} \times \frac{5280 \text{ft}}{\text{mile}} \times \frac{12 \text{in}}{1 \text{ft}} \times \frac{2.54 \text{cm}}{\text{in}} \times \frac{\text{m}}{100 \text{cm}} = 17 \text{ m/s}$$

$$\eta = \frac{c_{vacuum}}{c_{BEC}} = \frac{2.998 \times 10^{8} \text{ m/s}}{17 \text{ m/s}} = 1.8 \times 10^{7}$$

1.7 Molecules fluoresce from the lowest vibrational state of the lowest excited state because the rates of vibrational relaxation and internal conversion are much greater than the rate of fluorescence. Thus, hundreds of relaxation events can occur before fluorescence does, meaning that the molecules reach the lowest vibrational and excited state energy before fluorescing.

1.8 Fluorescence is a singlet-to-singlet transition (i.e., no electron spin change), whereas phosphorescence is a triplet-to-singlet transition requiring an electron spin flip.

1.9 First calculate the energy associated with 400 MHz EMR:

(a) $E = h\upsilon = (6.626 \times 10^{-34} \text{ Js})(400.0 \times 10^{6} \text{ s}^{-1}) = 2.650 \times 10^{-25} \text{ J}$

Now used the Boltzmann distribution equation to calculate the relative populations:

$$\frac{N*}{N^{\circ}} = \frac{P*}{P^{\circ}} e^{\left(\frac{-\Delta E}{kT}\right)} = 1\left(2.718^{\frac{-2.650\times10^{-25}\,\text{J}}{\left(1.38\times10^{-23}\,\text{J/K}\right)\left(298\,\text{K}\right)}}\right) = 0.99993556$$

(b) Note that more significant figures than are warranted have been reported simply to show that while the ratio is very close to 1, meaning that the two populations are nearly equal, the lower energy state (spin up) is slightly more populated than the higher energy state (spin down).

1.10 Complete the table below. For simplicity, assume $P* = P^{\circ}$.

First, convert the wavelength, frequency, or wavenumber to energy in joules (values are shown in the table) and the temperatures to Kelvin. Then apply the Boltzmann equation to calculate $N*/N^{\circ}$.

Type	Typical radiation	Energy (J)	Temperature (°C)	$\dfrac{N*}{N^{\circ}}$	$\dfrac{\left(N*/N^{\circ}\right)_{300}}{\left(N*/N^{\circ}\right)_{25}}$
UV-visible	400 nm	4.966×10^{-19}	25	3.626×10^{-53}	1.472×10^{25}
			300	5.337×10^{-28}	
IR	1600 cm^{-1}	3.178×10^{-20}	25	4.404×10^{-4}	40.81
			300	1.797×10^{-2}	
NMR	400 MHz	2.650×10^{-25}	25	0.9999356	1.00003
			300	0.9999665	

(Note: Because of the exponential dependence, values of $N*/N^{\circ}$ are quite sensitive to the number of decimal places that are used in ΔE, k, and T, so your answers may be somewhat different than those presented here but should be of the same order of magnitude.) What we see from this is that increasing the temperature has a significant impact on the relative distribution of molecules in ground vs. excited states when the energy gap is large (the UV-visible example), but not much of an effect when the energy gap is small and the ratio is already close to 1.00 as it is in NMR. This impacts the sensitivity of measurements and how they change with temperature in specific techniques.

1.11 (a) $(N*/N^{\circ})_{\text{NMR}}/(N*/N^{\circ})_{\text{UV-vis}} = 0.9999356/3.626 \times 10^{-53} = 2.757 \times 10^{52}$

(b) $(N*/N^{\circ})_{\text{NMR}}/(N*/N^{\circ})_{\text{IR}} = 0.9999356/4.404 \times 10^{-4} = 2.270 \times 10^{3}$

(c) Because the energy gap is so large in UV-visible spectroscopy, the ratio of excited state to ground state molecules is very small, whereas for NMR, where the energy gap is much smaller, the two populations are nearly the same. So the ratio of the NMR ratio to the UV-visible ratio is tremendously large.

When considering the NMR/IR comparison, because IR spectroscopy involves lower energy transitions than UV-visible spectroscopy, the ratio of ratios is not nearly as dramatically different as it is in the NMR/UV-visible comparison. Looked at another way, the energy associated with IR spectroscopy is closer to that of NMR spectroscopy than is the energy associated with UV-visible spectroscopy.

1.12 $m\lambda = 2D\sin\theta$

(a) At $30°$:

$$\lambda = 2(1.5\times10^{-10}\text{ m})\sin(30°) = 2(1.5\times10^{-10}\text{ m})(0.50) = 1.5\times10^{-10}\text{ m}$$

Remember to convert angles to radians to calculate the sine function when using some calculators or certain spreadsheet commands.

(b) At $45°$:

$$\lambda = 2(1.5\times10^{-10}\text{ m})\sin(45°) = 2(1.5\times10^{-10}\text{ m})(0.707) = 2.1\times10^{-10}\text{ m}$$

(c) These are in the X-ray region of the spectrum.

1.13 $\dfrac{I_r}{I_o} = \dfrac{(\eta_{\text{diamond}} - \eta_{\text{benzene}})^2}{(\eta_{\text{diamond}} + \eta_{\text{benzene}})^2} = \dfrac{(2.417 - 1.501)^2}{(2.417 + 1.501)^2} = 0.05466$ or 5.466%

1.14 The numerator in the equation for reflection shows that the closer two materials are in refractive index, the lower the fraction of reflected light will be, diminishing to 0.000 when the two refractive indices are perfectly matched. In this case, the refractive indices of water and heptane are closer to each other than are those of water and benzene. Thus, the water/heptane interface reflects less light, and the water/benzene interface reflects more light.

1.15 $\dfrac{\sin\theta_{\text{water}}}{\sin\theta_{\text{diiodomethane}}} = \dfrac{\eta_{\text{diiodomethane}}}{\eta_{\text{water}}}$

$\dfrac{\sin(10°)}{\sin\theta_{\text{diiodomethane}}} = \dfrac{1.737}{1.333} = 1.303$

$\sin\theta_{\text{diiodomethane}} = \dfrac{\sin(10°)}{1.303} = \dfrac{0.1737}{1.303} = 0.1333$

$\theta_{\text{diiodomethane}} = 7.6°$

1.16 Rayleigh scattering

1.17 Tyndall effect/Mei scattering

CHAPTER 2

2.1 (a) $E = \dfrac{hc}{\lambda} = \dfrac{(6.626\times10^{-34}\text{ Js})(2.998\times10^8\text{ m/s})}{200\times10^{-9}\text{ m}} = 9.93\times10^{-19}\text{ J/photon}$

$\dfrac{9.93\times10^{-19}\text{ J}}{\text{photon}} \times \dfrac{6.022\times10^{23}\text{ photons}}{\text{mole}} = 5.98\times10^5\text{ J/mol} = 598\text{ kJ/mol}$

(b) Energy of a C—C bond ≈ 350 kJ/mol. Thus, the energy of a mole of 200 nm photons exceeds that of a mole of C—C bonds.

2.2 $\text{Abs} = \log\left(\dfrac{P_o}{P}\right) = -\log\left(\dfrac{P}{P_o}\right) = -\log T$

$T = \left(\dfrac{\%T}{100}\right)$

(a) $\text{Abs} = -\log\left(\dfrac{0.5\%}{100}\right) = 2.301$

(b) 2.00

(c) 0.82

(d) 0.52

(e) 0.30

(f) 0.12

(g) 0.046

(h) 4.3×10^{-4}

Note that the lower the percent transmittance, the greater the absorbance.

(i) The graph should resemble Figure 2.40 in the text.

2.3 (a) $\text{Abs} = \log\left(\dfrac{P_o}{P}\right) = -\log\left(\dfrac{P}{P_o}\right) = -\log T$

$T = \left(\dfrac{\%T}{100}\right)$

$0.003 = -\log T$

$T = 0.993$

$\%T = T \times 100$

$\%T = 99.3\%$

(b) 89.1%

(c) 66.1%

(d) 50.1%

(e) 15.8%

(f) 6.31%

(g) 1.0%

(h) 0.32%

(i) 0.10%

(j) 0.032%

2.4 (a) $A = \log\left(\dfrac{P_o}{P}\right) = -\log\left(\dfrac{500\ \mu A}{50\ \mu A}\right) = 1.00$

(b) $A = 2.20$

(c) $A = 3.00$

(d) $A = 3.11 \times 10^{-3}$

2.5 $A = \log\left(\dfrac{P_o}{P}\right) = \log\left(\dfrac{8000\ \mu A}{44\ \mu A}\right) = 2.26$

$A = \log\left(\dfrac{P_o + P_{stray}}{P + P_{stray}}\right) = \log\left(\dfrac{8000\ \mu A + 20\ \mu A}{44\ \mu A + 20\ \mu A}\right) = 2.10$

$\%\ \text{error} = \left|\dfrac{2.10 - 2.26}{2.26}\right| \times 100 = 7.08\%$

2.6 $A = \log\left(\dfrac{P_o + P_{stray}}{P + P_{stray}}\right)$

(a) $A = \log\left(\dfrac{100 + 0}{50 + 0}\right) = \log\left(\dfrac{100}{50}\right) = 0.301$

(b) $A = \log\left(\dfrac{100 + 2}{50 + 2}\right) = 0.293$

(c) $A = \log\left(\dfrac{100 + 0}{2 + 0}\right) = 1.70$

(d) $A = \log\left(\dfrac{100 + 2}{2 + 2}\right) = 1.41$

(e) For (a) and (b): $\%\ \text{error} = \left|\dfrac{0.293 - 0.301}{0.301}\right| \times 100 = 2.7\%$

(f) For (c) and (d): $\%\ \text{error} = \left|\dfrac{1.41 - 1.70}{1.70}\right| \times 100 = 17.1\%$

2.7 $A = \log\left(\dfrac{P_o}{P}\right) = \varepsilon bc$

$\log\left(\dfrac{85.4}{20.3}\right) = 0.624 = (\varepsilon_{500nm})(2\ \text{cm})(1.00 \times 10^{-4}\ \text{mol/L})$

$3120\ \text{L/mol}\cdot\text{cm} = \varepsilon_{510nm}$

2.8 $A = \log\left(\dfrac{P_o}{P}\right) = 0.185$ so $P_o/P = 1.53$

If P becomes P_o and P_o becomes P, then the new $\dfrac{P_o}{P} = \dfrac{1}{(P_o/P)_{initial}} = \dfrac{1}{1.53} = 0.653$

So Abs $= \log(0.653) = -0.185$

Thus, the sign of the absorbance would switch from positive to negative. After the switch, P would be greater than P_o, which does not make sense given the way they are typically defined.

2.9 Because $A = \varepsilon bc$, a negative absorbance implies a negative concentration, which does not make sense. Thus, while it is possible to obtain negative absorbance readings with spectrophotometers, their occurrence means that the reference/blank solution is actually absorbing more light than the solute solution, which suggests something may not be correct.

2.10 Greater sensitivity (i.e., more signal) and less chance for nonlinear behavior.

2.11 (a) Littrow

(b) Holographic grating with 1440 lines/mm

(c) Chopper

(d) Photomultiplier tube

(e) InGaAs and PbS detector

(f) ≤ 0.05 nm

(g) $0.00007\ \%T$

(h) Variable: 0.05–5.00 nm in 0.01 nm increments

(i) $\leq 0.0002\,A/h$

(j) National Institute of Standards and Technology, $\pm 0.003\,A$

2.12 $A = \log(P_o/P)$

Suppose that P_o is 100 and that if there is no drift, P would be some value, for example, 20. In this case, $A = \log(100/20) = 0.70$. If the lamp intensity is decreasing, then P will be lower than 20, say, 15. In this case, $A = \log(100/15) = 0.82$. This shows that the absorbance will be artificially high. Another way to think about this is that a decreasing lamp intensity will produce less light, making it seem like more was absorbed, leading to a falsely high absorbance.

2.13 (a) $\lambda = \dfrac{2t\eta}{m} = \dfrac{2(185\times10^{-9}\ \text{m})(1.35)}{1} = 499.5\ \text{nm}$

(b) $\lambda = \dfrac{499.5\ \text{nm}}{2} = 250\ \text{nm}, \dfrac{499.5\ \text{nm}}{3} = 166.5\ \text{nm}, \text{etc.}$

2.14 $\dfrac{1180\ \text{grooves}}{\text{mm}}$

$\dfrac{\text{mm}}{1180\ \text{grooves}} \times \dfrac{\text{m}}{10^3\ \text{mm}} \times \dfrac{10^9}{1\ \text{m}} = 847\ \text{nm/groove}$

2.15 $\Delta\lambda = 2WD^{-1}$

(a) $5.0\ \text{nm} = 2W(1.6\ \text{nm/mm})$

$$W = \dfrac{5.0\times10^{-9}\ \text{m}}{2(1.6\times10^{-9}\ \text{m}/1\times10^{-3}\ \text{m})} = 1.56\times10^{-3}\ \text{m} = 1.6\ \text{mm}$$

(b) $D^{-1} = \dfrac{b}{mF}$

$$b = D^{-1}mF = \left(\dfrac{1.6\ \text{nm}}{\text{mm}}\right)(1)(0.50\ \text{m}) = \left(\dfrac{1.6\times10^{-9}\ \text{m}}{1.0\times10^{-3}\ \text{m}}\right)(1)(0.50\ \text{m}) = 800\ \text{nm}$$

2.16 (a) Wide slits will let more light through and improve the precision of the absorbance readings, and because the peak is specified as being wide, there is little worry about nonlinearity due to varying molar absorptivities.

(b) Narrow slits are required to accurately measure the absorbance of fine spectral features. With narrow spectral features, nonlinearities arise if the slits are too wide.

2.17 (a) $R = \dfrac{\lambda}{\Delta\lambda} = \dfrac{589.3 \text{ nm}}{0.59 \text{ nm}} = 999$

(b) $R = mN = 999$

$N = 999$ lines in the first order ($m = 1$) and 250 lines in the fourth order ($m = 4$).

(c) Just under 0.59 nm bandpass.

(d) $\Delta\lambda = 2\,WD^{-1}$

$0.59 \text{ nm} = 2\,W\,(1.6 \text{ nm/mm})$

$\dfrac{0.59 \text{ nm}}{2(1.6 \text{ nm/mm})} = 0.18 \text{ mm} = W$

2.18 The sensitivity factor of each diode contributes to both P and P_0. Thus, when the ratio P/P_0 is taken, the sensitivity cancels. Because each diode is dedicated to a particular narrow wavelength range, the cancelation occurs for each diode regardless of wavelength and a spectrum that is free of distortion is obtained.

2.19 $A_\lambda = \varepsilon_\lambda bc$

$0.257 = \varepsilon_\lambda\,(2.00 \text{ cm})(5.643 \times 10^{-4} \text{ mol/L})$

$227.7 \text{ L/mol} \cdot \text{cm} = \varepsilon_\lambda$

$A_\lambda = \varepsilon_\lambda bc$

$A_\lambda = (227.7 \text{ L/mol} \cdot \text{cm})(1.00 \text{ cm})(2.74 \times 10^{-3} \text{ mol/L})$

$A_\lambda = 0.624$

2.20 $A_\lambda = \varepsilon_\lambda bc$

$1.289 = \varepsilon_\lambda\,(1.00 \text{ cm})(1.16 \times 10^{-4} \text{ mol/L})$

$11\,112 \text{ L/mol} \cdot \text{cm} = \varepsilon_\lambda$

$A_\lambda = \varepsilon_\lambda bc$

$0.500 = (11112 \text{ L/mol} \cdot \text{cm})(1.00 \text{ cm})(c)$

$4.50 \times 10^{-5} \text{ mol/L} = C$

$\dfrac{4.50 \times 10^{-5} \text{ mol complex}}{\text{L}} \times \dfrac{1 \text{ mol Fe}}{1 \text{ mol complex}} \times \dfrac{55.845 \text{ g Fe}}{1 \text{ mol Fe}} = 2.51 \times 10^{-3} \text{ g Fe} = 2.51 \text{ mg Fe}$

2.21 (a) Concentration of standards: 1.17×10^{-4} M, 9.35×10^{-5} M, 7.02×10^{-5} M, 4.68×10^{-5} M, 2.34×10^{-5} M

Calibration curve equation: Abs = $(15\,730 \pm 380 \text{ L/mol})(\text{Conc.}) - (0.024 \pm 0.029)$

Slope = εb

$15\,730 \text{ L/mol} = \varepsilon(1.00 \text{ cm})$

$15\,730 \text{ L/mol} \cdot \text{cm} = \varepsilon$

(b) $A_\lambda = \varepsilon_\lambda bc$

$0.681 = (15\,730 \text{ L/mol} \cdot \text{cm})(1.00 \text{ cm})(c) - 0.024$

4.48×10^{-5} M p-nitroaniline $= c$

(c) $\dfrac{4.48 \times 10^{-5} \text{ mol}}{\text{L}} \times \dfrac{25.00 \times 10^{-3} \text{ L}}{5.00 \times 10^{-3} \text{ L}} \times \dfrac{100.0 \times 10^{-3} \text{ L}}{5.00 \times 10^{-3} \text{ L}} = 4.48 \times 10^{-3} \text{ mol/L} = 4.48 \times 10^{-3} \text{ M}$

2.22 Because the molar absorptivity at 375 nm in cyclohexane is lower, the unknown solution does not absorb as much light as it would if methanol were the solvent for the same solute concentration. Thus, the measured absorbance is low compared to what it should be based on the calibration curve. Because absorbance in proportional to concentration, the concentration that is determined using the calibration curve will be low compared to the actual concentration.

2.23 (a)

(b) $A_\lambda = \varepsilon_{\lambda,1} b c_1 + \varepsilon_{\lambda,2} b c_2$

$A_{294\text{nm}} = (1028 \text{ L/mol·cm})(1.00 \text{ cm})(c_p) + (4211 \text{ L/mol·cm})(1.00 \text{ cm})(c_m)$

$A_{400\text{ nm}} = (18\,782 \text{ L/mol·cm})(1.00 \text{ cm})(c_p) + (1485 \text{ L/mol·cm})(1.00 \text{ cm})(c_m)$

Solve simultaneous equations by multiplying the first equation by 18782 L/mol·cm and the second equation by 1028 L/mol·cm. Then subtract the two resulting equations and solve for c_m. Repeat the procedure to solve for c_p, but multiply the top by 1485 L/mol·cm and the bottom by 4211 L/mol·cm.

$c_m = 1.25 \times 10^{-4} \text{ M}$

$c_p = 6.85 \times 10^{-5} \text{ M}$

2.24 625 nm is red light. So red light is absorbed and blue light is transmitted, making the dye appear blue.

525 nm is green light – so red is transmitted along with some blue, but our eyes are more sensitive to red light than to blue, so the overall appearance of the dye is red.

420 nm is blue light, meaning that green, yellow, and red light are transmitted. To our perception, however, red and green combine to make yellow, so the dye appears yellow.

2.25 (a) A sigmoidal plot is obtained by graphing absorbance versus pH for both sets of wavelength data. At low pH, *p*-nitrophenol is protonated and has a high molar absorptivity at 317 nm (and low molar absorptivity at 407 nm). As the pH is increased, the *p*-nitrophenol gets deprotonated. At high pH, all of the *p*-nitrophenol is deprotonated (i.e., exists as the phenolate ion). The molar absorptivity of phenolate is low at 317 nm and high at 407 nm. The inflection point of the plots occurs at the pH where half of the *p*-nitrophenol is protonated and half is deprotonated. In other words, the inflection point occurs when pH = pK_a.

Based on the plots generated from the data, the pK_a of p-nitrophenol is approximately 7.0.

(b) The nitro group in the *para* position withdraws electrons from the O—H bond via resonance and inductive effects. This weakens the O—H bond, making it energetically easier for the proton to dissociate from p-nitrophenol relative to p-methylphenol, which lacks this electron-withdrawing effect.

2.26 The high absorbance caused by the solvent means that almost all of the light from the lamp is being absorbed by the sample such that $P_0 \gg P$. Under these conditions, stray radiation and electrical noise in the circuitry compete with the actual signal arising from the minimal amount of light being transmitted through the sample.

2.27 (a) $\dfrac{0.874 \text{ g benzene}}{\text{mL}} \times \dfrac{1 \text{ mol benzene}}{78 \text{ g benzene}} \times \dfrac{1000 \text{ mL}}{L} = 11.2 \text{ mol/L} = 11.2 \text{ M}$

(b) $A_\lambda = \varepsilon_\lambda bc = (240 \text{ L/mol} \cdot \text{cm})(1.00 \text{ cm})(11.2 \text{ mol/L}) = 2688$ – obviously not a measurable absorbance

(c) Virtually none of the light of the specified wavelength makes it through the benzene.

(d) No – the absorbance due to the solute would be immeasurable compared to the absorbance due to the benzene.

CHAPTER 3

3.1 The conjugated portion of the molecule is responsible for fluorescence, as circled in the figure. This molecule, given the long alkyl chain, could be used to probe cell membranes and other nonpolar biological environments.

3.2 (a) Excite near 280 nm and observe emission near 360 nm.

(b) If a different wavelength is used for excitation, the fluorescence maximum will not shift to a different wavelength. The maximum wavelength of emission is dictated by the energy difference in the excited and ground states, which does not depend on the excitation wavelength. On the other hand, exciting at 275 rather than 260 will lead to an increase in the overall intensity of the emission. This occurs because more photons are absorbed at 275 than at 260, meaning that more molecules are excited at 275 nm. Because more molecules are excited, more also fluoresce, leading to greater fluorescence intensity at all wavelengths in the emission spectrum.

3.3 (a) Decreasing molar absorptivity decreases fluorescence because of lower likelihood that molecules absorb a photon. If fewer molecules are in the excited state, fewer fluoresce.

(b) Increasing slit widths increases fluorescence because more molecules will be excited. With more molecules in the excited state, more will fluoresce, all else being equal.

(c) Weaker sources lead to decreased fluorescence because fewer photons per unit time are going into the sample, meaning fewer molecules will be excited. If fewer molecules are excited, fewer fluoresce, ultimately decreasing the fluorescence intensity.

(d) Higher sample temperatures generally lead to decreased fluorescence. This occurs because the increased temperature means more thermal agitation/molecule motion. This allows molecules to lose their energy through collisions with other molecules and with the walls of the container. This leads to nonradiative relaxation of molecules and therefore decreases the fluorescence intensity.

(e) Decreasing solvent viscosity, like increasing temperature, allows molecules to move around more, which increases the probability of losing energy through collisions (i.e., nonradiative relaxation) rather than through fluorescence. This decreases the fluorescence intensity.

(f) Bromoethanol contains a bromine substituent, which tends to promote intersystem crossing. This decreases the number of excited state fluorescent molecules in the singlet state, thereby decreasing the likelihood of fluorescence, which in turn decreases the fluorescence intensity. Using ethanol instead of bromoethanol as the solvent would decrease the intersystem crossing that bromoethanol promotes, thus increasing fluorescence intensity.

(g) Exciting at a wavelength other than the wavelength of maximum absorbance leads to fewer molecules absorbing photons and thus fewer in the excited state. Fewer molecules in the excited state result in fewer molecules fluorescing and consequently decreased fluorescence intensity.

3.4 Lasers are intense sources of electromagnetic radiation (i.e., greater number of photons per unit time) – much greater in intensity than lamp sources. Putting more photons per unit time into a sample increases the number of molecules that are in the excited state at any point in time and thus increases the fluorescence intensity. This phenomenon is reflected in Eq. (3.5), which shows that fluorescence intensity (P_f) is proportional to the source intensity (P_o).

3.5 Increasing the temperature of a sample increases molecular motion, leading to increased collisions between molecules. These collisions transfer energy from an analyte molecule in the excited state to solvent molecules, which leads to nonradiationless relaxation, thus reducing the number of analyte molecules that decay by fluorescing. The increased thermal motion can also bring excited state molecules into contact with quenchers more quickly, which also decreases fluorescence.

3.6 (a)

The circled lone pair and conjugated electrons are responsible for the absorption and emission of di-8-ANEPPS.

(b) The wavelength of maximum emission of di-8-ANEPPS shifts toward shorter wavelengths (higher energies) when going from polar solvents to nonpolar solvents. This occurs because polar solvents like ethanol solvate (and thus reduce the energy of) the polar excited state more than they do the less polar ground state. This stabilization effect is greater in polar solvents than in nonpolar solvents. Thus, the excited-to-ground state energy gap is smaller in polar solvents than it is in nonpolar solvents. The smaller energy gap means that the fluorescent photons di-8-ANEPPS emits in polar solvents are of longer wavelength (lower energy) than in nonpolar solvents like benzene. Therefore, in benzene, which does not solvate the polar excited state (or the less polar ground state for that matter) as well as ethanol does, the wavelength of maximum emission shifts toward shorter wavelengths (higher energy). See Figure 3.2 and the accompanying explanation in Section 3.1 for more information.

3.7 Toluene leads to the smallest difference between the wavelengths of maximum absorbance and maximum fluorescence, while ethanol leads to the largest difference in these wavelengths. After the bithiophene absorbs a photon, its excited state structure is more polar than its ground state. Ethanol, being polar, can reorient around the excited state of the fluorescent molecule prior to the fluorescence event. This reorientation maximizes dipole–dipole and any potential hydrogen-bonding interactions. This lowers the energy of the excited state prior to emission. This means that the bithiophene derivative shown in Figure 3.3 needs to emit a photon with lower energy to return to the ground state when in ethanol compared to the higher energy photon it needs to emit when dissolved in toluene (where no substantial stabilization of the excites state occurs because toluene is nonpolar). Photons of lower energy have longer wavelengths. Thus, we see fluorescence shifted to longer wavelengths as the solvent polarity increases. Hydrogen bonding likely also plays a role given the specific structure of this molecule, as evidenced by the fact that dimethyl sulfoxide, which has a larger dipole (3.96 D) than ethanol (1.69 D), but which does not donate hydrogen bonds, produces a slightly shorter wavelength of maximum fluorescence than ethanol.

3.8 Intersystem crossing involves the molecule changing from a singlet state to a triplet state via an electron spin flip, whereas internal conversion occurs between two different excited state electronic distributions but does not involve a change in the spin state – the molecule remains in the singlet state during internal conversion.

3.9   (a) Excitation above approximately 315 nm produces emission from phenanthrene only. Excitation at 275 nm maximizes the naphthalene fluorescence, and monitoring the fluorescence at 315 nm is nearly free of emission from phenanthrene. Phenanthrene emission can be monitored between 375 and 400 nm with little or no emission from naphthalene.

(b) Due to the high degree of overlap at nearly all wavelengths, phosphorescence offers no advantages over fluorescence.

3.10 (a) An emission spectrum is a plot of the intensity of fluorescence as a function of emission wavelength, typically collected by holding the excitation wavelength constant and scanning through the emission wavelengths by rotating the emission grating. An excitation spectrum is a plot of fluorescence intensity as a function of the excitation wavelength. Excitation spectra are collected by holding the emission grating at a fixed wavelength and rotating the excitation grating.

(b) The excitation spectrum of a molecule more closely resembles its absorbance spectrum because a molecule does not fluoresce unless it absorbs light. Therefore, wavelengths at which a molecule strongly absorbs also tend to be wavelengths that result in greater fluorescence, so excitation spectra more closely resemble absorbance spectra than do emission spectra.

3.11 One of the main advantages of fluorescence is that the measurements are generally made against a "dark" background, meaning there is little or no background light being emitted at the wavelength of interest. So the measurement, at its simplest, relies on measuring the amount of light emitted from a sample compared to no light being emitted. In contrast, absorbance measurements rely on measuring a *decrease* in light intensity, meaning the detector has to differentiate between two different light levels. Additionally, because fluorescence intensity is proportional to source intensity, bright sources can be used to produce fluorescence in very dilute solutions. Brighter sources do not improve sensitivity in absorbance measurements because absorbance fundamentally depends on the *ratio* of light intensity with and without the sample. Increasing the source intensity does not increase the absorbance value. In contrast, increasing the source intensity in fluorescence measurements, which do not depend on a ratio but rather just on the total amount of fluorescence observed, increases the fluorescence intensity.

3.12 Answer: 0.0530 μg of 3,4-benzopyrene in 1 L of air.

Explanation: Because a blank produced a reading of 3.5, that value is subtracted from all of the readings. The first standard has a concentration of 0.0750 μg/mL and the second has concentration of 0.125 μg/mL, with signals of 21.0 and 35.1, respectively (after subtracting the blank reading). Next, a two-point calibration curve is created, and the equation for the line connecting the two points is determined, yielding the result

Fluorescence = 282 (concentration in μg/mL) − 0.15

The fluorescence of the sample solution was 29.8 (33.3 − 3.5). Entering that into the equation for the line yields

29.8 = 282 (concentration in μg/mL) − 0.15

Solving for the concentration yields

0.106 μg/mL = concentration of 3,4-benzopyrene in the sample that was analyzed

Therefore, in the 10 mL solution through which the air was drawn, there is 1.06 µg of 3,4-benzopyrene (0.106 µg/mL × 10 mL).

Those 1.06 µg got into the 10 mL of solution because 20 L of air was drawn through the solution. Therefore, there is 1.06 µg in 20 L of air, or 0.0530 µg in one liter of air (1.06 µg/20 L).

3.13 (a) $13\,425 = 1.35 \times 10^{11}$ (quinine sulfate (M)) $- 1.52 \times 10^3$

$13\,425 + 1.52 \times 10^3 = 1.35 \times 10^{11}$ (quinine sulfate (M))

$14\,945 = 1.35 \times 10^{11}$ (quinine sulfate (M))

$14\,925/1.35 \times 10^{11} =$ (quinine sulfate (M))

1.11×10^{-7} M = concentration of quinine sulfate

(b) The fluorescence intensity will decrease proportionally with the decrease in the lamp intensity, so the same concentration will now yield a signal of $0.837 \times 13\,425 = 11\,237$. Using this value in the equation for the calibration curve yields

$11\,237 = 1.35 \times 10^{11}$ (quinine sulfate (M)) $- 1.52 \times 10^3$

$11\,237 + 1.52 \times 10^3 = 1.35 \times 10^{11}$ (quinine sulfate (M))

$12\,757 = 1.35 \times 10^{11}$ (quinine sulfate (M))

$12\,757/1.35 \times 10^{11} =$ (quinine sulfate (M))

9.45×10^{-8} M = concentration of quinine sulfate

(c) Percent error $= 100 \times \dfrac{1.11 \times 10^{-7} - 9.45 \times 10^{-8}}{1.11 \times 10^{-7}} = 14.9\%$ error

3.14 Equation (3.5) shows that for solutions with low absorbance, $P_f \approx 2.303\Phi_f P_o \varepsilon bc$. Based on this, at 280 nm, we can write $3.69 \times 10^5 \approx 2.303\Phi_f P_o(1280\ M^{-1}\ cm^{-1} \times b \times c)$. The concentration does not change at the two wavelengths. So even though we cannot solve numerically for the concentration, we can write

$$c = \frac{3.69 \times 10^5}{2.303\Phi_f P_o \left(1280\,M^{-1}cm^{-1} \times b\right)}$$

But this must equal the concentration at the different wavelength, for which $e = 930\,M^{-1}cm^{-1}$ and the source intensity $= 2.6 \times P_o$. This leads to

$$c = \frac{P_{f,\text{new wavelength}}}{2.303\Phi_f\, 2.6 \times P_o \left(930\,M^{-1}cm^{-1} \times b\right)}$$

Setting the two equations equal to one another results in

$$\frac{P_{f,\text{new wavelength}}}{2.303\Phi_f\, 2.6 \times P_o \left(930\,M^{-1}cm^{-1} \times b\right)} = \frac{3.69 \times 10^5}{2.303\Phi_f P_o \left(1280\,M^{-1}cm^{-1} \times b\right)}$$

such that

$$P_{f,\text{new wavelength}} = \frac{\left(3.69 \times 10^5\right) \times 2.303\Phi_f\, 2.6 \times P_o \left(930\,M^{-1}cm^{-1} \times b\right)}{2.303\Phi_f P_o \left(1280\,M^{-1}cm^{-1} \times b\right)}$$

After canceling the factor of 2.303, the quantum yield (which was assumed constant), the path length, and the original source intensity, P_o, the result is

$$P_{f,\text{new wavelength}} = \left(3.69 \times 10^5\right) \times \frac{2.6 \times \left(930\,\text{M}^{-1}\text{cm}^{-1}\right)}{\left(1280\,\text{M}^{-1}\text{cm}^{-1}\right)} = 6.97 \times 10^5$$

So, the new fluorescence intensity is affected by the increase in the source intensity (the increased factor of 2.6) and affected to a small extent by the decrease in molar absorptivity at the new excitation wavelength relative to the original wavelength.

3.15 (a) A careful examination of the data, and plotting it, shows that the measurement of 51 500 corresponds to two different concentrations – approximately 70 and 177 μM – as indicated by the dashed lines in the plot below. It is therefore impossible to tell which, if either of them, is correct.

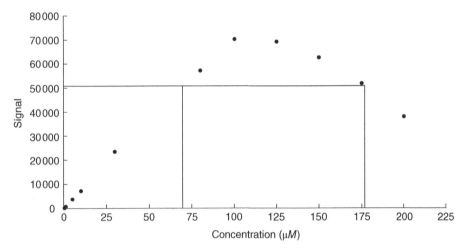

(b) The 10-fold dilution puts the signal squarely in the linear portion of the curve. Using the first seven data points to establish a correlation yields

Signal = 771 (concentration) − 79

Plugging in 13 520 for the signal leads to

(13 520 + 79)/771 = 17.6 μM

Because a 10-fold dilution was done on the sample to get this result, the original concentration is 10 times this value, or 176 μM.

Note: In part (a) we raised the possibility that the concentration might be around 70 μM. This can be ruled out, not only because of the result above but also by thinking about the calibration curve. 70 μM is in the linear portion of the curve, so a 10-fold dilution would result in a 10-fold decrease in fluorescence. Thus, if the concentration really were 70 μM, the 10-fold dilution would have led to a signal of approximately 5 150, instead of the 13 520 stated in the problem. Because the dilution did not result in a proportional reduction in signal, we know the original concentration was not in the linear range and thus not 70 μM.

3.16 $3 \times 10^9 \text{ base pairs} \times \dfrac{13 \text{ hours}}{200 \text{ base pairs}} \times \dfrac{\text{day}}{24 \text{ hours}} \times \dfrac{\text{year}}{365 \text{ days}} \approx 20\,000 \text{ years}$

We can conclude from this that thousands of capillary electrophoresis instruments were simultaneously running in parallel (and used capillary arrays), working on small sections of the genome that were eventually "stitched" together to produce the entire genetic code.

3.17 In some cases, particularly when analyzing opaque solutions, there is little to no fluorescence emitted at 90° from the excitation beam because the beam cannot penetrate into the solution. The advantage of front-face collection, then, is that fluorescence from the "layer" of solution where the excitation beam first encounters the solution can be collected and analyzed. In other words, it allows for signal to be collected from the "face" of the solution that would otherwise be weak or impossible to measure at 90°. Naturally, then, front-face collection is useful in situations where the excitation beam gets significantly attenuated as it propagates through the sample solution. This occurs with opaque/cloudy samples, highly concentrated samples, and solid samples.

3.18 The spikes in the fluorescence emission at 610 are caused by spikes in the intensity of the mercury lamp. Mercury lamps emit radiation at the discrete wavelengths stated in the problem (the reader is encouraged to search the Internet for an image of the emission of a mercury lamp). Because fluorescence intensity is proportional to source intensity, spikes in the source intensity can create spikes in fluorescence. For this reason, the *ratio* of fluorescence intensity to source intensity is often plotted, rather than just the raw fluorescence intensity.

3.19 (a) Inner filter effect – At high concentrations, the first "layer" of analytes absorbs so much of the excitation beam that analyte molecules that are several layers into the solution "see" less light and are therefore less likely to be excited. They are thus less likely to fluoresce, leading to negative deviations from linearity.

(b) Self-absorption – Fluorescence emitted by one molecule is absorbed by another, which then decays through a nonradiative mode. This causes negative deviations from linearity.

3.20 $\text{Number of molecules excited} = 0.1 \times \left(1 \times 10^{-15} \text{ mol}\right) \times \dfrac{6.022 \times 10^{23} \text{ molecules}}{\text{mol}} = 6.022 \times 10^7$

$\Phi = \dfrac{\text{Number of molecules that fluoresce}}{\text{Number of molecules that absorbed a photon}} = \dfrac{4.3 \times 10^7}{6.022 \times 10^7} = 0.71$

3.21 Very few molecules fluoresce, and even fewer phosphoresce because phosphorescence requires the spin-forbidden transition from a singlet to a triplet state, followed by emission of a photon and a return to a single state. The spin-forbidden transition is highly unlikely, plus there are other radiative and nonradiative relaxation pathways through which an excited state molecule can relax back down to the ground state before phosphorescence occurs.

3.22 Because chemiluminescence is generated by a reaction between molecules, it is not necessary to have a source of electromagnetic radiation. It is also not necessary to have an excitation monochromator because there is no excitation source. Further, if one is only interested in the total amount of electromagnetic radiation produced by a given reaction, it would not be necessary to have an emission monochromator. An emission monochromator would only be needed to select a specific wavelength of light being emitted by the chemiluminescent reaction. Whether or not an emission monochromator is used, some method of collecting the emitted light and focusing it onto a detector would likely be necessary.

CHAPTER 4

4.1 (a) $\text{Abs} = -\log T$

10.0% transmittance means transmittance $= 0.100$

$\text{Abs} = -\log T = -\log 0.100 = 1.000$

(b) 1.00% transmittance means transmittance $= 0.0100$

$\text{Abs} = -\log T = -\log 0.0100 = 2.000$

4.2 $\text{Mass of }^1\text{H} = \dfrac{1.008\,\text{g}}{\text{mol}} \times \dfrac{1\,\text{mol}}{6.022 \times 10^{23}} \times \dfrac{1\,\text{kg}}{1000\,\text{g}} = 1.674 \times 10^{-27}\,\text{kg}$

$\text{Mass of }^{35}\text{Cl} = \dfrac{34.969\,\text{g}}{\text{mol}} \times \dfrac{1\,\text{mol}}{6.022 \times 10^{23}} \times \dfrac{1\,\text{kg}}{1000\,\text{g}} = 5.807 \times 10^{-26}\,\text{kg}$

$\text{Mass of }^{37}\text{Cl} = \dfrac{36.966\,\text{g}}{\text{mol}} \times \dfrac{1\,\text{mol}}{6.022 \times 10^{23}} \times \dfrac{1\,\text{kg}}{1000\,\text{g}} = 6.138 \times 10^{-26}\,\text{kg}$

For $^1\text{H}-^{35}\text{Cl}$

$$\mu = \frac{m_1 m_2}{m_1 + m_2} = \left(\frac{1.674 \times 10^{-27}\,\text{kg} \times 5.807 \times 10^{-26}\,\text{kg}}{1.674 \times 10^{-27}\,\text{kg} + 5.807 \times 10^{-26}\,\text{kg}} \right) = 1.627 \times 10^{-27}\,\text{kg}$$

$$\upsilon = \frac{1}{2\pi} \sqrt{\frac{k}{\mu}} = \frac{1}{2\pi} \sqrt{\frac{481\,\text{N/m}}{1.627 \times 10^{-27}\,\text{kg}}} = \frac{1}{2\pi} \sqrt{\frac{481\,\text{kg} \cdot \text{m/s}^2\text{m}}{1.627 \times 10^{-27}\,\text{kg}}} = 8.6536 \times 10^{13}\,\text{s}^{-1}\,(\text{or Hz})$$

Convert frequency to wavenumber $(E = h\upsilon = hc\bar{\nu})$ so

$$\bar{\nu} = \frac{\upsilon}{c} = \frac{8.6536 \times 10^{13}\,\text{s}^{-1}}{2.998 \times 10^8\,\text{m/s}} = \frac{2.88646 \times 10^5}{\text{m}} \times \frac{1\,\text{m}}{100\,\text{cm}} = 2886.5\,\text{cm}^{-1}$$

$^1\text{H}-^{37}\text{Cl}$

$$\mu = \frac{m_1 m_2}{m_1 + m_2} = \left(\frac{1.674 \times 10^{-27}\,\text{kg} \times 6.138 \times 10^{-26}\,\text{kg}}{1.674 \times 10^{-27}\,\text{kg} + 6.138 \times 10^{-26}\,\text{kg}} \right) = 1.630 \times 10^{-27}\,\text{kg}$$

$$\upsilon = \frac{1}{2\pi}\sqrt{\frac{k}{\mu}} = \frac{1}{2\pi}\sqrt{\frac{481\,\text{N}/\text{m}}{1.627\times10^{-27}\,\text{kg}}} = \frac{1}{2\pi}\sqrt{\frac{481\,\text{kg}\cdot\text{m}/\text{s}^2\text{m}}{1.630\times10^{-27}\,\text{kg}}} = 8.6456\times10^{13}\,\text{s}^{-1}\left(\text{or Hz}\right)$$

$$\bar{v} = \frac{\upsilon}{c} = \frac{8.6456\times10^{13}\,\text{s}^{-1}}{2.998\times10^8\,\text{m}/\text{s}} = \frac{2.88379\times10^5}{\text{m}}\times\frac{1\text{m}}{100\,\text{cm}} = 2883.8\,\text{cm}^{-1}$$

The difference in predicted wavenumbers is only 2.7 cm⁻¹, so they would not be fully resolved on an instrument with a 4 cm⁻¹ resolution.

Note: We have admittedly carried a few more significant figures through these calculations than are allowed for the sake of making the comparison.

4.3 (a) Because a C—C bond in an aromatic ring is neither a single nor a double bond, but rather acts somewhere in between, we expect the force constant to be between 450 N/m (single bond character) and 900 N/m (double bond character). So a reasonable estimate is midway between them, or approximately 675 N/m.

(b) $c\bar{v} = \upsilon$ so $\dfrac{2.998\times10^8\,\text{m}}{\text{s}}\times\dfrac{1480}{\text{cm}}\times\dfrac{100\,\text{cm}}{\text{m}} = 4.437\times10^{13}\,\text{s}^{-1}$

$$\upsilon = \frac{1}{2\pi}\sqrt{\frac{k}{\mu}}$$

$$4.437\times10^{13}\,\text{s}^{-1} = \frac{1}{2\pi}\sqrt{\frac{k}{9.963\times10^{-27}\,\text{kg}}}$$

$$7.772\times10^{28}\,\text{s}^{-2} = \frac{k}{9.963\times10^{-27}\,\text{kg}}$$

$$774\,\text{kg}/\text{s}^2 = k = 774\,\text{N}/\text{m}\,\left(\text{because 1 N} = \text{kg}\cdot\text{m}/\text{s}^2\right)$$

in reasonable agreement with the estimate of 675 made above.

4.4 The strength of the bond is modeled by the force constant, k, in Hooke's law. Lower force constants result in lower frequencies. Therefore, C—C single bonds, which have the weakest bond of the three listed, have the lowest stretching frequency, and C—C triple bonds have the highest.

4.5 (a) 51 peaks expected ($3N-6$ with $N = 19$ $C_8H_8O_3$), approximately 45–50, are present.

(b) Overlapping peaks, weak peaks, peaks outside of typical IR range, and vibrations that do not produce a change in dipole moment are all possible reasons for fewer peaks than expected.

(c) Overtone bands (near multiples of expected wavenumbers due to $\Delta v = \pm2$ or ±3) and combination bands (sum of two fundamental frequencies).

4.6 The mirror has to move a greater distance for greater resolution. Longer distances require more time given an equal velocity.

4.7 (a) $3N - 6 = 3$ normal modes predicted.

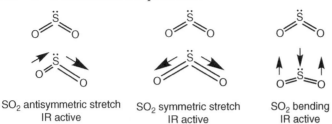

SO$_2$ antisymmetric stretch   SO$_2$ symmetric stretch   SO$_2$ bending
            IR active   IR active   IR active

(b) All modes of SO$_2$ are active because all create a change in the dipole moment of the molecule.

(c) SO$_2$ contrasts with CO$_2$ in that all modes are active, whereas the symmetric stretch in CO$_2$ is IR inactive. This arises because CO$_2$ is linear, whereas SO$_2$ has a bent geometry due to the lone pair of electrons on the sulfur.

4.8 A cell with a known path length is used when quantitative information is desired. A fixed path length is important for creating linear calibration curves based on the Beer–Lambert law: $A = \varepsilon bc$.

4.9 (a) $\Delta E = hc\bar{v} = 6.626 \times 10^{-34} \text{ Js} \left(\dfrac{2.998 \times 10^8 \text{ m}}{\text{s}} \right) \left(\dfrac{2143}{\text{cm}} \right) \left(\dfrac{100 \text{ cm}}{\text{m}} \right) = 4.257 \times 10^{-20} \text{ J}$

$$\frac{N^*}{N^\circ} = \frac{P^*}{P^\circ} e^{\frac{-\Delta E}{kT}}$$

$$\frac{-\Delta E}{kT} = \frac{-4.257 \times 10^{-20} \text{ J}}{\left(1.381 \times 10^{-23} \text{ JK}^{-1} \right) \left(298.15 \text{ K} \right)} = -10.34$$

$$\frac{N^*}{N^\circ} = e^{-10.34} = 2.718^{-10.34} = 3.235 \times 10^{-5}$$

This is the ratio of the number of molecules in the higher vibrational state relative to the lower, but we have been asked for the ratio of the lower population relative to the upper population, so we need to take the inverse, which yields

$$\frac{N^\circ}{N^*} = 3.091 \times 10^4$$

So for every 1 molecule in the upper vibrational state, 30910 are in the ground vibrational state at 298 K.

(b) Higher temperatures reflect the fact that molecules have more energy. So at higher temperatures, more molecules will have enough energy to be in a higher vibrational state. Therefore, this ratio decreases as temperature increases.

(c) The ratio of ground to excited state populations is so great in electronic transitions because electronic transitions have much larger energy gaps. In the Boltzmann equation, the energy gap is in the exponential term, meaning that there is an exponential dependence of the ratio on the energy gap. For this reason, the much larger energy gaps associated with electronic transitions relative to the smaller energy gaps associated with vibrational levels give rise to population distributions that significantly favor the ground state, roughly 10^{52} to 1 (!) for electronic transitions relative to 10^4 to 1 for vibrational transitions.

4.10 Fourier transform instruments offer two primary advantages over dispersive instruments, namely, the Fellgett (or multiplex) advantage and the Jacquinot (or throughput) advantage. The Fellgett advantage arises because an entire spectrum can be recorded in seconds by simply displacing a mirror less than a few centimeters. This allows multiple spectra to be collected and signal averaged in a very short time. By taking averages, noise, which is random, tends to cancel, while signal remains the same. This improves the signal-to-noise ratio over dispersive instruments, with which typically only one spectrum is recorded, making averaging impossible.

The throughput advantage arises because in FTIR instruments, the optics allow a significant portion of the source electromagnetic radiation through the optical system all at the same time. This allows a greater intensity of radiation to reach the detector compared to dispersive instruments, where most of the radiation is blocked by slits in order to select a narrow wavelength range to pass through the sample and on to the detector. The greater intensity of the radiation in FTIR instruments compared to dispersive instruments, along with the ability to signal average multiple spectra, improves S/N ratios in FTIR spectrometers.

4.11 (a) The ratio of $5/2.7 = 1.85$. Because S/N depends on the square root of the number of scans, we square this result and obtain 3.43. So 3.42 times as many scans as produced the S/N = 2.7 are needed; in other words, 10×3.43, or 34.3 scans. Rounding up, we would collect at least 35 scans.

(b) 35 scans is 3.5 times more than the original 10 scans, so it will take 3.5 times longer, or 3.5 minutes, to collect the new set of scans. S/N improves with the square root of the scan number, but analysis time is linear with the scan number.

4.12 (a) 4534 cm^{-1} (first overtone) and 6801 cm^{-1} (second overtone).

(b) These are located in the near-IR portion of the spectrum.

4.13 (a) The very bottom spectrum has the best resolution.

(b) The very top spectrum has the smallest retardation (associated with the worst resolution and highest R), and the very bottom spectrum has the largest retardation and best resolution. Specifically, the retardations from top to bottom are 0.0625, 0.125, 0.25, 0.5, 1.0, and the maximum retardation of the instrument.

4.14 NIR is better for calibration curves because broad peaks mean that molar absorptivities at all the wavelengths near the λ_{max} are approximately equal – a requirement for linear Beer's law plots as detailed in the UV-visible spectroscopy chapter. Spectral features in the mid-IR are sharp, meaning that molar absorptivities vary dramatically from one wavelength (wavenumber) to the next, making nonlinear calibration curves more likely. This is not to say that calibration curves cannot be used in the mid-IR – they can, and they are, but they require attention to details that can produce nonlinearities as discussed in Chapter 2.

4.15 Multiple reflection ATR cells provide more opportunities for electromagnetic radiation to be absorbed than single reflection cells. This is because each reflection presents an opportunity for the sample to absorb radiation. Greater absorption makes it easier to distinguish the intensity of the electromagnetic radiation in the presence (P) and absence (P_0) of the sample and thus easier to measure transmittance.

4.16 Chemometric techniques are needed because spectral features often overlap in NIR spectra, making it difficult to attribute all of the observed absorbance at any wavelength to just a single cause (i.e., just one molecule or just one particular stretch or overtone band). Chemometrics provides a means for deconvolving the signals and obtaining quantitative information.

4.17 Salt plates are transparent to IR radiation. Plastics and glass contain chemical bonds (Si–O, C–H, etc.) that absorb IR radiation. Thus, if plastics or glass are used as cell windows in the IR region, very little of the radiation makes it through to the sample and very little reaches the detector, making it impossible to measure a spectrum of the sample.

4.18 (a) MCT detectors are more sensitive and respond faster than DTGS detectors. They are therefore best suited for applications in which signals are weak (i.e., low light levels reaching the detector) and for fast analyses such as tracking rapid kinetic processes or chromatographic peaks that elute over a short period of time.

(b) MCT detectors can saturate more easily than DTGS detectors, and they require liquid nitrogen for their operation. This adds expense and labor to operating them. For these reasons, they are typically not the standard detector in commercial instruments.

CHAPTER 5

5.1 The wavelength of the laser can be calculated with Eq. (5.1).

$$\Delta \bar{v} = \frac{1}{\lambda_0} - \frac{1}{\lambda}$$

$$\frac{1}{\lambda_0} = \Delta \bar{v} + \frac{1}{\lambda}$$

$$\lambda_0 = \left(\Delta \bar{v} + \frac{1}{\lambda} \right)^{-1}$$

$$\lambda_0 = \left(990 \times 10^{-7} \text{ nm}^{-1} + \frac{1}{850 \text{ nm}} \right)^{-1} \approx 785 \text{ nm}$$

5.2 When polyaniline is fully deuterated, the vibrational peak shifts downward because deuterium is heavier than hydrogen. The Raman shift of the C–D bending mode $\bar{v}_{(C-D)}$ can be estimated using Hooke's law given the masses of carbon ($m_C = 12\mu$), hydrogen ($m_H = 1\,\mu$), and deuterium ($m_D = 2\,\mu$), the energy of the C—H bending mode $\bar{\nu}_{(C-H)} = 1192\text{ cm}^{-1}$, and the relationship that $\bar{\nu} = c\upsilon$.

$$\bar{\nu}_{(C-D)} = \bar{\nu}_{(C-D)} \times \frac{\bar{\nu}_{(C-H)}}{\bar{\nu}_{(C-H)}} = \frac{\upsilon_{(C-D)}/c}{\upsilon_{(C-H)}/c} \times \bar{\nu}_{(C-H)}$$

$$= \left[\frac{1}{2\pi}\sqrt{\frac{k(m_C + m_D)}{m_C m_D}}\right] \Bigg/ \left[\frac{1}{2\pi}\sqrt{\frac{k(m_C + m_H)}{m_C m_H}}\right] \times \bar{\nu}_{(C-H)}$$

$$= \sqrt{\frac{m_H(m_C + m_D)}{m_D(m_C + m_H)}} \times \bar{\nu}_{(C-H)}$$

$$= \sqrt{\frac{14}{26}} \times 1192\text{ cm}^{-1} \approx 875\text{ cm}^{-1}$$

According to Hooke's law, the C—H bending mode near 1192 cm^{-1} shifts down to approximately 875 cm^{-1} after deuteration. This agrees with the experimentally measured vibrational frequency (Quillard, S. et al. *Synthetic Met.* 1997, *84*, 805–806).

Alternative Solution
The location of the peak can also be determined by explicitly solving for the force constant k using Hooke's law.

$$\upsilon_{(C-H)} = c \times \bar{\nu}_{(C-H)} = \frac{1}{2\pi}\sqrt{\frac{k(m_C + m_H)}{m_C m_H}}$$

$$k = c^2 \times \bar{\nu}_{(C-H)}^2 \times 4\pi^2 \times \frac{m_C m_H}{m_C + m_H}$$

$$= \left(3 \times 10^{10}\text{ cm/s}\right)^2 \times \left(1192\text{ cm}^{-1}\right)^2 \times 4\pi^2 \times \frac{(12\,\mu)(1\,\mu)}{12\,\mu + 1\,\mu}$$

$$= 4.66 \times 10^{28}\;\mu/\text{s}^2 = 77.35\text{ kg/s}^2$$

The Raman shift of the C–D bending mode can then be determined by plugging the force constant k and masses of carbon and deuterium into Hooke's law.

$$\bar{\nu}_{(C-D)} = \frac{1}{2\pi c}\sqrt{\frac{k(m_C + m_D)}{m_C m_D}}$$

$$= \frac{1}{2\pi\left(3 \times 10^{10}\text{ cm/s}\right)}\sqrt{\frac{4.66 \times 10^{28}\;\mu/\text{s}^2 \times (12\,\mu + 2\,\mu)}{(12\,\mu)(2\,\mu)}}$$

$$= 5.31 \times 10^{-12}\text{ s/cm} \times \sqrt{2.72 \times 10^{28}\text{ s}^{-2}}$$

$$\approx 875\text{ cm}^{-1}$$

5.3 Because CO_2 is a linear molecule, it has $3N - 5$ or $3(3) - 5 = 4$ vibrational degrees of freedom. CO_2 has a center of symmetry so the mutual exclusion principle holds. This condition means that the symmetric vibrations are Raman active but not IR active, while the antisymmetric ones are IR active but not Raman active.

5.4 The symmetric stretching mode of H_2O is IR active because of the change in dipole moment and Raman active because of the change in polarizability near the molecular equilibrium position. The mutual exclusion principle does not apply because H_2O does not have a center of symmetry.

5.5 The required incident power can be determined using Eq. (5.8) and equating the Raman intensities produced by the two sources.

$$I_0(\upsilon_2)\upsilon_2^4 Nf(\alpha)^2 = I_0(\upsilon_1)\upsilon_1^4 Nf(\alpha)^2$$
$$I_0(\upsilon_2)(c/\lambda_2)^4 = I_0(\upsilon_1)(c/\lambda_1)^4$$
$$I_0(\upsilon_2) = I_0(\upsilon_1)(\lambda_2/\lambda_1)^4$$
$$I_0(\upsilon_2) = 10 \text{ mW} \left(633 \text{ nm}/375 \text{ nm}\right)^4 \approx 80 \text{ mW}$$

5.6 At room temperature, Stokes Raman scattering is more intense because most molecules are in the ground vibrational state. The molecule gains energy during the Stokes Raman scattering interaction.

5.7 Given that the Stokes signal is 10 times greater than the anti-Stokes signal, the population ratio is $N_e/N_0 = 1/10$. The temperature can be determined by plugging this ratio and the energy difference between the states ($E_e - E_0 = 500$ cm^{-1}) into Eq. (5.9).

$$\frac{N_e}{N_0} = e^{-(E_e - E_0)/kT}$$
$$\ln\left(\frac{N_e}{N_0}\right) = -\frac{E_e - E_0}{kT}$$
$$T = -\frac{E_e - E_0}{k \ln(N_e/N_0)}$$
$$T = -\frac{500 \text{ cm}^{-1}}{0.695 \text{ cm}^{-1}/\text{K} \times \ln(1/10)} \approx 312 \text{ K} \approx 39\,^{\circ}\text{C} \approx 100\,^{\circ}\text{F}$$

5.8 A nonzero depolarization ratio means that a component of the Raman-scattered intensity is polarized perpendicular to the excitation polarization. The excitation and scattered light therefore have different polarizations, and the off-diagonal terms in the polarizability matrix must be nonzero.

5.9 The depolarization ratio $\rho = I_{\text{perpendicular}}/I_{\text{parallel}} = 1/10 = 0.1$. The vibration is likely symmetric because ρ is less than 0.75.

5.10 Scanning monochromators, acousto-optic tunable filters, angle-tuned interference filters, and birefringent filters are examples of variable-wavelength filters. Acousto-optic filters use transducers to generate acoustic waves in a crystal that operates like a diffraction grating. Rapid wavelength tuning can be achieved by changing the frequency of the transducer signal, which affects the period of the grating.

5.11 The minimum angle of incidence can be determined by solving Eq. (5.11) for θ.

$$\Delta\lambda = \lambda_0 \left[1 - \sqrt{1 - (\sin\theta / n_{\mathrm{eff}})^2} \right]$$

$$\Delta\lambda / \lambda_0 = 1 - \sqrt{1 - (\sin\theta / n_{\mathrm{eff}})^2}$$

$$\sqrt{1 - (\sin\theta / n_{\mathrm{eff}})^2} = 1 - \Delta\lambda / \lambda_0$$

$$(\sin\theta / n_{\mathrm{eff}})^2 = 1 - (1 - \Delta\lambda / \lambda_0)^2$$

$$\sin\theta = n_{\mathrm{eff}} \sqrt{1 - (1 - \Delta\lambda / \lambda_0)^2}$$

$$\theta = \arcsin\left[n_{\mathrm{eff}} \sqrt{1 - (1 - \Delta\lambda / \lambda_0)^2} \right]$$

$$\theta = \arcsin\left[1.5\sqrt{1 - (1 - 2\,\mathrm{nm}/635\,\mathrm{nm})^2} \right] \approx 7°$$

5.12 Czerny–Turner and axial transmissive spectrometers are two dispersive spectrometers frequently used for Raman spectroscopy. These instruments are often coupled with charge-coupled device (CCD) camera arrays for detection.

5.13 Confocal instruments achieve high spatial resolution by placing a pinhole in a confocal plane (a plane that is conjugate to the object and images planes) to block out-of-focus light. This enables data acquisition from planes below the sample surface and the generation of three-dimensional Raman images.

5.14 (a) Quantitative analysis methods often take advantage of the linear relationship between concentration and Raman intensity (see Eq. 5.8). (b) Principal component regression and partial least squares are two implicit analytical methods. Unlike explicit methods, these techniques do not require prior knowledge of the concentrations or Raman spectra of all chemical constituents in the sample.

5.15 (a) The drawbacks of resonance Raman spectroscopy include increased risks of absorption, fluorescence, and photodegradation. (b) The resonant oscillation of conduction electrons at the sample surface is called a surface plasmon. (c) The Raman shift is $\bar{v}_m = \bar{v}_1 - \bar{v}_2 = 1/\lambda_1 - 1/\lambda_2 = 1/850\,\mathrm{nm} - 1/1150\,\mathrm{nm} \approx 3070\,\mathrm{cm}^{-1}$.

The CARS signal is emitted at a new frequency $v_1 + v_m = 2v_1 - v_2$.
The corresponding wavelength is

$$\lambda = c / (2v_1 - v_2) = (2\bar{v}_1 - \bar{v}_2)^{-1}$$

$$\lambda = (2/850\,\mathrm{nm} - 1/1150\,\mathrm{nm})^{-1} \approx 675\,\mathrm{nm}.$$

INDEX

Spectroscopy: Principles and Instrumentation, First Edition. Mark F. Vitha.
© 2019 John Wiley & Sons, Inc. Published 2019 by John Wiley & Sons, Inc.

Printed and bound by CPI Group (UK) Ltd, Croydon, CR0 4YY

16/04/2025

14658421-0001